GENERAL PLANT BREEDING

By the Same Author

Elements of Biometrical Genetics

About the Author

Arvind R. Dabholkar (*b.* 1937) obtained his Master's degree in Agriculture, with specialization in Horticulture from Agra University, in 1958. In 1963, he obtained M.Sc. (Ag.) degree in Agricultural Botany from Vikram University. From University of Indore, Indore, he achieved his Ph.D, in 1979.

From 1963 to 1997, he was engaged in sorghum breeding work in Jawaharlal Nehru Krishi Vishwa Vidyalaya. He worked in various capacities starting from Research Assistant to Professor (Plant Breeding)/Senior Scientist. He taught Advanced Genetics, Quantitative Genetics and Plant Breeding to postgraduate students for over 20 years. He guided several students for their M.Sc. (Ag.) thesis work.

He is author of *Elements of Biometrical Genetics*, published by Concept Publishing Company, New Delhi, in 1992. The second revised and enlarged edition of this book was brought out in 1999.

GENERAL PLANT BREEDING

A.R. Dabholkar

CONCEPT PUBLISHING COMPANY, NEW DELHI-110059

Cataloging in Publication Data—DK
Courtesy: D.K. Agencies (P) Ltd. <docinfo@dkagencies.com>

Dabholkar, A. R. (Arvind Rangrao), 1937-
General plant breeding / A.R. Dabholkar.
p. cm.
Includes bibliographical references (p.)
Includes index.
ISBN 8180692426

1. Plant breeding. I. Title.

DDC 631.52 21

ISBN 81-8069-242-6 (HB)

First Published 2006

Published and Printed by
Ashok Kumar Mittal
Concept Publishing Company
A/15-16, Commercial Block, Mohan Garden
New Delhi-110059 (INDIA)

Phones : 25351460, 25351794
Fax : 091-11-25357103
Email : publishing@conceptpub.com

Preface

I was teaching a course on Plant Breeding to M.Sc. (Ag.) students specializing in Plant Breeding and Genetics at the College of Agriculture, Indore. Post-graduate students who had offered horticulture also joined my class subsequently. A thought peeped in my mind regarding the justification about what I was teaching to both the groups of students. I should have talked about vegetables, ornamentals and fruits to students of Horticulture. As the things are, traditional students of Plant Breeding would learn about cereals, pulses, oilseeds, fibre crops etc. Thus there seems to be a compartment between courses taught. Why should the traditional M.Sc. (Ag.) students of Plant Breeding not learn about vegetable breeding? The science of Plant Breeding is same for both the horticultural crops and agricultural crops.

There are exquisite, knowledgeable books on plant breeding such as those by R.W. Allard, Briggs and Knowles, Poehlman, B.D. Singh and book edited by V.L. Chopra. Several symposiums are available which provide adequate knowledge about various aspects of plant breeding. Several books are published on vegetable breeding also. In my opinion there should be books on plant breeding which deal with both agricultural and horticultural crops.

There are diverse products which are of commercial importance in case of vegetables, from leaves to stems to flowers to roots. Each has its quality requirement. Moreover, they are perishable. There is tremendous opportunity in the International market for cut flowers. Vigorous and intelligent breeding efforts are required in these crops. The concepts developed in cereals, pulses, oilseeds etc. for evolving excellent varieties/hybrids could be employed for betterment of horticultural crops also.

Although maize and castor, dealt with by traditional plant breeding students, are monoecious the sex forms in monoecious cucurbits are interesting. The traditional students of plant breeding should know about various sex forms found in cucurbits. Self-incompatibility is regarded as a hurdle in crop production. But vegetable breeders have utilized it for hybrid seed production.

Bridging species have been employed for improvement in potatoes. This knowledge could benefit traditional students of plant breeding. Hence, the present effort was made to include agricultural and horticultural crops while talking about plant breeding.

I admit that I am not an expert in vegetable breeding. Despite being engaged in sorghum breeding for 33 years I do not claim that I am proficient plant breeder. But being a post-graduate teacher as well as a plant breeder I have endeavoured to write a book on plant breeding which also refers to vegetable breeding in addition to cereals, pulses, oilseeds, and fibre crops.

I invite healthy and concrete suggestions from teachers as well as students to rectify any errors or misinterpretations presented in the text.

During the preparation of the manuscript I required help from several eminent research scientists and my friend students. None of them hesitated to provide me the necessary literature. The Director and Scientists of the Central Potato Research Institute, Shimla, Dr. Bellum V.S. Reddy and Dr. H.C. Sharma of the International Crops Research Institute for Semi-Arid Tropics (ICRISAT), Hyderabad, Dr. K.R.M. Swamy of the Indian Institute of Horticultural Research, Bangalore and Dr. (Mrs.) S. Sardana from the National Bureau of Plant Genetic Resources, New Delhi have very kindly and in time supplied me the literature that I required. I am obliged to them. Dr. H.C. Sharma from ICRISAT permitted me to utilize data and photographs from the literature supplied. I sincerely acknowledge this kindness. Dr. K.P. Singh, of the Indian Institute of Vegetable Research, Varanasi, shared some information about vegetable breeding. Dr. Swamy of IIHR, Bangalore was another scientist to educate me about vegetables.

I am thankful to Dr. A.N. Mishra from the Regional Station, Indian Agricultural Research Institute, Indore and Dr. A.N. Sharma of the National Research Centre for Soybean, Indore for their valuable suggestions while preparing chapters on disease and insect resistance, respectively.

My sincere acknowledgements to my students and colleagues Dr. (Mrs.) Usha Saxena and Dr. Mukesh Saxena are an essential requirement. Both of them provided me necessary help during preparation of the manuscript. I should not fail to comment about the library of College of Agriculture, Indore. It is full of treasure of very early literature on plant breeding and genetics. This has also been the source of knowledge that helped me while drafting this manuscript. I should not fail to thank Dr. (Mrs.) Mrudula Billore, for freely providing me the literature on plant breeding.

Mr. R.C. Jain "Badnasib" typed the manuscript very meticulously. I am responsible for any typographical errors, if at all they are still found in the text.

Lastly, I acknowledge very sincerely encouragement always extended to me by my wife, Mrs. Kalpana Dabholkar. For last so many months she tolerated my occupancy with reading, taking notes and writing the manuscript. She never expressed displeasure about being neglected by me, although unknowingly. This is also an important factor for refraining from household responsibilities and remaining busy with this creation.

I hope this work will be useful to the students of plant breeding and will not be only 'Swant Sukhaya' (own's pleasure)!

Indore
July, 22, 2004

A.R. Dabholkar

Contents

List of Tables

List of Figures

1

Plant Breeding — Scope and Achievements

Various kinds of plants that we observe today in different parts of the world are nature's precious gift to the mankind. The food material that a man predominantly consumes is either a plant material or derived from plant material. No doubt, plants are the major source of food, feed, clothing, fuel, shelter and drugs. Fruits and vegetables are nourishing and protective foods. It has been recognized that fruits are a rich source of antioxidants, which protect us from dreadful diseases. Vegetables and fruits contain vitamin *E*, *K* and β carotene in addition to minerals like zinc, copper and manganese. Considering the usefulness of plants, it is not surprising that the man has long been concerned with developing superior forms of the plants useful to him. Improvement of crop plants is an ancient accomplishment of man.

Plant breeding, it can be said, began when the primitive man selected useful plants from vegetation growing around him. Early plant breeding was a very slow process. It remained an art only, based on skill and eye judgement for the choice of superior plant forms. The early plant breeders were not aware about the principles of inheritance of various trait differences found among the plants. It was only after the rediscovery of Mendel's laws of inheritance and further developments in the field of genetics that plant breeding was placed on sound scientific footings.

Genetics is the science that is primarily concerned with the precise understanding of the biological properties that are transmitted from parents to offspring. This science provides information about the manner in which a given character is inherited. It is genetics which provides understanding about the physical basis of heredity, the genetic material that governs inheritance of living organisms, organization and mode of action of genetic material, linkage relationships, recombination, mutational effect, polyploidy, regulation of action of genetic material and so many interesting facts concerned with the heredity.

Plant breeding is the art and science of improving genetic pattern of plants in relation to their economic use (Smith, 1967). Present day plant breeding is more of a science than an art. However, the choice of parents for hybridization, selection of superior plants from segregating population and the process of hybridization itself requires keen observation of plants and skill. Hence, the role of art in plant breeding cannot be entirely denied.

Modern science of plant breeding rests heavily upon the knowledge provided by sciences such as agronomy, plant physiology, horticulture, entomology, plant pathology, microbiology, statistics, plant biochemistry, food technology, genetics and cytogenetics, molecular genetics, plant genetic engineering, etc.

It is primarily plant breeding which has been instrumental for maximization of productivity of various crop plants. Genetic potential for high yield coupled with quality attributes of commercial products, resistance to various biotic and abiotic stresses could be incorporated in new varieties via plant breeding techniques.

1.1 AUGMENTATION OF PRODUCTION OF VARIOUS CROP SPECIES

Reverend Thomas Robert Malthus postulated in 1798 that unless human populations are checked by the wars and/or disasters, they will continue to increase until hunger takes control. In a way Malthus was right because there is lack of correspondence between increase in agricultural production and rate of increase of human population. The human population follows geometric progression for multiplication while the agricultural production is increased with great difficulty.

Plant breeding has averted the doomsday predicted by Malthus, but some believe that the day has only been postpone. The reason for such thinking is not baseless. There is no control over the population advancement. A plant breeder is, therefore, endowed with the responsibility of evolving, through the techniques of plant breeding, higher and still higher yielding varieties of various crop species. He has to work ceaselessly for enhancing the productivity of varieties of various crops.

Production of food grains, oilseeds, vegetables and fruits could be increased by enhancing area under cultivation. But there is little chance to do so in India because most of the land is already under cultivation. The per capita availability of land and water is going down

(Swaminathan, 1995). Many scientists have called attention to the potential food supplies in the ocean, in the form of both animal and plant life. But plant breeding has the potential to enhance the productivity of various crop species. Along with plant genetic engineering it may not be difficult for plant breeder to attain a still another quantum jump in productivity.

In 1950, a few years after the independence of India, the total food grain production was 50 million metric tonnes. The population was about 35 crores. The health organizations recommend consumption of at least 400 g food grains per capita per day. In other words, the food grains produced fell tremendously short to feed the population then inhabiting India. Vegetable production was merely 15 million tonnes. This quantity was also insufficient to meet the standard dietary recommendation of about 300 g per capita per day for a balanced diet (Singh and Kalloo, 2000). Average productivity of most of the cereals, like wheat and rice, was merely 663 and 668 kilogrammes per hectare, respectively in 1950. Food grains were regularly imported from other countries to meet the demand of population and consequently India was regarded as a "begging bowl" by other countries (Swaminathan, 1996). Efforts to enhance the productivity and production of wheat and rice brought home the fact that the indigenous tall, late maturing varieties of these crops are unable to efficiently make use of higher doses of plant nutrients applied, especially the nitrogen and irrigation levels. Wheat crop, which received more than 40 kilogrammes of nitrogen per hectare, coupled with 4 to 5 irrigations, lodged. Lodging tends to reduce yield. Rice varieties were also unable to respond favourably to higher doses of nitrogen and irrigations. Productivity of indigenous varieties of sorghum, pearl millet, maize, pulses, oilseeds and cotton was also very low (Table 1.1).

Till independence, India was a major cotton producer with surplus for export. But after the partition in 1947, most of the promising irrigated area of cotton was ceded to Pakistan and hence was left with 60 per cent area, which was mostly rainfed (Kairon *et al.*, 2000). Average productivity during 1950 was only 88 kilogrammes of lint per hectare.

The varieties of vegetables were poor yielder and responded poorly to higher doses of nutrients. Moreover, many varieties of most of the vegetable crops were season bound. Hence, these vegetables were available in certain seasons. Shelflife of many vegetables was short and many perished during long distance transportation.

Table 1.1: Productivity of some crops in India (*kg/ha*)

Crop	*1951*	*1995*
Wheat	663	2493
Rice	668	1855
Maize	538	1570
Sorghum	353	1014
Pearl millet	300	707
Gram	510	810
Cotton (lint)	88	266
Potato	6590	18526*

*(1999)

Dwarf, stiff strawed, disease resistant, fertilizer responsive varieties of wheat were bred at CIMMYT (Centro Internacional de Mejora Miento de Maize y Trigo), Mexico, under the leadership of Nobel laurate Dr. N.E. Borlaug, during sixties. These wheat varieties possessed genes, which reduced the height of plants (by shortening the length of internodes), *viz.*, rht_1 and rht_2. These genes are known as 'Norin'. Some varieties like Sonora 63, Sonora 64, Lerma rojo, PV 18, were evaluated in 1964 and 1965 on farmers fields, on large scale in India. Their performance was excellent under high fertility levels and adequate irrigations. In 1967, about 18,000 tonnes seed of these varieties was imported to India and distributed to farmers. Farmers accepted the package of practices recommended for cultivation of these dwarf wheat varieties. Consequently, there was a rise in productivity and production of wheat. This was the harbinger of green revolution in India (The term 'Green Revolution' was coined by Dr. William Gaud of United States Department of Agriculture, in 1968). Indian plant breeders evolved dwarf, stiff strawed, fertilizer responsive and disease resistant varieties of wheat in subsequent years. Wheat productivity and production in India has recorded big leaps in following years. In 1965, 12 million metric tonnes of wheat was produced from 14 million hectares at the rate of about 865 kilogrammes per hectare. But in 1998, 70.78 million metric tonnes of wheat was produced from just 27.40 million hectares. At the 1965 rate of productivity, about 82 million hectares of land would have been required to produce 70.78 million tonnes of wheat. Because of high yielding varieties and new production technology, 55 million hectares of land was saved. This technology for maximizing the production, without utilizing more land, was known as "Land Saving Agriculture"

Fig. 1.1: High yielding variety of wheat
(*Courtesy* : Dr. P.K. Verma, I.A.R.I., Indore)

(Swaminathan, 1990). It was primarily plant breeding which was instrumental for land saving.

Around the same time when dwarf Mexican wheat varieties were introduced in India, dwarf, non-photosensitive, erect leaved, fertilizer responsive varieties of rice were introduced from the International Rice Research Institute (IRRI), Philippines. These varieties, viz., IR 8, Taichung Native 1, etc., possessed "Dee Geo Woo Gen" genes responsible for shortening of the internodes of rice plants. Dwarf, fertilizer responsive varieties were later on bred in India also. Due to the adoption of new varieties along with cultivation practices recommended for their production, the production and productivity of rice exhibited strides. Productivity of rice and wheat at the national level is currently 1905 and 2700 kilogrammes per hectare (1998-99), respectively.

In 1961, four maize hybrids, viz., Ganga 1, Ganga 101, Deccan and Ranjit, were developed in India and released for cultivation. Sorghum hybrid *CSH* 1 was released in 1964 and pearl millet hybrid *HB* 1 in 1965. The sorghum and pearl millet breeders utilized cytoplasmic-genetic male-sterility (CMS) for evolving hybrid varieties. As a result of the availability of high yielding varieties (*HYVs*) the food grain production in 1998 was about 200 million metric tonnes.

The first manually produced hybrid of cotton, viz., Hybrid 4, was bred in India and released in 1970. Several hybrid varieties were subsequently bred. The average productivity of cotton is presently 310 kilogrammes per hectare (Kairon *et al.*, 2000).

Recently, rice hybrids, based on cytoplasmic-genetic male sterile, *indica* types, viz., IR-58025A and IR-62829A, have been developed and released for general cultivation. These have yield potential of about 6 to 8 tonnes per hectare. Hybrid varieties of sunflower (such as *BSH*-1, *APSH* 11, *LDMRSH* 3), pigeonpea (*e.g. ICPH* 8) were also released. Hybrid varieties of castor are also available to Indian farmers.

Despite severe droughts in India, after sixties, no death has been reported due to famines. Between 1870 and 1900, famines claimed a toll of about 30 million lives.

In the earlier days the per capita per day availability of vegetables was 95 grammes. The low availability of vegetables was obviously due to low productivity. Relatively less attention was given in India to horticulture in the first six five year plans (Reddy, 2000). All India Coordinated Vegetable Improvement Project (*AICVIP*) was established by the Indian Council of Agricultural Research, (I.C.A.R.), New Delhi in 1970-71, which was upgraded in 1987 to the Project Directorate of Vegetable Research. The Indian Institute of Horticultural Research (IIHR) was established at Bangalore in 1967. Presently there are 10 National Research Centres (*NRCs*) and 16 All India Coordinated Research Projects engaged in horticulture research. More than 70 types of vegetables are grown on a national level in the country. Maximum emphasis is on vegetables like tomatoes, brinjal, chillies, cauliflower, cabbage and peas. Not much effort is made to enlarge the area under minor and exotic vegetables, even though they hold the key to export markets.

Before the launching of research projects for improvement of vegetable crops, introductions were made from various countries. For example, varieties Arkel and Bonneville of garden pea were introduced from United Kingdom and United States of America, respectively. Variety Sioux of tomato (from USA), Sugarbaby, a variety of watermelon, and Early Grano of onion from USA, variety Pusa Chetki (improvement from material introduced from Denmark) of radish have become popular in India. Late Dr. H.B. Singh of the Plant Introduction Division deserves credit for his contribution to the wealth of vegetable varieties such as Pusa Sawani (Okra), Pusa Purple Long, Pusa Purple Round and Pusa Cluster of brinjals. Production of vegetables and also fruits has increased in India after independence (Table 1.2).

Table 1.2: Production of fruits and vegetables in India

(*Lakh tonnes*)*

Year	*Fruits*	*Vegetables*
1984-85	237.60	454.00
1989-90	268.00	486.00
1990-91	282.00	538.00
1991-92	286.31	583.30
1992-93	329.60	710.00
1993-94	394.70	650.90
1994-95	386.00	672.90
1995-96	415.00	715.90
1996-97	469.70	808.00

**Source*: Indian Agriculture, 1999.
— Indian Economic Data Research Centre, New Delhi. Fertilizer Statistics, 1999-2000.
— Fertilizer Association of India, New Delhi.

The real impact of vegetable research on enhanced productivity was seen with the release of first tomato hybrid "Karnataka" in 1973 and first capsicum hybrid 'Bharat' by the Indo-American Hybrid Seed Company, based at Bangalore (Kalloo, 1995). Several hybrids of tomato, cabbage, cauliflower, capsicum, okra, chilli, brinjal (eggplant), gourd-melons and cucumbers have been developed and released for cultivation in subsequent years by both public and private sector research organizations. Hybrids of vegetable crops which carry heat tolerance, for example of cabbage, have been developed which enabled off-season cultivation of these vegetables. Watermelons, which were earlier available only during hot summers in India, are now available in winter months of November, December. Cauliflower is raised during March to May in north-western plains of the country enabling availability in hot seasons (Attavar, 2000). About 30 per cent area under tomato and cabbage in India is covered under hybrid varieties. There is a steady increase in coverage under hybrids of brinjal, okra and chillies. As a result, the production of vegetables is about 90.8 million tonnes (Negi, 2002) while it was only 20 million tonnes during 1947 (Kalloo and Pandey, 2002). In cauliflower, onion and cabbage production, India now stands first, second and third, receptively in the world (Table 1.3). Productivity of tomato is about 17.7 tonnes per hectare, on average. This progress could be attributed largely to plant breeding.

Fig. 1.2: Tomato hybrid (*Courtesy*: Indo-American Hybrid Seeds (India) Pvt. Ltd., Bangalore)

1.2 GOALS OF PLANT BREEDING

Basically, most if not all, the breeding programmes aim at evolving high yielding varieties of the crop plants they are working with. Because of the ever increasing population there is a pressure on agriculture to enhance production. Except during the periods of plenty human populations have suffered from hunger. Each period of plenty brought a jump in population, followed by hunger and its after effects, *viz.*, pestilence, high infant mortality and sickly lives. Hence, the plant breeder is ceaselessly engaged in breeding of high yielding varieties. However, yield alone is not the plant attribute for which breeding efforts are required. There are several characters, which demand attention of the plant breeder.

Table 1.3: Average productivity of some vegetable crops

(*quintals per hectare*)*

Vegetable	*Average yield in*	
	World	*India*
Brinjal	Not available	130.80
Cabbage	216.20	156.80
Cauliflower	136.24	147.00
Onion	137.96	141.90
Chilli (green)	82.92	8.80
Okra	Not available	85.00
Green peas	60.83	72.90
Tomato	250.96	156.00

**Source*: Kalloo and Pandey (2002)

The breeding methods employed for genetic improvement of vegetable crops differ in no way from those used for cereals, oilseeds, pulses and fibre crops. Nevertheless, certain peculiarities of vegetable crops need specific attention while planning the improvement programme. In certain vegetable crops the edible part that is consumed is not the seed but other parts as in case of cauliflower, cabbage, radish, carrot, onion. Hence, seed production of these crops requires different methods. Some vegetables are consumed both as fresh or processed products. Consequently, while breeding, quality requirement in both the cases is a separate breeding objective. This certainly is the case in other crops also like wheat. Wheat quality for bread, biscuits, semolina, suji, *etc.* is different. Even the species are different. Most important aspect is perishable nature of vegetable crops, suitability for long transportation and the shelflife. It may be pointed out that biotechnological tools are now available, which shall enable the breeders in future to breed varieties of tomato, which ripen late and stand long transportation. Bruise resistant tomatoes have been developed which express antisense RNA against polygalacturonase (PG), which attacks pectin in cell walls of ripening fruits and softens the skin. These are called 'Flavr Savr'. These tomatoes stay long on plant giving more time for accumulation of sugars and acids for improving flavour (Radhakrishnan, 2002).

High Yield

High yielding varieties have been bred in India in the past few years.

High yielding varieties could be evolved by breeding for yield *per se*. Improving physiological efficiency can result in evolution of high yielding varieties. Most of the present day varieties of several crops owe their high yielding ability to superior harvest index (Jain, 1993). Harvest index is the ratio of economic yield to the total biological yield produced by a plant. In case of wheat, tomato and cotton, economic yield is represented by kernel yield, yield of fruits and yield of lint, respectively. Efficient plants partition relatively more dry matter to the economic part of plants, while the inefficient varieties are unable to do so. For instance, the indigenous varieties of sorghum produce tall plants and accumulate about 350 to 400 g dry matter per plant in 140 to 150 days. But these are not physiologically efficient because only 30 per cent dry matter is found in grains, the economic yield. On the other hand, newly bred hybrids partition about 50 per cent dry matter towards grains, *i.e.* their harvest index is 50 per cent (Rao and Venkateswarlu, 1971).

High yielding types could be evolved by exploiting the phenomenon of hybrid vigour. Hybrids of sorghum, maize, pearl millet, rice, cotton, castor, tomato, sunflower, several vegetable crops yield significantly more than earlier grown varieties of these crop species. For instance, with average management of tomato hybrids, 50 to 60 tonnes of yield could be achieved per hectare while the varieties yield around 20 to 25 tonnes per hectare (Chhabra, 1992).

High yield could be obtained by breeding varieties, which respond favourably to the higher doses of nutrients applied. For instance, indigenous varieties of wheat were unable to respond to application of nitrogen beyond 80 kilogrammes per hectare. On the other hand, the recently bred dwarf, semidwarf varieties can utilize efficiently upto 120 kilogrammes of nitrogen per hectare.

Breeding for Quality and Nutrition

Besides yield, consumer acceptance also determines the success of a variety. The consumer preferences may be varied. For example, tomato may be consumed as vegetable or may be used for manufacture of sauce, ketchup, puree, paste, etc. Similarly, potato may be consumed as vegetable or used by industry for manufacturing chips, French fries, *etc*. The ultimate quality of either tomato or potato will differ according to its utilization. For example, for preparation of chips those potato varieties are preferred, which produce tubers with high specific gravity and lower sugar contents, especially the reducing sugars *viz*., glucose, fructose, etc. Chips containing larger amount of sugars turn brown and black upon exposure to air and hence potato tubers containing high sugars are unacceptable (Talburt and Smith, 1987).

Uniformly red coloured, soft and juicy tomatoes are usually preferred for fresh consumption. But tomatoes with firm fruits are desirable for long distance transportation. For processing purposes, the quality of tomato is determined by attributes such as high total soluble solids (4.5 to 7%) and viscosity. But those fruits are desirable which posses fewer locules, have more thicker pericarp, increased firmness and resistant to cracking.

A breeder is equally concerned about the nutritional aspects of the plant products. The main source of food in developing countries of the world is cereals, which provide 45 to 89 per cent of the total calories and 50 to 80 per cent of the total protein requirements. Animal protein is nutritionally well balanced but is costly. On the other hand, plant proteins are cheaper to produce and easier to transport and store. Proteins are made up of about 20 amino acids. For a man, eight amino acids are considered essential in a well balanced diet. These are valine, methionine, leucine, isoleucine, threonine, phenylalanine, lysine and tryptophan. The cereals are generally deficient in lysine, tryptophan and/or threonine, while pulses are deficient in sulphur containing amino acids (Bansal and Mehta, 1989). Hence, plant breeder aims at increasing protein contents and improving amino acid pattern.

Chapatti making quality of wheat flour, cooking quality of rice are the attributes, which are priority items while making improvements in wheat and rice, respectively.

Elimination of Undesirable Substances

Economically important produce of several crop species contains undesirable substances. Some times these substances are toxic to man or animal after consumption. Hence, during breeding for high yield, or even otherwise exclusive efforts may have to be made to reduce or to eliminate these substances from the economic product.

Lathyrus sativus, for example, contains in their grains a neurotoxin, BOAA (β-N-oxalyl-amino-L-alanine). Prolonged consumption of seeds of lathyrus may lead to paralysis of lower limbs of man due to this neurotoxin. Indigenous varieties of sorghum contain in their leaves and stalks cyanogenic glucoside, an alkaloid also called as dhurrin. Consumption of such a fodder is toxic to ruminants.

Rapeseed and mustard contain very high amounts (45-60%) of erucic acid in its fatty acid profile. The oil also contains high glucosinolates. Glucosinolates in the seed meal or cake break into toxic sulphur compounds in the intestines of farm animals and impair their growth and reproductive ability. Breeding efforts have been successful in reducing such substances.

Disease and Insect Resistance

It is estimated that more than 50 per cent of the world's crop production is lost each year through the activities of pests and diseases of plants (Russell, 1978). Genetic yield potential of a variety is diluted if it is susceptible to a disease or an insect pest. Biffen (1905) showed that resistance to disease is an inherited attribute of plants. Interaction between genes for resistance in plants and genes for virulence/avirulence in pathogen determines the resistance reaction of plants (Flor, 1955, 1956). Resistance against insect pests, plant parasites are also heritable in nature. Breeding programmes have been successful to incorporate genes for resistance in plants against insect pests or fungal pathogen, plant parasites and bacteria. Varieties resistant against viral infections have also been evolved.

Yellow Vein Mosaic (YVM) is most serious disease of okra in tropics and subtropics. Imported hybrids of okra were not successful in India because they did not carry tolerance to yellow vein mosaic virus. Hybrids of okra, Varsha, Vijay and Vishal were first virus tolerant hybrids developed in India. Parbhani Kranti, a variety of okra resistant to YVM was bred at Marathwada Agricultural University, Parbhani. Punjab Padmini was evolved by Punjab Agricultural University using *Abelmoschus manihot* spp. *Manihot*, a cultivar from African continent.

Black rot is a serious disease of both cabbage and cauliflower. Variety Pusa Shubhra of cauliflower has been bred which possesses resistance against this disease.

Painter (1951) proposed three mechanisms for resistance of plants to insect attack, *viz.*, non-preference, antibiosis and tolerance. These are heritable traits of plants. Rice varieties possessing resistance against brown plant hopper (*Nilaparvatha lugens*) and gall midge (*Orseolia oryza*), have been evolved. Surakasha and Mansarovar are such varieties (Rao *et al.*, 1995).

Sorghum variety SAR-1 was evolved at the International Crops Research Institute for Semi-Arid Tropics (ICRISAT), Hyderabad, which possesses resistance to a semi-root parasite *Striga asiatica*.

Resistance or Tolerance to Abiotic Stress

Insects and diseases caused by fungi, bacteria, virus, and plant parasites comprise of biotic stress to the plant. Plants may experience abiotic stress due to several factors. Consequently, yield is considerably reduced. Excess accumulation of salts in the soil solution, such as sodium carbonate, sodium bicarbonate, magnesium chloride etc., in toxic concentrations, is detrimental for normal plant growth and performance. In extreme cases even the germination is adversely affected. Reclamation of such soils is possible through addition of certain amendments and irrigation. But this is costly and not a permanent or long lasting solution. Varieties could be bred which tolerate certain concentrations of these accumulated salts in soil solution. Varieties of rice such as CSR 10, USAR 1, USAR 2 and Narendra have been bred which can tolerate salt stress. These have 20 to 80 per cent yield advantage in stress conditions over the traditional varieties (Siddiq, 2000). Wheat variety KRL 1-4, suitable for saline soil conditions has been evolved at the Central Soil Salinity Research Institute, Karnal (Singh, 1995).

About 70 per cent agriculture in India is rainfed. The amount, distribution and the intensity of rainfall is unpredictable. Consequently rainfed crops experience moisture stress at various stages of their growth and development. Productivity is severely influenced by such stress. Varieties have to be bred which can tolerate water stress at critical stages of their growth.

Besides the traits mentioned in earlier paragraphs for which breeders seek improvements in plants, there are several characters, which require their attention. Plant height is one character, which has been corrected in the past. For example, the tropical sorghum varieties introduced in United States of America had to be corrected for photoperiod and thermosensitivity in order to make the varieties suitable for this country. Several varieties did not even flower because of differences in photoperiod. Non-photosensitive varieties were developed by backcross method of breeding. Since the varieties were tall and not suitable for combine harvesting medium to dwarf forms were developed.

The indigenous varieties of most of the cereal crops were adapted to specific regions. When shifted to other geographical area, their performance was not impressive. Widely adapted varieties were evolved so that they could be grown over a wider geographical region.

Shattering of pods is observed in some crops like mung, urd, soybean, etc. Shattering results in considerable loss of yield. Breeding programmes have successfully incorporated genes for non-shattering/ less shattering in such crops. Synchronous flowering is also one attribute that needs breeder's attention.

Vegetable crops are mostly season bound. Increase in production of vegetables could be achieved by developing varieties, which are not very strongly season bound. Varieties have been developed in some crops, which could be successfully cultivated in off season. Bruise resistant and varieties with delayed ripening are desirable attributes to enhance the transportability and shelflife of some vegetables.

Improvement for any desirable character of a crop plant by conventional methods of plant breeding is restricted and has sexual barriers. Intervarietal, inter- and intra-specific hybridization has been exploited successfully for improvement. In a few cases inter-generic crosses have also led to fruitful results. But through the techniques developed in plant genetic engineering, such as recombinant DNA technology, transgenic plants have been developed, which have shattered sexual barriers for gene exchange. A desirable gene could be integrated into a plant from any microorganism, unrelated plant species and animal. Very recently transgenic varieties of cotton, potato and maize have been developed in United States of America, which carry a gene, called Bt gene, from a bacteria, *Bacillus thuringiensis*. This gene is responsible for production of a crystalline protein, which is toxic to lepidopterous insect pests, such as bollworm of cotton. Transgenic cotton varieties possessing Bt gene are resistant to bollworm.

Breeding of superior forms of fruits has lagged behind. This could be attributed to a few factors like, (i) more time requirement, (ii) laborious task, (iii) mode of inheritance of several fruit crops is not yet perfectly known, (iv) varying degrees of self-incompatibility mechanism is observed among fruit crops (v) long juvenile phase, (vi) long time is required to identify promising hybrid combinations (Majumdar, 1989). Despite these hurdles, breeders have made considerable contribution to fruit crops. Regular bearing habit has been incorporated in mango through breeding. For example, Mallika, Amrapalli, Ratna and Sindhu are recently bred regular bearing mango varieties. In Punjab, "Kinnow", a hybrid between King and Willow leaf has replaced the traditional citrus fruits, such as sweet orange and local mandarin. High yielding varieties of papaya, e.g., Honey Dew, Pusa Dwarf, Pusa Giant, Pusa Delicious; hybrids of grape such as Arka Hans, Arka Kanchan, Arka Shyam have been released for cultivation. Hybrids of apple, Lal Ambri (Red delicious × Ambri) and Sunehari (Ambri × Golden delicious) have also been evolved and are available for cultivation. Ambri is a local type of apple grown in Jammu and Kashmir.

Plant breeding has also played significant role in the evolution of novel varieties of gladiolus, roses, chrysanthemum, etc.

These few examples are adequate to convince about the usefulness of plant breeding in the improvement of crop plants. This science is an effective tool to overcome the danger of food deficit, if executed properly.

References

Allard, R.W. (1960). *Principles of Plant Breeding*. John Wiley & Sons, Inc. New York, p. 485.

Attawar, M.M. (2000). Tool for productivity gains. In *The Hindu-survey of Indian Agriculture*, Kasturi and Sons Ltd., Chennai, pp. 145-49.

Bansal, H.C. and S.L. Mehta (1989). Breeding for protein quantity and quality. In *Plant Breeding—Theory and Practice*. Ed. Chopra, V.L. Oxford and IBH Publ. Co., New Delhi, pp. 309-32.

Bhatia, C.R. (1996). Biotechnology—Many application areas. In *The Hindu-survey of Indian Agriculture*, Kasturi and Sons Ltd., Chennai, pp. 173-75.

Biffen, R.H. (1905). Mendel's law of inheritance and wheat breeding. *Jour. Agric.Sci.*, 1: 4-48.

Briggs, F.N. and P.F. Knowles (1967). *Introduction to plant breeding*. Reinhold Publishing Corp., New York, p. 426.

Chhabra, N. (1992). The tomato. *Indian Horticulture*. April-June, 1992, pp. 11-20.

Chopra, V.L. (1995). Biotechnology. Potential for crop improvement. In *The Hindu-survey of Indian Agriculture*. Kasturi and Sons Ltd., Chennai, pp. 171-73.

Flor, H.H. (1955). Host-parasite interaction in flax rust—its genetics and other implications. *Phytopath*, 45: 680-85.

Flor, H.H. (1956). The complementary genic systems in flax and flax rust. *Adv.Genet*, 8: 29-54.

Jain, H.K. (1993). Plant genetic resources and policy implications for a changing agriculture. *Indian J. Genet.*, 53(3): 223-237.

Kairon, M.S.; P. Ramasundaram and M.V. Venugopalan (2000). Agenda for new millennium. In *The Hindu-Survey of Indian Agriculture*, Kasturi and Sons Ltd., Chennai, pp. 109-112.

Kalloo, G. (1995). Heterosis and breeding in vegetable crops: Present status and future prospects. In *Hybrid Research and Development* Ed. M. Rai and S. Mauria. Proceedings of Seminar. Indian Society of Seed Technology, New Delhi, pp. 165-172.

Kalloo, G. and A.K. Pandey (2002). Commendable progress in research. In *The Hindu-Survey of Indian Agriculture,* Kasturi and Sons Ltd., Chennai, pp. 159-163.

Negi, J.P. (2002). Preventing a near collapse. In *The Hindu-Survey of Indian Agriculture*. Publ. Kasturi and Sons Ltd., Chennai, 600-002 pp. 175-178.

Painter, R.H. (1951). *Insect resistance in crop plants.* Publ. McMillan, New York.

Reddy, P.P. (2000). Rich export potential. In *The Hindu Survey of Indian Agriculture*. Publ. Kasturi and Sons Ltd., Chennai, pp. 141-144.

Ram, H.H. (1998). *Vegetable breeding: Principles and Practices.* Kalyani Publishers, Ludhiana.

Rao, N.G.P. and J. Venkateswarlu (1971). Genetic analysis of some exotic x Indian crosses in sorghum. III. Heterosis in relation to dry matter production and nutrient uptake. *Indian J.Genet*. 31: 156-176.

Rao, U.P.; M.V.S. Sastry; J.R.K. Rao, G.S.V. Prasad and M.B. Kalode (1995). Breeding varieties resistant to major insect pests in rice. In "Genetic Research and Education: Current Trends in Next Fifty Years": Ed. B. Sharma *et al.*, *Indian Soc. Genetics and Plant Breeding*, New Delhi, pp. 223-227.

Radhakrishnan, G. (2002). Genetically modified plants from understanding to applications. In proceedings of National Seed Seminar on Quality Seed to Enhance Agricultural Profitability. Editors Shekhargouda, M. and L. Krishna Naik, organized by Indian Society of Seed Technology, New Delhi and University of Agricultural Sciences, Dharwad, pp. 34-46.

Russell, G.E. (1978). *Plant breeding for pest and disease resistance*. Publ. Butterworth, London.

Singh, H.P. (2000). Harnessing the potential of horticulture. *Fertilizer News*, 45(2): 41-46.

Singh, K.N. (1995). Recent approaches to breeding for salt tolerance in crop plants. In "Genetic Research and Education: Currrent Trends in Next Fifty Years" Ed. Sharma, B. *et al. Indian Society of Genetics and Plant Breeding*, New Delhi, pp. 490-499.

Singh, K.P. and G. Kalloo (2000). Nutrient management in vegetable crops. *Fertilizer News*, 45(4): 77-81.

Smith, D.C. (1967). Plant breeding—development and success. In *Plant breeding* Ed. Frey, K.J. Iowa State University Press, Ames, Iowa, pp. 3-54.

Swaminathan, M.S. (1990). Making agricultural progress sustainable—role of new technologies. In "Agricultural Development Policy Adjustments and Reorientation". *Indian Soc. Agric. Eco.* Oxford and IBH Publishing Co., New Delhi, pp. 108-132.

Swaminathan, M.S. (1995). Rationale of integrated systems. In *The Hindu—Survey of Indian Agriculture,* Publ. Kasturi and Sons, Chennai, pp. 7-15.

Swaminathan, M.S. (1996). Hunger free area programme. In *The Hindu—Survey of Indian Agriculture.* Publ. Kasturi and Sons, Chennai, pp. 9-15.

Talburt, W.F. and O. Smith (1987). *Potato processing.* 4th edition. AVI Publishing Co., West Port Conn. U.S.A.

2

Reproductive Systems in Crop Plants

The genetic composition of a population of a crop species depends, primarily, upon reproductive system. The choice of an efficient breeding programme, employed for improvement of a given character, also depends upon the reproductive system of the crop species. The precise steps to be executed depend upon the floral morphology, flowering habits and floral biology. Breeding methodology and handling of segregating material is also determined by the extent of natural cross-pollination. Although alfalfa and maize are both cross-pollinated, both selfing and crossing is easy in maize because of its floral morphology. A tobacco plant when subjected to crossing, yields numerous crossed seeds from single emasculation and pollination. On the other hand, in wheat one emasculated floret yields only one seed upon pollination by an appropriate male plant.

The crop plants of horticultural and agricultural value, exhibit two predominant kinds of reproductive mechanisms, *viz*., asexual and sexual. In asexual reproduction, union of sexual gametes is not involved. Instead the progeny plants develop from vegetative parts of the plant (vegetative propagation) or may develop from unfertilized gametes, cells from various parts of the ovule and modified cells (apomixis). Vegetative propagation includes various modes of reproduction, e.g., rhizomes (e.g. ginger, *Zingiber officinale*), bulbs (e.g. onion, *Allium cepa*), corms (e.g. gladiolus, *Gladiolus grandiflorus*), tubers (e.g. potato, *Solanum tuberosum*), runners or stolens (e.g. strawberry, *Frageria* sp.), bulbils (e.g. lilies, *Lilium* sp.), suckers (e.g. banana, *Musa acuminata*), root cuttings (e.g. poplar, *Populus alba*), stem cuttings (e.g. sugarcane, *Saccharum officinarum*) and leaves (e.g. bryophyllum, *Bryophyllum pinnatum*). Because the offsprings produced by any of these vegetative means do not involve nuclear fusion, but result from mitotic cell division, progeny of an individual is genetically identical. These are known as clones. A clone is a group of individuals descended by mitosis from a common ancestor. All individuals descending in this way from an individual plant are identical genetically with the mother plant.

Sexually reproducing plants can be categorized into two broad groups, viz., predominantly self-pollinating and cross-pollinating.

General steps involved in the execution of any breeding scheme are influenced by reproductive system. The kind of pollination system that prevails in a crop species is very much important for decision taking. In recent days crossing between deliberately selected plants of two (or more) varieties is a regular feature of breeding programme to generate more genetic variability. In cross-pollinated crops, population improvement schemes, which involve cyclical selection and crossing of selected plants, is practiced to increase the frequency of superior genes and maintain genetic variability. The ease with which controlled crosses could be executed and also selfed enables employment of various breeding schemes.

Occurrence of male-sterility permits application of population improvement schemes in certain self and often cross-pollinated crops. Hybrid seed production has become much easier on commercial scale, in certain crops, where male-sterility or self-incompatibility occurs.

There is another asexual reproductive process, known as apomixis in plants. Apomixis is the substitution for sexual reproduction of another asexual reproductive process that does not involve nuclear or cellular fusion (Stebbins, 1941). In this mode of reproduction embryo develops without fertilization. Consequently, the progeny is genetically identical with mother plant.

In apomictic species, sexual reproduction is either entirely absent or suppressed. When sexual reproduction also occurs, the apomixis is classified as facultative. But when apomixis is the only method of reproduction apomixis is known as obligate. Apomixis is reported in mango, citrus, cotton, maize, wheat, tobacco, kentuky-blue grass, dallies grass (*Paspalum* sp.). Apomixis was reported in sorghum also (Rao and Narayana, 1968).

In vegetative apomixis, sexual reproduction is substituted by vegetative reproduction. For example, small seed like bulbils replace seeds in the flowering panicles of bulbous blue grass (*Poa bulbosa*) and onion. Carmils in gladiolus also represent vegetative apomixis.

In agamospermy seeds are produced. This form of apomixis is sub-divided into three groups (Hays *et al.*, 1955), viz., adventitious embryony, agamogony and non-recurrent apomixis. Origin of embryos from some cells in the nucellus (Figure 2.1) or adjacent tissues involves adventitious embryony. It is a type of sporophytic budding. Adventitious embryony is commonly observed in oranges and roses. Parthenogenesis is the development of a sporophyte from female gamete without syngamy. Such an individual could be haploid or diploid. This form of reproduction is observed in maize, wheat and tobacco. Agamogony refers to the kind of apomixis in which the daughter sporophyte develops from diploid gametophyte. Non-recurrent apomixis involves development of a haploid sporophyte from haploid gametophyte via normal meiosis. Such haploid plants are being produced in maize and converted into homozygous diploid plants, because sectors of diploids are produced. Apogamy is the development of embryo from nucleus other than the egg cell. Commonly apogamy results from fusion of two cells of embryosac, either synergids or antipodals (Poehlman, 1986). Polyembryony is the occurrence of more than one embryo in seed. It is common in mango, citrus and some cassia. In some species large number of embryos may be present, but usually only one or few will produce seedlings. Twin seedlings may occur in many plants and they may be genetically identical or different. This will depend upon the manner of origin.

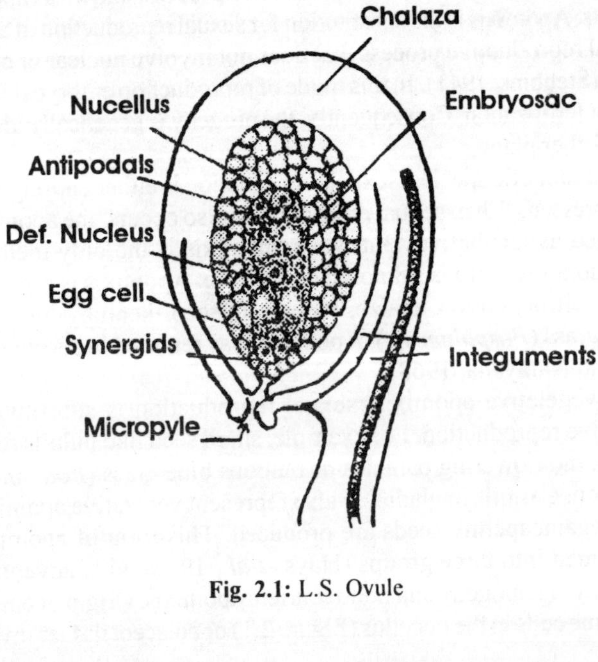

Fig. 2.1: L.S. Ovule

It is important to recognize that apomixis is due to modification of the normal meiotic and fertilization processes. It may be advantageous because apomixis inhibits genetic recombination and segregation. And if proportion of apomictic progeny is high, good phenotypes may remain relatively stable after isolation.

Because asexually propagated plants do not involve nuclear fusion of male and female gametes, as in case of sexual reproduction, there is no opportunity for segregation and recombination in successive clonal generations. This may be an advantage or disadvantage. Asexual reproduction may be beneficial in fruit and ornamental plants where it is desirable to retain flavour (as in fruits), or colour of foliage or flowers (as in many ornamental plants). But disadvantageous because (i) they gradually accumulate naturally occurring mutations leading to increased heterozygosity, (ii) they progressively accumulate deleterious genes, and (iii) the constant disuse of sex leads to progressively higher levels of male and female sterility.

2.1 SEXUAL REPRODUCTION

A typical perfect flower of an angiospermic plant comprises of gynoecium or pistil (which is composed of one or more carpels), androecium (composed of a number of stamens. Each stamen consists of filament, anther and connective), calyx consisting of a number of sepals and corolla consisting of a number of petals.

The flower is said to be complete when all the four whorls, viz., calyx, corolla, androecium and gynoecium are present. Incomplete flowers are those in which any of the four whorls is absent. Flowers possessing both gynoecium and androecium in addition to calyx and corolla are called hermaphrodite. On the other hand, those flowers, which bear only androecium, calyx and corolla, are known as androecious while those carrying only gynoecium along with calyx and corolla are called gynoecious flowers.

The dominant phase of a plant is known as sporophyte because some cells in anthers (microsporocytes) and ovary (megasporocytes) undergo sporogenesis leading to production of pollen grains and embryosac or female gametophyte, respectively. Pollen grains are the male gametophytes or known as microspores.

2.1.1 Microsporogenesis

Formation of pollen grains (microspores) from microsporocyte is a process called "microsporogenesis". One of the cell in anther, the pollen mother cell or microsporocyte, undergoes meiotic cell division and forms four haploid nuclei—a tetrad of microspores (pollen grains). The number of chromosomes in gametes, such as the sperm or an egg, is half as compared to the somatic number of that plant. This is called as the haploid chromosome number. Haploid is a cell having gametic chromosome number. During maturation of pollen, haploid set of chromosomes divides mitotically to produce a generative nucleus and a vegetative (or tube) nucleus. Pollen is usually released at this binucleate stage. The transfer of pollen grains from an anther of a flower to the stigmatic surface of the same or another flower of the same (or sometimes allied) species is called as pollination. Pollination taking place within a single flower (evidently hermaphrodite) or between two flowers (bisexual or unisexual) borne by same plant is self-pollination. On the other hand, pollination taking place between two flowers borne by two separate parent plants of same or allied species is called cross-pollination or allogamy. Self-pollination is also known as autogamy. After the pollen grains alight on the stigma or during the growth of pollen tube (the intine of pollen grows out into a slender tube, called pollen tube), through the stylar tissues downwards towards the ovary, generative nucleus divides by mitosis resulting in production of two sperm nuclei.

2.1.2 Megasporogenesis

One of the cells in the ovule, the megasporocyte, divides meiotically and gives rise to an axial row of four haploid megaspores. These are not the female gametes. Three of the megaspores, usually those away from micropyle (Figure 2.1), disintegrate. Remaining megaspore enlarges to form an embryosac or female gametophyte. Haploid nucleus within the enlarged megaspore undergoes three successive mitotic divisions and results in eight haploid nuclei. These eight nuclei are arranged in two groups of four each at each end of the developing female gametophyte. During the last mitosis cell plates are formed across the spindle in such a fashion that three cells at each end are cut off and one nucleus from each of the two groups moves to the centre. These two polar nuclei fuse together and form definitive nucleus, also called the fusion nucleus. It is the only cell with two nuclei in the embryosac. Thus seven celled and eight nucleate structures are formed within the ovule. Three of the cells at micropylar end form an egg apparatus, comprising of an egg and two synergids. The three cells at opposite end are called antipodals.

2.1.3 Fertilization or Syngamy

After the pollen grain alights on the stigmatic surface, it germinates, in compatible combinations and produces pollen tube. The pollen tube grows down through the style towards ovary. The pollen tube enters the ovule through the micropyle. The sperms, two in number, are then discharged into the embryosac. One of the sperms fertilizes the egg cell and another fuses with the definitive nucleus, thus resulting in double fertilization. The fertilized definitive nucleus now carries three sets of chromosomes, i.e. becomes triploid, while the egg cell after fertilization forms a diploid zygote. The zygote develops into an embryo and the fertilized definitive nucleus into endosperm.

A few other changes take place in ovule after fertilization. The two integuments develop into two seed coats; the outer called the testa and inner as tegmen. Embryo is the only part in seed, which serves to continue the hereditary stream of plant characteristic of species. Endosperm, when present, may exhibit influence traceable to the pollen. The effect of pollen on embryo and endosperm is known as xenia. This phenomenon was noted by Focke in 1881 (Hays *et al.*, 1955). Example of xenia can be seen in maize. The endosperm may be sugary or starchy. Starchy is determined by one dominant gene Su and sugary is recessive to it (su). When a variety with sugary endosperm is pollinated by a variety with starchy endosperm, the endosperm of sugary variety is starchy. In other words, two sugary genes (su su) in endosperm are suppressed in phenotypic effect by the single Su gene. This is xenia. In some cases, other maternal tissues of the fruit are influenced by the pollen, a phenomenon known as metaxenia, described by Swingle in 1926 (Hays *et al.*, 1955). In case of date palm the quality of fruit is influenced by metaxenia (Allard, 1960). Hence, one has to be careful about the pollen parent. Reader should note that date palm is dioecious and hence male and female flowers are present on separate plants. Although female plants are valuable commercially, male plants have to be maintained as pollen source.

2.2 SELF- AND CROSS-POLLINATED CROPS

In general, efficient self-pollination schemes are not difficult to secure as compared to efficient cross-pollination schemes. Genetic differentiation is not involved for promoting self-pollination. Cleistogamy and chasmogamy ensure self-pollination in the crop plants.

It is essential for self-pollination to take place that anthers and stigma of hermaphrodite flower mature at same time. Cleistogamy refers to the situation where the flowers never open and hence pollen grains may only pollinate the stigma of same flower. Cleistogamy is reported for *Comelina bengalensis*, some species of pansy (*Viola*), balsam (*Impatiens*) and wood sorrel (*Oxalis*). In California oat grass the inflorescence is borne between the culm and leaf sheath, and never emerges out. Hence, it exhibits cleistogamy. In case of *Festuca megalura*, the inflorescence emerges through the boot but flowers never open. In case of chickpea anthers dehisce long before the flowers open and hence result in self-pollination (Poehlman and Borthakur, 1969). Cleistogamy has been recorded in mung (*Vigna radiata*), urd (*Vigna mungo*) and chickpea (*Cicer arietinum*) by Ayyar and Balasubrahmanyan (1935), Chandrasekharan and Parthasarathy (1965). Bowden and Neve (1953) reported cleistogamy in Nunaba variety of sorghum.

Chasmogamy is the act of fertilization at the time of flower opening, when pollen is shed and stigma becomes receptive. The flowers open and pollen is available for transport by wind (anemophily) or insects (entomophily), the receptive stigma is pollinated before or just as the flowers open as in lettuce, wheat and barley. In tomato, pollination follows opening of flowers but the stamens form a cone around pistil in such a way that self-pollination is almost always ensured.

Stigmatic surface of tobacco is sticky and hence pollen grains adhare to the stigma after the anthers of same flower dehisce. In cotton, the pollen grains are shed directly on the stigma after dehiscence of anthers. Since pollen grains of cotton are sticky very less amount of pollen is drifted away by wind.

Complete self-pollination imposes restrictions on the population, e.g. it prohibits introduction of new genes and production of new forms. Moreover, after hybridization of two genetically diverse forms (whether natural or artificial) the segregating material approaches rapidly towards fixation of homozygous genotypes. However, there is sufficient evidence that some cross-pollination takes place in self-pollinated crops.

Mather (1943) pointed out that the genetic system evolved by a species is a compromise between its need for immediate fitness to a particular environment, which surrounds it, and for flexibility, expressed in the ability of the population to produce new gene combinations adapted to a changing environment. This compromise can be established at any level, and we may find difficult to divide plants sharply into those, which are obligately cross-pollinated and exclusively self-fertilized. All degrees of intermediacy may exist. A survey of distribution of types of genetic system among the higher plants shows that self-fertilization is particularly common among annual species (Stebbins, 1950). But according to Mather (1943) genetic organization of self-pollinated crops is less flexible as compared to that of cross-pollinated crops. Out breeding species are, therefore, better adapted to long range of environmental fluctuations. Inbreeding, according to him may be suicidal in the long run because it drastically restricts genetic variability among individuals in population, which is essential to meet the requirements of changes in the environment.

All the plants in the population of cross-fertilized crops are highly heterozygous. Heterozygosity is characteristic of the genetic organization of cross-fertilized crops. The populations of these crops are genetically variable and hence flexible. Consequently they can quickly develop optimum gene combinations for a given situation and get adapted. Without exception, almost all cross-pollinated crops (except cucurbits) deteriorate upon inbreeding. On the other hand, cross-fertilization restores vigour.

Because heterozygosity appears important for efficient functioning of genetic systems of cross-pollinated populations, nature has bestowed means of encouraging or enforcing cross-pollination in these crop plants. Monoecy, protrandry and protogyny are mechanisms which encourage cross-pollination but some self-pollination can occur. On the other hand, dioecy, self-incompatibility and male- or female-sterility enforce cross-pollination. Genetic differentiation is involved in dioiecy, self-incompatibility and either kind of sterility. Genetic differentiation is not involved or noticed for protandry, protogyny and monoecy.

Dichogamy, i.e., differences in the time of maturity of anthers and stigma within the same flower (or flowers of same plant) is known in several crop species. This mechanism may be responsible for promotion of cross-pollination as mentioned in previous paragraph. In protogynous form of plants the stigma attains maturity earlier than the anthers within the same flower or same plant if male and female flowers are separated in space. Protogyny has been reported in okra, chillis (Peter, 1998), cabbage (Ram, 1998) among vegetable crops. Ficus (Banyan), Four

O'clock plant, Magnolia, custard apple, walnut, and avocado also show protogyny (Dutta, 1980). Pearl millet among cereals is protogynous. In protandry, the anthers mature earlier than the stigma of same flower (or male and female flowers of same plant if separated in space). This phenomenon has been reported for carrot, cotton, chicory, muskmelon, onion, pointed gourd (*Trichosanthes dioica*), marigold, sunflower, maize, raspberry and several other plants.

When the male and female flowers are borne by same plant but separated in space, the situation is called as monoecy. In maize the androecious flowers are borne in the tassel while the gynoecious flowers are located in the cob. In castor, pistillate flowers are borne in the upper portion of an inflorescence while the staminate flowers occupy lower position on same inflorescence. Nevertheless, proportion of staminate and pistillate flowers on the same raceme varies a great deal within and among the genotypes. Amaranth is also monoecious. Grapes, pecan, walnut, blackberry and strawberry among fruits also exhibit monoecy. Cucurbits also exhibit monoecy. But because there are various forms, cucurbits will be described separately.

When the pistillate and staminate flowers are borne on different plants of same species the situation is called as dioecy. Monoecious plants and dioecious plants represent two forms of dicliny. Dioecious nature is not very common among the plant species. This type of reproductive system is wasteful of gametes because plants are stationary. The two individual plants of opposite sex cannot approach each other for mating. Dioecy is known for date palm, hemp, hops, asparagus, spinach, papaya, kundroo (*Coccinia indica*) and parwal (*Trichosanthes dioica*).

Dioecism in *Spinacea oleracea* is controlled primarily by a pair of sex-chromosomes, *X* and *Y* (Bemis and Wilson, 1953). The homozygote, *XX* is a female plant while the heterozygote, *XY*, is a male plant. This is similar to *Drosophila melanogaster*.

The general form of flowering in cucurbits is monoecious, i.e. staminate and pistillate flowers are borne on same plant but are separated in space. Kaloo (1988) described various forms observed in cucurbits viz., trimonoecious, monoecious, androecious, andromonoecious, gynoecious, gynomonoecious and hermaphrodite.

The trimonoecious form of plant bears three kinds of flowers on same vine, viz., staminate, pistillate and hermaphrodite (Figure 2.2). This is a rare form and found in breeding lines of cucumber, muskmelon, and ridge gourd. Monoecious plants bear both staminate and pistillate flowers, but separated in space. It is a predominant form observed in cucumber, muskmelon, watermelon, bitter gourd, bottle gourd, pumpkin, ridge gourd, squash. In monoecious plants staminate flowers appear first and subsequently pistillate flowers are formed. Androecious plants bear only staminate flowers. Androecious plants are observed in cucumber. Both staminate and hermaphrodite flowers appear in andromonoecious forms. Andromonoecy is predominantly observed in muskmelon, but they are found in cucumber also. Gynoecious plants carry only pistillate flowers. Gynoecious plants are of great interest to the plant breeder because they can be easily utilized in hybrid seed production as seed parent. It is observed in cucumbers. Gynomonoecious forms of plants produce pistillate as well as hermaphrodite flowers. This form is sometimes observed in the breeding material of ridgegourd, muskmelon and cucumber. Hermaphrodite plant forms bear hermaphrodite flowers only. Hermaphrodite plants are observed in cucumbers.

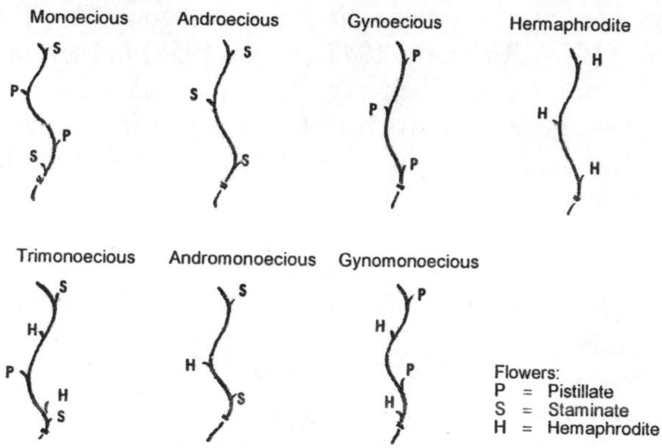

Fig. 2.2: Some sex forms in cucurbits

Kalloo (1988), Peter (1998) and Ram (1998) presented the various aspects of inheritance of sex expression in different cucurbits. A brief information is presented here for general information.

In muskmelon (*Cucumis melo*) andromonoecious, gynoecious and monoecious forms are observed. Monoecious and andromonoecious forms are most common. Gene *A* promotes formation of staminate flowers while gene *G* leads to formation of pistillate flowers. But genes *A* and *G* are dominant to a and g, respectively. A cross of monoecious (*AAGG*) with hermaphrodite (aagg) plant produces F_1, which is monoecious (*AaGg*). In F_2 generation, the proportions of four phenotypes are:

Monoecious	$A-G-\frac{9}{16}$
Andromonoecious	$aaG-\frac{3}{16}$
Gynomonoecious	$A-gg\ \frac{3}{16}$
Hermaphrodite	$aagg\ \frac{1}{16}$

However, environmental agencies also influence the sex expression.

Sex forms in ridgedgourd (*Luffa acutangula*) are monoecious, androecious, gynoecious, gynomonoecious andromonoecious and hermaphrodite, but mostly monoecious in smoothgourd (*L. cylindrica*). Singh *et al.* (1948), Richharia (1948), Singh (1958) and several others studied the inheritance of sex expression in *Luffa acutangula* and principally two gene pairs have been found responsible (Peter, 1998). Accordingly monoecious forms are AAGG, gynoecious and others are as shown below:

Monoecious	*AAGG*
Gynoecious	*A–gg*
Androecious	*aaG–*
Andromonoecious	*a'a'GG*
Gynomonoecious	*Aag'g'*
Hermaphrodite	*aagg, a'a'gg, aag'g', a'a'g'g'*.

A scientific report from *IARI* (1963) was quoted by Peter (1998), which suggests that two gene pairs govern the inheritance of sex in *L. acutangula*. In F_2 generation of a cross between monoecious × hermaphrodite individuals, 9/16 are monoecious, 3/16 androecious, 3/16 gynoecious and 1/16 hermaphrodite.

Three major gene pairs viz., Acr/acr, M/m and A/a are responsible for sex expression of cucumber (*Cucumis sativus*). Gene Acr is responsible for gynoecy (Tatlioglu, 1993). Monoecious plants are recessive at this locus, *i.e.* are acr acr, but possess genes *A* and *M* responsible for monoecism. Gene Acr promotes female sex expression.

Development of hermaphrodite form is due to recessive gene m (Acr -, *A*-). When both Acr and *M* are dominant gynoecious forms are produced. Gene a is responsible for androecy. This gene is hypostatic to Acr and hence has an effect in acr/acr plants only (acr acr aa mm).

Coccinia indica (Kundru) and parwal (*Trichosanthes dioica*) are dioecious forms among the cucurbits. The male plants are heteromorphic in both the crop species, i.e. 22*A* + *XY* in *Coccinia* and 20A + *XY* in *Trichosanthes* (A denotes autosomes).

A partial list of self-pollinated and cross-pollinated crops is presented in Table 2.1. Unfruitfulness sometimes results in crop plants. It could be due to sterility (either female or male) and incompatibility. Because self-incompatibility is used for hybrid seed production in some vegetables, indication has been given in this table about self-incompatibility observed in some crops. Cross-pollination resulting from various mechanisms, such as monoecism, dioecism, protandry, protogyny has also been mentioned. But some mistakes may be possible.

Mating systems can change in response to external requirements. For example, tomato was originally cross-pollinated South American crop species. But after its introduction in Europe, it was modified to almost exclusively self-pollinated form (Rick and Butler, 1956). Likewise intense selection in originally self-incompatible *Brassica oleracea* var. botrytis (L.) resulted in compatible annual form of cauliflower (Watts, 1963). Cucurbits, in general, are cross-pollinated. But these crops do not or suffer very negligible inbreeding depression upon selfing. Cucurbit plants are large and occupy more land. The farmers in the back yards of their houses plant cucurbits and a few plants are grown to satisfy the family requirement. Because population size is restricted inbreeding cannot be avoided. Hence, cucurbits got adapted to inbreeding.

The classification of crop species as self-pollinated and cross-pollinated is not absolute. For instance upto 50 per cent cross-pollination may be observed in nature, in crops like sorghum and cotton. Proportion of out crossing is also relatively larger in brinjal, chillies and okra. These crops need special care during multiplication of genetically pure seed. As a matter of fact, very few species are completely self-pollinated. Usually upto 5 per cent out-crossing may be a satisfactory limit to classify a crop as self-pollinated. Species, which manifest more than 15 to 20 per cent natural out-crossing, could be regarded as cross-pollinated. But again some self-pollination may take place in cross-pollinated crops also.

Table 2.1: A partial list of self-pollinated and cross-pollinated plants

(A) Self-pollinated plants		
Barley	Lettuce (partial cleistogamy)	Ragi
Brinjal* (Heterostyly)	Lentil	Sesamum
Chickpea	Limabean*	Sorghum*
Chilli/pepper*[d]	Linseed	Soybean
Cotton*[e]	Mung	Sunhemp
Cowpea	Oat	Tomato
Finger millet	Okra*[d]	Tobacco
Guar	Pea	Urd
Jute	Potato	Wheat
Kodo millet	Rice	

*Often cross-pollinated

(B) Cross pollinated plants		
Alfalfa[c]	Celery	Papaya[a]
Almond[c]	Chenopodium	Pear[c]
Amaranth	Cherry[c]	Pearl millet[d]
Apple	Chicory[e]	Pecan[b]
Artichoke	Citrus	Pistachio[a]
Ash gourd[b]	Coccinia[a]	Plum[c]
Asparagus[a]	Coriender	Pumpkin[b]
Avocado[c,d]	Cucumber	Radish[c]
Banana[b]	Date[a]	Rasbberry[e]
Beet[c]	Fig	Rye[c]
Bitter gourd[b]	Grapes[b]	Safflower
Bottle gourd[b]	Hemp[a]	Spinach[a]
Broccoli[c]	Hops[a]	Strawberry[b]
Brussels sprout[c]	Knolkhol[c]	Sugarcane
Cabbage[c,d]	Maize[b,e]	Sunflower
Carrot[e]	Mango[c]	Sweet potato[c]
Castor[b]	Muskmelon[b,e]	Turnip[c]
Cauliflower[c,e] (Some varieties protandrous)	Mustard[c]	Walnut[b,d]
	Olive	
	Onion[e]	

a = dioecious, b = monoecious, c = self-incompatible, d = protogynous
e = protandrous

In order to produce genetically pure seed of self-and cross-pollinated crops isolation of space and time is essential. But in case of self-and some cross-pollinated crops, limited genetically pure seed could be produced by enclosing the flowers/inflorescences within paper bags or muslin cloth bags, and if required pollinated by the pollen collected from same plant or the sibs.

2.3 DETERMINATION OF THE NATURE OF POLLINATION MECHANISM

For most of the crop plants the mode of reproduction and the nature of pollination mechanism that prevails is very well known. However, a plant breeder may have to deal with a crop with which he is not acquainted and/or information about nature of pollination mechanism is not available, at hand. The mode of pollination that exists is to be essentially known, before a plant breeder launches a plant improvement programme. Without this information he can not decide about the efficient breeding plan.

A close examination of the floral morphology may, at times, suggest the mode of pollination mechanism for that crop. For example, those crop species, which are monoecious and dioecious, are sure to be allogamous in nature. On the other hand, cleistogamous nature of flower suggests self-pollination. Protogyny and protandry also encourage out-crossing. In order to understand whether a crop is enforced for cross-pollination through self-incompatibility, raising a plant in isolation or enclosing the flower (or inflorescence) inside paper bag would be desirable. If cross-pollination is essential for a crop, an isolated plant would not set seed. However, enclosing an inflorescence/flower inside a paper bag changes the microenvironment around it.

Loss of vigour upon continuous self-pollination (in case of self-compatible species) will also suggest that the crop is naturally cross-pollinated.

In order to determine the magnitude of cross-pollination that can take place an experiment has to be planned. Two varieties of the same crop are chosen, one of which bears a dominant marker gene. It will be desirable if marker gene can be identified in seed or seedling stage. This will enable quick conclusion about the extent of cross-pollination.

For determination of extent of cross-pollination the variety with recessive gene is planted in the field, as per the commercial planting rates of course, in rows. Among the lines of this variety the variety with dominant marker gene is planted at regular intervals. At maturity each row of the variety with recessive gene is harvested separately. If the dominant marker can be identified in the harvested seed out crossing can be known immediately after harvest. For example, the maize endosperm may be sugary (governed by gene su, and hence endosperm shall be su su su), or starchy (*Su Su Su*). When starchy variety is used as dominant marker line among sugary rows, because of xenia (immediate effect of pollen on endosperm, due to the phenomenon of double fertilization in seed), pollination of sugary plants by the pollen from starchy plants will lead to formation of starchy endosperm. The sugary endosperm, because of high sugar, is wrinkled and translucent when dry. Starchy types shall not be wrinkled and translucent, at least as much as sugary types. One Su gene in endosperm can dominate two su genes in definitive nucleus and hence endosperm of sugary plant shall not be sugary.

In sorghum purple coleoptile is dominant to green (Ayyangar and Reddy, 1942). This character can also be used for studying extent of natural crossing. Similarly red purple or black purple stem colour of sorghum is dominant to tan plant colour. Such traits could be studied in seedling stage.

One may be interested to find out the distance upto which natural cross-pollination can take place. A variety with dominant marker gene is planted in centre of a large and isolated field. Another variety, which is recessive for the marker gene is planted in rows on all four sides of dominant marker variety at regular intervals, say at an interval of one metre. The recessive variety may be sown upto 600 to 800 metres. If the marker gene has xenia effect, as in case of starchy/sugary endosperm of maize, each plant in recessive rows is closely observed before harvest. The distance of rows upto which dominant marker is observed in recessive rows indicates the distance upto which natural cross-pollination has taken place. When xenia effect is not available for such studies, each row is separately harvested. In the next season, seed from each row is sown and observed for dominant marker. The recessive lines in proximity of dominant marker lines shall have more cross-pollination. But as the distance increases extent of cross-pollination shall be smaller. Lines farthest from dominant marker line may not have cross-pollination at all.

Experiments carried out earlier revealed that crops differ with respect to natural crossing. For example, about 4 to 5 per cent, maximum out-crossing is observed in crops like groundnut, cowpea, tomatoes, linseed, wheat, soybean, etc. But out-crossing is enormous in crops like cabbage and cauliflower. And this is the reason why isolation distance for production of seed of wheat and soybean is 3 metres as against 1600 metres in cabbage and cauliflowers (Agrawal, 1980). Even a single undesirable plant within the isolation limits can cause considerable contamination. Ayyangar and Rao (1931) estimated that a single inflorescence of sorghum may be made up of as many as 2000-4000 sessile spikelets. A single panicle of sorghum may produce from 24 to 100 million pollen grains (Poehlman and Borthakur, 1969). Because pollen grains of sorghum are light and pollen is easily drifted away by wind one can imagine how far the pollen can travel and how many plants a single undesirable plant can contaminate. But the average natural cross-pollination of sorghum is around 6 per cent (Graham, 1916).

References

Allard, R.W. (1960). *Principles of plant breeding*. Publ. John Wiley and Sons, Inc. New York.

Agrawal, R.L. (1980). *Seed technology*. Oxford and IBH Publishing Co., New Delhi.

Ayyangar, G.N.R. and T.V. Reddy (1942). Seedling — adult colour relationships and inheritance in sorghum. *Ind.J. Agric. Sci.*, 12:341.

Ayyangar, G.N.R. and V.P. Rao (1931). Studies in sorghum: 1 Anthesis and Pollination. *Ind.J.Agric.Sci.*, 1: 445.

Ayyar, V.R. and R. Balasubrahmanyan (1935). Anthesis and pollination in Bengal Gram (*Cicer arietinum*). *Madras Agricultural Journal*, 23: 170-178.

Bemis, W.P. and G.B. Wilson (1953). A new hypothesis explaining the genetics of sex determination in *Spinacea oleracea* L. *J.Hered.*, 44: 90-95.

Bowden, J.E. and R.A. Neve (1953). Sorghum midge and resistant varieties in Gold coast. *Nature*, London, 172: 551.

Briggs, F.N. and P.F. Knowles (1967). *Introduction to plant breeding*. Reinhold Publishing Corporation, New York.

Chandrasekharan, S.N. and S.V. Parthasarathy (1965). *Cytogenetics and plant breeding*. Publ. P. Varadachary and Co., Madras. 2nd Edition.

Dutta, A.C. (1980). *A class book of botany*. Publ. Oxford University Press, Delhi.

Graham, R.J.D. (1916). Pollination and cross-fertilization in jowar plant (*Andropogon sorghum*, Rot) Memoirs, Department of Agriculture, *India (Bot.Ser.)*: 8: 201-216.

Hays, H.K., F.R. Immer and D.C. Smith (1955). *Methods of plant breeding.* Publ. Mc Graw Hill Book Co. Inc., New York.

Kalloo, G. (1988). *Vegetable Breeding.* Vol.I. C.R.C. Press, Florida, pp. 1-30.

Mather, K. (1943). Polygenic inheritance and natural selection. *Biol.Rev.Camb.Phil.Soc.*, 18: 32-64.

Peter, K.V. (1998). *Genetics and breeding of vegetables.* Indian Council of Agricultural Research, New Delhi.

Poehlman, J.M. and D. Borthakur (1969). *Breeding Asian Field Crops.* Oxford and IBH Publishing Co., New Delhi.

Ram, H.H. (1998). *Vegetable Breeding: Principles and practices.* Kalyani Publishers, Ludhiana.

Rao, N.G.P. and L.L. Narayana (1968). Apomixis in grain sorghum. *Indian J.Genet.*, 28: 121-127.

Richharia, R.H. (1948). Sex inheritance in *Luffa acutangula*. *Current Science*, 17: 359.

Rick, C.M. and L. Butler (1956). Cytogenetics of tomato. *Adv. Genetics*, 8: 267-382.

Seshadri, V.S. (1989). Vegetable crops. In *Plant Breeding*. Ed. V.L. Chopra. Oxford and IBH Publishing Co., New Delhi, pp. 391-408.

Singh, H.B.; S. Ramanujam and B.P. Pal (1948). Inheritance of sex forms in *Luffa acutangula*. *Nature*, 161: 775.

Singh, S.N. (1958). Studies in the sex expression and sex ratio in *Luffa* species. *Indian Journal of Horticulture*, 15: 66-75.

Stebbins, G.L. (1941). Apomixis in Angiosperms. *Bot.Rev.*, 7: 507-542.

Stebbins, G.L. (1950). *Variation and evolution in plants*. Columbia University Press, New York, p. 643.

Tatlioglu, T. (1993). Cucumber—*Cucumis sativus*, L. In "Improvement of vegetable crops". Eds. Kallo, G. and B.O. Bergh. Pergamon Press, Oxford, New York, pp. 197-233.

3

Self-Incompatibility and Male-Sterility

A passing reference was made in chapter 2, about self-incompatibility and male-sterility, while describing the mechanisms, which enforce cross-pollination in crop plants. Crane and Lawrence (1934) made a distinction between incompatibility and sterility, which lead to unfruitfulness. Self-incompatibility is the inability of a plant to set seed when self-pollinated despite the fact that both the male and female gametes are functional. The same plant sets seed when stigma receives compatible pollen. In nature, stigma receives a variety of air borne and/or insect carried pollen but not all the pollen grains that alight on the stigmatic surface succeed in effecting fertilization. The pistil is adequately equipped with devices, which allow the pollen of only the right mating type to function normally and others to be discarded. If a pistil carrying functional female gamete fails to set seed following pollination with viable and fertile pollen, capable of bringing about fertilization in another pistil, the two are said to be incompatible and the phenomenon is called as sexual incompatibility. Self-incompatibility enforces out-breeding. Unfruitfulness resulting from non-functional gametes is known as sterility. Sterility may result due to chromosomal aberrations (such as translocation and inversion), gene action or cytoplasmic influences, which lead to abortion or modification of entire flower, stamens or pistils. Sterility may be caused by disturbances in the development of pollen grains, embryo, embryo-sac or endosperm. External environmental agencies can also induce sterility. When the stamens or male gametophytes are not functional, the sterility is known as male-sterility on the other hand, when the female gametes or pistils are non-functional, the sterility is referred to as female-sterility.

Male-sterility has been employed as a device to exploit the phenomenon of hybrid vigour by plant breeders in several crops. This kind of sterility has been used to produce hybrid seed on commercial scale. Male-sterility has been advocated to be used to effect random mating in self-pollinated crops for population improvement programmes. Genetic male-sterile line was used in sorghum to develop random mating populations (Webster, 1965, Ross *et al.*, 1971).

In some cole crops, self-incompatibility has been advocated for production of single, three-way and double-cross hybrids. It is for this reason that self-incompatibility and male-sterility have been assigned to a separate chapter.

Male-sterility, resulting from inability of anther to dehisce and disperse normal and viable pollen grains, has been reported in *Solanum melongena* (Sneep *et al.*, 1979). This is truly speaking not male-sterility and hence was termed "functional male-sterility". Inheritance of this sterility is determined by a single pair of recessive genes (Nuttal, 1963). Female-sterility was reported in sorghum (Casady *et al.*, 1960) and some other crops.

3.1 SELF-INCOMPATIBILITY

According to East (1940) incompatibility occurs in more than 3000 species spread among 20 families of flowering plants. Subsequently, many additional cases were found which led Lewis (1954) to the conclusion that East's estimate was too low. Breeding tests are required for detection of incompatibility. Among the crop and ornamental plants, most perennial forage grasses (from *Gramineae*), forage legumes (such as from Leguminoseae), members from family Crucifereae (e.g. *Brassica oleracea*, *Raphanus sativus*), Compositeae (e.g., sunflower and cosmos) and Solanaceae (e.g. *Lycopersicon* sp., potato, petunia) show self-incompatibility of varying kinds and degrees of effectiveness (Duvick, 1967). Self-incompatibility is also reported in *Beta vulgaris* and *Ipomea batatas* (Kalloo, 1988), Coconut (Nair and Rekha, 1996) and Mango (Sharma and Singh, 1970).

3.1.1 Classification of Incompatibility

Lewis (1954) classified incompatibility systems into two main groups, namely heteromorphic and homomorphic. The homomorphic incompatibility is further divided into gametophytic and sporophytic depending upon the control of incompatibility by the genotype of pollen itself or that of the sporophyte.

(a) Heteromorphic

This kind of self-incompatibility depends upon the relative lengths of the stamens and pistils in the flowers. Morphologically two (distyly) or three (tristyly) distinct mating types have been observed in plants. Heterostyly is a common feature in brinjal (Ram, 1998). The styles may be longer, medium or shorter in length than the stamens. Fruit setting in long styled flowers varies from 70 to 85 percent but only 12-55% in medium styled flowers. In many species of *Primula* the thrum types have shorter styles and highly placed anthers while the pin forms have flowers with long styles and short anthers. (Figure 3.1). Tristyly has been reported in *Lythrum*. In addition to differences in the heights of stamens and pistils within the flowers and incompatibility, the thrum and pin plants also manifest differences for some other characters, such as pollen size and size of stigmatic cells. The inheritance of self-incompatibility in *Primula* is controlled by a single locus with two alleles, viz., S and s. The thrum plants are genotypically Ss and the pin ss. In tristyly two loci each with two alleles govern the inheritance of different forms. Gene S is completely dominant to s. The incompatibility reaction of the pollen, in heterostyled plants is controlled by the genotype of sporophyte. In other words, the self-incompatibility is sporophytic. The genotype of plant, which produces the pollen grains, impresses the incompatibility reaction on the pollen. Thus all the pollen of thrum plants behaves as if they were all S in genetic constitution.

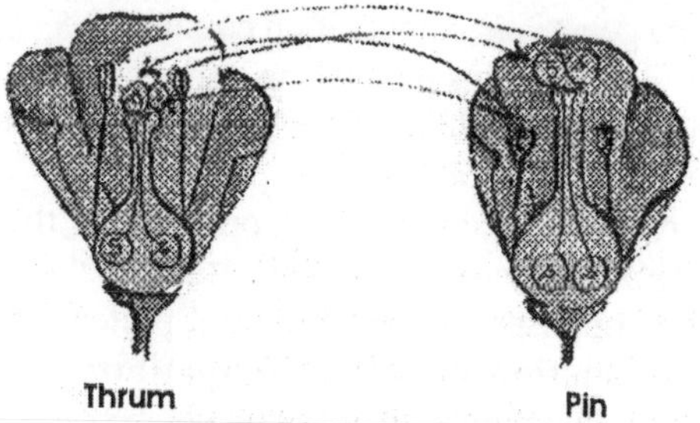

Fig. 3.1: Heteromorphic incompatibility

The compatible mating are:

Pin (*ss*) × thrum (*Ss*), and
Thrum (*Ss*) × pin (*ss*).

That is pollen from thrum type flowers can fertilize the pin type flowers and pollen from pin can fertilize the thrum flowers. (Figure 3.1). The pollen of genotype ss will grow in the style of plant of genotype Ss and *vice versa*.

(b) Homomorphic

In this category all the mating types within the species are morphologically indistinguishable and require a breeding test for their recognition. The self-incompatibility reaction of pollen, in homomorphic types, may be controlled either by the genotype of pollen itself or by the genotype of plant which produced pollen grains. The former is called as gametophytic while the latter is known as sporophytic incompatibility.

(i) Gametophytic Incompatibility

The gametophytic incompatibility or the 'oppositional factor system' was first observed by East and Mangelsdorf (1925) in *Nicotiana sanderae*. Since then it has been reported in several plants, including red clover, white clover, alsike clover and yellow sweet clover. According to East and Mangelsdorf the self-incompatibility reaction is governed by a series of multiple alleles. The allelic series is designated as S_1, S_2, S_3, S_4 ......S_n. The diploid stylar tissues will normally bear any two alleles of the series, but never the same two. For example, the genotypic composition could be S_1S_2, S_1S_3, S_1S_4, S_2S_3, S_2S_4 and so on. Pollen grains possessing genes similar to one present in the stylar tissues is unable to effect fertilization. In this system of self-incompatibility the hinderence to pollen growth is localized in stylar tissues. A pollen grain will have only one of the alleles of the series (because pollen grains are haploid). Because the incompatibility allele in the style opposes the penetration of pollen tubes, with same allele, this explanation of self-incompatibiltiy was called as oppositional factor hypothesis. In *Solanum phureja* and *S. stenotomum* bifactorial control of gametophytic self-incompatibility has been reported. Whereas in *Solanum pinnatisectum* genes at two unlinked loci viz., *S* and *R*, govern the inheritance of self-incompatibility. In *Beta vulgaris*, four gene pairs control this character (Kalloo, 1988). Further, Kalloo (1988) reported self-incompatibility in *Lycopersicon peruvianum*, *L.hirsutum* and *L.chilense*. Pollen tube growth, in self-incompatible matings was arrested in upper part of style. The self-incompatibility reaction exhibited by *L. peruvianum* was influenced by temperature. It was noticed at 25°C but not at 10 to 20°C.

The gametophytic self-incompatibility results in three kinds of pollinations, viz., fully incompatible, in which both the alleles of stylar tissues and in pollen grains are common (for example, $S_3S_4 \times S_3S_4$), half the pollen grains compatible ($S_1S_2 \times S_2S_3$) and fully compatible ($S_1S_2 \times S_3S_4$). This is shown in Figure 3.2. East and Yarnell in 1929 reported a self-compatibility allele Sf at S locus. Plants with genotype *Sx Sf*, where *Sx* is any allele at this locus, is self-compatible.

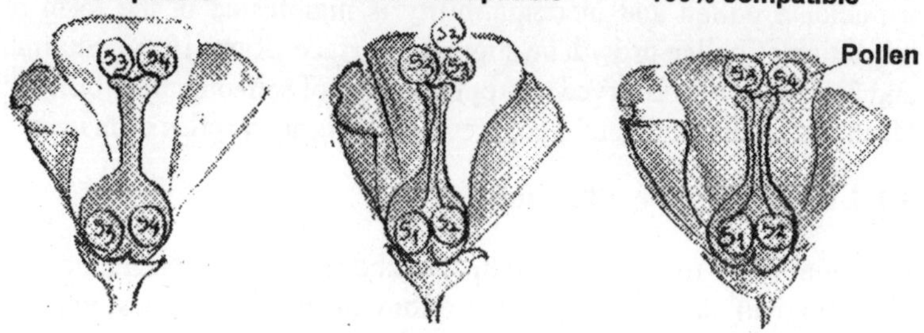

Fig. 3.2: Various kinds of pollinations in gametophytic incompatibility

(ii) Sporophytic Incompatibility

The incompatibility reaction of the pollen grains in this type is controlled by the genotype of the plant that produces them. In this respect self-incompatibility reaction is similar to that observed in heterostyly. Nevertheless, the inheritance is governed by a single locus with an allelic series. But unlike the gametophytic incompatibility, the alleles display dominance in sporophytic system. The dominance is determined by the plant, which produces the pollen. For instance, the genotype of plant is S_1S_2 and S_1 is dominant to S_2, then all the pollens produced by S_1S_2 plant will behave as if they were S_1. Hence the S_1 and S_2 pollen will be incompatible in S_1 style but compatible in S_2 style. In this type of self-incompatibiltiy hindrance to germination of pollen is localized in stigma.

Self-incompatibility observed in all the species of Crucifereae is sporophytic (Batman, 1955). According to Ram (1998) the incompatibility is controlled by one locus, *S* with multiple alleles (50 to 70 in *B.oleracea*). The self-incompatibility alleles may display dominance. Further all the pollen grains produced by a plant exhibit similar incompatibility reaction. For instance in *Primula*, the thrum forms of plants are genotypically *Ss*. Nevertheless, regarding incompatibility reaction both genes, i.e. *S* and *s*, behave similarly. They shall be incompatible when pollinated on thrum forms but completely compatible on pin forms. Sporophytic incompatibility differs from gametophytic system in that the plants may be produced that are homozygous for *S* allele. This feature has been made use of in production of hybrids in case of self-incompatible crop species. This type of self-incompatibility in cabbage is usually associated with trinucleate pollen and incompatibility is manifested in the form of inhibition of pollen growth on stigmatic surface (Ram, 1998). Sharma and Singh (1970) observed sporophytic type of self-incompatibility in Dashahari, Chausa, Bombay Green and Langra varieties of mango.

3.1.2 Barriers to Fertilization

Incompatibility reaction of pollen could be ascribed to biochemical or physiological causes. It can occur at any of the three stages, viz., on the stigmatic surface after the pollen alights, in the stylar tissues while pollen is travelling downwards towards ovary and during the fertilization of egg cell (Sears, 1937).

Gametophytic incompatibility is common with 2 celled pollen whereas the sporophytic incompatibility is associated with species shedding pollen at 3 celled stage (Bhojwani and Bhatnagar, 1988). Barriers to pollination at stage 1 are more of morphological and ecological nature. However, physiological barriers appear to play role at stages 2 and 3. In the gametophytic system the pollens germinate normally, the pollen tubes penetrate through the stigma but fail to reach the base of style. Thus in gametophytic type of incompatibility the rejection of incompatible male gametophyte occurs in the style. On the other hand, in the sporophytic system the rejection reaction occurs on the stigma. In this case either the pollen grains do not germinate or the pollen tubes fail to penetrate the stigma. However, if the stigma is bypassed, and the pollen grains are directly introduced into the style the pollen tubes grow normally and lead to fertilization.

Two reactions have been recognised in incompatibility process (Bhojwani and Bhatnagar, 1988), viz., recognition reaction and rejection reaction.

Non germination of pollen gains on the stigma was reported for cabbage, radish and *Capsella.* If the pollen germinated it produced tiny abortive pollen tube that was unable to accomplish fertilization in case of incompatible pollinations. The pollen grains of incompatible strains of *Nicotiana* germinated on stigma after self-pollination but the pollen tube was inhibited as it penetrated stigma, it then grew slowly down through the stylar tissues until flower dropped due to lack of fertilization in time. In genus *Lycopersicon*, *L. peruvianum*, *L. hirsutum* and *L. chilense* have been reported to be self-incompatible. After germination of pollen the growth of pollen tube was arrested in the self-incompatible matings in case of red clover (Sears, 1937). In *Theobroma cacao* the male gamete failed to pair with polar nucleus to form endosperm.

(a) Recognition Reaction

Recognition reaction occurs at the molecular level and by this the pistil recognises whether a particular pollen grain is to be accepted or rejected (incompatible). In the sporophytic incompatibility the recognition reaction sets in almost immediately after the pollen comes in contact with stigma. In *Prunus avium* the presence of stigma is necessary for recognition of pollen. Pollen deposited on artificial stigma, prepared on cut stump of the style, after removing original stigma, are able to germinate. The pollen tubes of the pollen placed on artificial stigma grow through the style and effect fertilization, even in case of incompatible pollinations.

(b) Rejection Reaction

The rejection reaction initiates physiological and biochemical processes in the pistil specific to the type of pollen received. In the incompatible pollination the phenotypic expressions of these changes bring about rejection reaction (Bhojwani and Bhatnagar, 1988). It may occur right on stigma, inhibiting pollen germination or prevent the entry of pollen tube into the stigma in case of sporophytic incompatibility system. In case of gametophytic system the pollen tube growth may be inhibited in style leading to bursting of the tube in style. As early as 1894, Green reported that intact pollen upon hydration releases various hydrolytic enzymes. These enzymes, according to him, play role in pollen stigma interaction. Pollen wall has been shown to be the storage site for mobile proteins, which may be responsible for incompatibility reaction. It has been suggested that in gametophytic incompatibility these proteins are present in the intine and in case of sporophytic incompatibility in the exine (Heslop-Harrison, 1975).

In the sporophytic incompatibility (except the heteromorphic forms) and some gametophytic systems the recognition and rejection reactions occur on the stigma. The pollen fail to germinate or malformed tubes are unable to penetrate the stigma. The fact that, in such cases, bypassing the stigma and introducing the pollen directly into the style can bring about normal fertilization, which clearly points to the fact that barrier to fertilization, is restricted in stigma. In *Arabis* and *Brassica* the stigma is covered with a cuticular layer and in case of compatible pollinations the pollen tubes are able to penetrate the cuticular layer, on the other hand, in case of incompatible pollinations the pollens are unable to accomplish this. However, if the pollen grains are left on compatible stigma for some time before shifting them on an incompatible stigma, the pollen tubes are able to penetrate the cuticular layer. It is suggested that an enzyme-cutinase essential to erode the cuticle is activated in compatible stigma, and once activated it is effective on incompatible stigma also.

Many species, where rejection occurs on the surface of stigma, are characterized by a "dry stigma", which lacks copious exudate. However, the dry stigma possesses a hydrated layer, called pellicle, over the cuticle. It consists of a lipid film in which float several proteins. Some of these proteins possess esterase activity. Soon after receiving pollen the stigmatic surface becomes moist due to water exuded by the papillate cells of stigma. Consequently, the exine held proteins flow out within seconds and bind with proteins of the pellicle. The pellicle has been suggested as the receptor site for exine proteins on the stigma. Interaction between the two protein fractions is responsible for incompatibility reaction.

In incompatible pollinations several types of abnormalities in behaviour of male gametophyte have been recorded, viz., the pollen may fail to germinate, if germination takes place the pollen tube is very short, the inhibition of one tube leads to emergence of another tube and appressoria like swellings develop at the tip of pollen tubes. The most distinct response of stigma to an incompatible pollen is the development of callosic plug between plasma membrane and pectocellulosic layer of stigmatic papillae, just below the point of contact with pollen. The callose plug also appears at the tip of the pollen tube and consequently growth of tube ceases (Bhojwani and Bhatnagar, 1988).

In the gametophytic incompatibility system the recognition and rejection reactions occur in the style after the pollen tube has grown to about two thirds the length of the style. The proteins incorporated in the intine during pollen development are possibly involved in these reactions. However, because the rejection reaction is considerably delayed in this system, as compared to the sporophytic incompatibility system, it is equally possible that the synthesis or activation and release of the recognition substances is delayed in gametophytic system (Heslop-Harrison, 1975). In the gametophytic system the incompatibility reactions occur due to production of proteins specified by the S alleles and secreted by the stylar cells.

In *Petunia hybrida* the incompatibility reaction is accompanied by a thickening of the pollen tube wall and degeneration of the tube cytoplasm.

The heteromorphic systems show some differences in incompatibility reactions as compared to the homomorphic system discussed above. In the heteromorphic system, pollen wall proteins, derived from tapetum do not have any role to play in the incompatibility reaction. Moreover, in the heteromorphic system there is no callose formation in the stigmatic papillae following self-pollination (Stevens and Murray, 1982). In *Primula obconia* no difference was observed in the rate of pollen germination between the pin and thrum plants or compatible and incompatible matings. The self-incompatibility reaction occurs only after pollen germination, and the sites of rejection reactions in the two forms are different. In the thrum forms the recognition and rejection reaction takes place on the stigma. Hence, pollen tubes grow only over the stigma surface and very few tubes penetrate stigma. On the other hand, in the pin flowers the pollen tubes penetrate the stigma equally rapidly following self and cross-pollination. Nevertheless, the number of tubes reaching the top of style in self-pollination is considerably low as compared to those in inter-morphological pollinations. In the thrum flowers, according to Stevens and Murrary (1982), the recognition reaction occurs on the stigma surface but the major rejection reactions occur within the pistil tissue. For more details on this topic the reader may refer to Bhojwani and Bhatnagar (1988).

3.1.2 Methods to Overcome Incompatibility Barriers

Self-incompatibility promotes cross-pollination and thus contributes to genetic diversity, which is required for survival of a species in nature. However, it is an impediment to achieve self-pollination to develop inbred lines and effect crosses between self-incompatible forms within the same species. Various methods have been suggested to overcome the incompatibility barrier.

(a) Mixed Pollination

A mixture of live incompatible pollen and killed (by chemicals or irradiation) compatible pollen, which is called as 'Mentor' or 'recognition pollen' is employed for pollination. The treatments used to kill the pollen do not disturb the wall held proteins. When the killed compatible pollen along with incompatible pollen is used for pollination, the proteins emitted from the recognition pollen mask the inhibition reaction at the surface of the stigma. Chatterjee and More (1991) employed technique of mentor pollen, in addition to use of benzyl adenine (1%) to overcome pre-fertilization barrier during inter-specific hybridization of *C.melo* with *C.figarei*, *C. meeusii*, etc.

(b) Bud Pollination

In some plant species pollination at an early bud stage has proved very effective in overcoming the intra-specific incompatibility barrier. In *Petunia axillaries* delayed pollination or pollination with stored pollen could not shatter the self-incompatibility barrier, but self-pollination of buds, two days before the anthesis resulted in seed set (Shivanna and Rangaswamy, 1969). At the bud stage stigma probably lacks the exudate, which appears only at anthesis. Probably protein secretion covers the stigmatic surface, which acts as barrier to penetration of stigma by germinating pollen grains. Protein secretion is probably laid down just before anthesis. If buds are opened and pollen is placed on stigmatic surface, i.e. before protein barrier is formed, seed set may be obtained. For production of inbred lines of cabbage, bud pollination has been recommended. With self-incompatible plants of cauliflower, bud pollination has been found to be effective in self-fertilization. In this system pollination is carried out in buds before 2 to 4 hours of opening with or without emasculation (Ram, 1988).

(c) Stub-Pollination

In those cases where incompatibility reaction is confined to stigma or in those cases where the length of style of the female parent is relatively longer than the pollen parent, removal of the stigma and part of the style has been helpful to overcome the incompatibility barrier. In *Ipomea trichocarpa* the primary site of self-incompatibility is the stigmatic surface, which inhibits pollen germination. Removal of the stigmatic lobes or trimming the terminal half of the style and placing the pollen on cut surface of style, results is good pollen germination and pollen tube growth down upto the ovary, without any inhibition.

(d) Intra-ovarian Pollination

Where the zone of incompatibility resides on stigma or in the style, direct introduction of pollen suspension into the ovary may be helpful to attain self-pollination. In this technique the ovary surface is sterilized with ethanol. Two holes are made on the ovary wall—one for the introduction of pollen suspension and another for escape of the air present in ovarian cavity. For details the reader may refer to Bhojwani and Bhatnagar (1988).

(e) Test-Tube Pollination

In this method the stigmatic, stylar and ovary wall tissues are completely removed from the path of pollen tube and the exposed ovule is directly dusted with pollen grains from same plant. The pollinated ovules are then cultivated in a suitable nutrient medium, which favours pollen germination as well as development of fertilized ovule into the seed.

(f) Heat Treatment

Moderately high temperatures (upto 50°C) are known to lessen the effect of self-incompatibility reaction in certain plants. In *Lilium longiflorum* self-incompatibility could be suppressed by heat treatment of the style before pollination. The style was exposed to 50°C for 6 minutes. Lower temperatures were not effective and higher temperatures were injurious to pistil tissues. Pandey (1973) suggested that self-compatibility induced in otherwise self-incompatible species through heat treatment of the style may be due to selective inactivation of or denaturization of specific isoenzymes involved in the self-incompatibility reaction.

Heat treatment of pistil upto 55 to 60°C reduced the self-incompatibility in *Lycopersicon peruvianum*. High temperatures reduced the incompatibility in cole crops also (Kalloo, 1988). High temperature has been utilized to break incompatibility in *Brassica oleracea* and *Raphanus sativus*.

(g) Irradiation

It has been reported to be useful for overcoming incompatibility. It induces mutation in the incompatibility alleles. Gametophytic incompatibility system has been much easier to change through mutation than the sporophytic system. By application of X-rays, P^{32} and chemical mutagenes to pollen mother cells, Lewis (1954) was able to obtain mutations of S alleles to self-fertile types (*Sf*) in *Prunus avium*. Mutations were easy to identify on selfed plants because the mutated pollen grains accomplished fertilization and led to seed production. According to Kalloo (1988) irradiation of pollen is effective in overcoming the self-incompatibility reaction of pollen.

(h) Repeated Pollinations

Repeated pollination with the incompatible pollen can sometimes lead to selfed seed formation. Repeated pollinations in self-incompatible chicory led to some selfed seeds (Kalloo, 1988). Further CO_2 treatment was found effective to overcome self-incompatibility barrier in *Brassica*.

(i) Late Season Pollination

A tobacco plant self-incompatible throughout the season was self-compatible at end of the season (Allard, 1960).

(j) Polyploidy

It was noticed that the self-incompatibility system observed in diploid of a species was often weakened, or disappeared, at the polyploid level (Briggs and Knowles, 1967). This is particularly true for gametophytic system. However, where incompatibility is determined sporophytically, it is not altered greatly due to doubling of chromosome number. Pear was self-incompatible when diploid but was compatible as autotetraploid. Briggs and Knowles (1967) were of opinion that when pollen grain carries two different alleles in a tetraploid, they might interact to neutralize the incompatibility reaction. Thus pollen grain carrying two different alleles may be functional on stigma whose style carries same two alleles for self-incompatibility. Thus self-pollination could be possible. But when pollen grain carries same two alleles, interaction is not possible and hence the pollen may not be functional on stigma with style carrying same two alleles as in the pollen.

3.2 MALE-STERILITY

Male-sterility has been observed in plants under natural conditions, although its frequency is not large. Male-sterility may take various forms (Gableman, 1956). The plant may not produce pollen or extremely sparse pollen is produced. The phenomenon is called as 'pollen sterility'. The male flowers or stamens may be severely malformed and hence no pollen is produced. This is 'staminal sterility'. The plant may produce normal pollen but anthers fail to dehisce. This is known as 'structural sterility'.

On the basis of nature of mode of inheritance the male-sterility may be of three kinds (Sears, 1947), viz., genetic, cytoplasmic and cytoplasmic-genetic. Among these, genetic and cytoplasmic-genetic (CMS) male-sterility has been employed to produce hybrid seed in various crops.

3.2.1 Genetic Male-Sterility

Virtually all the diploid species of plants, whether domesticated or wild, if investigated properly, show that they possess at least one male-sterility locus (Beadle, 1932). In this kind of male-sterility nuclear genes are entirely responsible for male-sterility. Usually the inheritance is governed by recessive genes (Lewis and Crowe, 1956). Genetic male-sterility may result from pollen abortion as observed in sorghum, (Singh and Hadley, 1961) or due to failure of anthers to dehisce as in tomato (Mital and Thomas, 1961) or abortion of anthers, pistillody of anthers or several other reasons.

Propagation of genetic male-sterile plant requires special efforts (Figure 3.3, *a*). Assume that the genotypic constitution of a male-sterile plant is AAmsms, where AA denotes genes other than those responsible for rest of the characters, while ms ms are responsible for male-sterility. Because usually the male-sterility is recessive, male-sterile line is homozygous for this gene. The male-sterile line does not produce pollen grains. Hence, a plant is developed which is AAMsms, i.e., heterozygous for sterility locus. For rest of the traits the plant is isogenic, i.e. possesses similar genes like the male-sterile plant. Pollination of the male-sterile plant will produce two kinds of progeny, viz., half will be genotypically AAmsms and other half AAMsms. The later will be fertile. Since the maintenance of genetic male-sterile line is by a plant heterozygous at sterility locus, male-fertile plants are produced in offspring. During hybrid seed production male-sterile line (used as seed parent) is planted alternating with a desirable pollen parent. But for pure hybrid seed production it is essential that male-fertile plants arising in sterile female line be rouged out before the anthesis of male-sterile plants. It would be desirable for identification, of fertile plants, therefore, to use a juvenile marker gene.

Genetic male-sterile lines have been identified and bred in pigeonpea (Reddy *et al.*, 1978), sunflower (Vranceanu, 1974) and safflower (Deshpande, 1940). In cotton, 11 loci have been identified of which 10 are in *Gossypium hirsutum* and one in *G.barbadense*. 'Gregg' line is the only good basic source of genetic male-sterility ($ms_5 ms_6$) for developing stable genetic male-sterile female parents. A number of genetic male-sterility sources are available in *Brassica* (Banga and Labana, 1983). Environment sensitive genetic male-sterile (*EGMS*) lines have been reported in rice (Siddiq *et al.*, 1995). Min Shong Shi in Hubei, China in 1973 discovered a photoperiod sensitive male-sterile mutant in rice, viz., Nongken 58*s* (*Shi*, 1981).

Genetic male-sterility could be employed to produce a random mating population in a bid to execute population improvement programme. It was used successfully in sorghum (Doggett, 1972).

3.2.2 Cytoplasmic Male-Sterility

This kind of male-sterility is governed by heritable factors in the cytoplasm and nuclear borne genes probably do not play any role in its determination. Plants carrying certain types of cytoplasm are male-sterile. Such sterile plants set seed when pollen is provided. But the offspring of such a pollination will be male-sterile (Figure 3.3, *b*). The reason is simple. The egg contributes a major share of cytoplasm to the zygote.

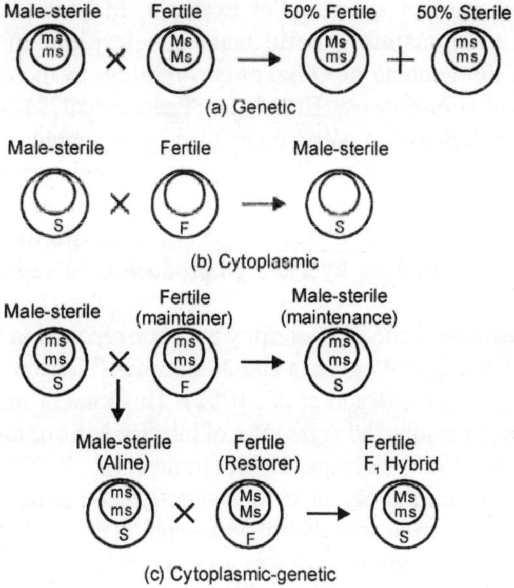

Fig. 3.3: Kinds of male-sterility

Maintenance of cytoplasmic male-sterile plant is straightforward. Pollination of the male-sterile plant by an isogenic plant (but fertile cytoplasm) will lead to seed formation. The progeny will be sterile, like its maternal parent. It is very simple to convert any desirable plant to cytoplasmic male-sterile form. Five to six backcrosses, using the desirable line as pollen parent and male-sterile as a seed parent, will result in male-sterile with desirable genotype. Cytoplasmic male-sterility can not be employed for production of hybrid seed, in those cases where commercially important product is seed. However, this male-sterility could be profitably used for ornamental plants. The male-sterile plants bloom for relatively longer period than the fertile plants.

3.2.3 Cytoplasmic-Genetic Male-Sterility

The male-sterility resulting from interaction of nuclear genes for sterility with heritable factors in the organelles of cytoplasm is known as cytoplasmic genetic male-sterility. This kind of male-sterility is usually observed when the nucleus from one species is placed into the cytoplasm of another species. For example, Meyers (1973, 1975) obtained a cytoplasmic-genetic male-sterile plant of cotton by transferring the genome of *Gossypium hirsutum* to the cytoplasmic background of *G.harknessii*, Brandagee. Pearson (1972) could induce male-sterility in *Brassica oleracea* by placing its genome into *B.nigra* cytoplasm. Wilson and Ross (1962) reported a male-sterility system in wheat based on *Triticum timopheevi* cytoplasm and bred wheat nucleus. Unfortunately cytoplasmic-genetic male-sterility system is not being widely used for hybrid seed production of vegetable crops in India.

Cytoplasmic-genetic male-sterility has been reported in 140 species of plants belonging to 47 genera and 20 families (Hanson and Conde, 1985; Levings, 1990; Braun *et al.*, 1992). This kind of male-sterility has been noticed among the segregates of inter-specific or intra-specific crosses (Frank, 1989; Levings, 1990; Braun *et al.*, 1992).

First report on the use of cytoplasmic-genetic male-sterility for hybrid seed production was that of Jones and Davis (1944) in onion. Following self-pollination there was no seed set in variety Italian red. Plant produced a large number of bulbils in the flower head. The male-sterile line was maintained through bulbils. Inheritance of male-sterility was investigated by Jones and Clarke (1943). The male-sterile condition resulted from interaction of recessive genes (msms) in the nucleus with factor (s) in cytoplasm.

Cytoplasmic-genetic-male sterility is being used to produce hybrid seed in many crops including sorghum, pearl millet, sunflower, rice, maize, etc.

The cytoplasmic-genetic male-sterility differs from cytoplasmic male-sterility only in that the progeny of male-sterile plants is not necessarily male-sterile. The cytoplasmic-genetic male-sterile plant can produce a fertile progeny when certain plants are used as pollen parents. For example, in sorghum male-sterility results from interaction of msc msc genes with factor in 'milo' cytoplasm (called as A_1 cytoplasm). But such a male-sterile could be multiplied by using an isogenic line with fertile cytoplasm as pollen parent (such lines are called maintainer lines or '*B* lines'). However, when a plant possessing Msc Msc genes (or even Msc msc genes) is used as pollen parent, the progeny is fertile. Such lines are called as restorer or '*R* lines' (Figure 3.3, *c*).

In several crop species diverse forms of cytoplasms have been identified which can produce male-sterility as a result of interaction with specific nuclear genes. For example, in sorghum four kinds of male-sterility inducing cytoplasms have been reported (Schertz and Pring, 1982). A_1 cytoplasm of milo was reported by Stephens and Holland (1954), A_2 isolated from IS-12662c a race caudatum type from Ethiopia (Schertz, 1977), A_3 isolated from IS-1112c or converted Nilwa, a race durra type from India (Quinby, 1980) and A_4 (a cytoplasm from *IS* 7920*c*, a race guinea from Nigeria (Worstell *et al.*, 1984). Several other cytoplasmic sources inducing male-sterility are reported. In maize Texas or *T*-type produces male-sterile plants after interaction with nuclear genes rf_1, rf_2 (Duvick, 1965), the Charrua or *C*-type and S-type cytoplasm induce male-sterility following interaction with specific nuclear genes. In pearl millet A_1 cytoplasm (Tift 23*A* of Burton, 1965) was largely used in India for hybrid seed production. Subsequently four more male-sterility inducing cytoplasms have been reported (Smith and Choudhary, 1989; Sujata *et al.*, 1994).

Although cytoplasmic-genetic male-sterile lines are being employed in various crops for production of hybrid seed on commercial scale, there are some problems associated with their use. A plant breeder should be aware of these problems. Some of the problems are:

(a) Modifier genes may influence the expression of male-sterility, which may break down during backcross programme while developing male-sterile lines.
(b) A small amount of cytoplasm is contributed to the offspring by the male gamete and eventually it may influence the expression of male-sterility.
(c) Male-sterility breaks down in some environmental conditions. For example, at temperatures higher than 40°C the male-sterility of sorghum shows partial break down. A few seeds can be observed in selfed earheads of male-sterile plants when temperatures are higher.
(d) In many crops use of a single source of cytoplasm for male-sterility leads to high degree of uniformity in cytoplasmic genomes in the hybrids, which even in the presence of nuclear diversity can make a crop vulnerable.

The glaring examples are those of maize and pearl millet. In maize the male-sterility of cms *T* is characterized by failure of anther protrusion and by pollen abortion. For two decades, the 1950*s* and

1960*s*, the cms *T* was used in hybrid production in order to avoid hand emasculation in United States of America. In 1969 and 1970, a fungal disease known as Southern corn blight, which is caused by *Cochliobolus heterostrophus* (Formerly known as *Helminthosporium maydis*, race *T*), occurred in epidemic proportions in south and Corn Belt regions of United States (Ullstrup, 1972). This fungal pathogen severely blighted maize carrying cms T, which at that time constituted more than 85 per pent of the maize acreage in United States. The pathogen has only mild effects on maize plants carrying normal cytoplasm. Another fungal pathogen, *Phyllosticta maydis*, is also specifically virulent to maize carrying cms *T*. The *Cochliobolus heterostrophus* and *Phyllosticta maydis* produce pathotoxins, Bm *T* and Pm, respectively, which are host specific to cms *T* maize (Lim and Hooker, 1972). These pathotoxins adversely affect the mitochondria from cms T maize but not the mitochondria from other maize cytoplasms or other plant species. The male-sterility and disease susceptibility of cms-*T* are inseparable traits and are associated with mitochondrial gene, *T*-urf13, which encodes a 13-kilodalton polypeptide (URF 13). An interaction between fungal toxins with URF 13 results in permibilization of inner mitochondrial membrane, and leads to specific susceptibility to fungal pathogens (Levings, 1990). Similarly pearl millet hybrids based on Tift 23A, male-sterile line used as seed parent, were found to be highly susceptible to downy mildew caused by *Sclerospora graminicola*. During seventies hybrids based on this male-sterile line were destroyed in various parts of India. Disease susceptibility of tobacco associated with genome variants is also reported (Durbin and Uchytil, 1977).

Because the cytoplasmic-genetic male-sterility has been so extensively employed for hybrid seed production of various crops, scientists have made efforts to understand the mechanisms underlaying this sterility. The light and electron microscopic studies revealed important differences between the fertile and sterile anthers. The wall of anthers, of maize for example, is usually composed of four layers. The inner most layer is called tapetum. The tapetum surrounds the pollen sac early in anther development and degenerates during later stages of development. Consequently, it is not present as an organised tissue in mature anthers. The tapetal cells serve to nourish the developing pollen by exporting nutrients and other molecules needed for pollen formation.

The cytoplasmic organelle, mitochondrion, is the powerhouse in eukaryotic cells. It synthesizes ATP from stored energy in the form of fats, carbohydrates and proteins. Pollen development requires higher levels of mitochondrial activity. Even a slight amount of mitochondrial dysfunction could result in pollen abortion. A twenty to fortyfold increase in mitochondrial number occurs in tapetal cells during microsporogenesis. This indicates an increased requirement for mitochondria in pollen formation. Under these conditions a mitochondrial gene mutation or rearrangement may interrupt mitochondrial biogenesis in anther cells and result in cytoplasmic male-sterility. First signs of abnormality in the tapetum of sterile anthers are the mitochondrial degeneration leading to pollen abortion (Nair, 1993). Tapetal degeneration in the anthers of male-sterile plants was reported in sorghum (Singh and Hadley, 1961) and cotton (Murthi and Weaver, 1974). Nevertheless, degeneration of tapetal cells was not observed in the male-sterile pigeonpea (Reddy *et al.*, 1978).

Plant mitochondrial genome encodes polypeptides, which are components of electron transport chain or $F_0 \rightarrow F_1$ ATPase. These are essential for respiration. Several mitochondrial genes have been identified in plant genomes, for example, Cox I, Cox II and Cox III, the sub-units of cytochrome c oxidase enzyme (Fox and Leaver, 1981), *atp* 6, *atp* 8, *atp* 9 and *atp*A sub-units of $F_0 \rightarrow F_1$ ATPase complex (Boutry *et al.*, 1983).

Restriction endonuclease fragment analyses of the mitochondrial DNAs (mt DNAs) from normal and T cytoplasm of maize have shown that the DNAs are different and maternally inherited (Levings and Pring, 1976). Subsequently, Pring and Levings (1976) found that the *C* and *S* mitochondrial genomes each have a distinct fragmentation pattern. According to their data male-sterile cytoplasms possess a unique mtDNA and thus implicate mitochondria as a carrier of male-sterility factor. Seventeen sorghum cytoplasms, 16 of which were male-sterile, were compared by restriction endonuclease fragment analysis of mitochondrial and chloroplast DNA (*ct* DNA). Seven groups of male-sterile cytoplasms were differentiated by mtDNA restriction fragment differences while only three were differentiated by *ct* DNA (Pring, Conde and Schertz, 1982). In maize evidence suggests that the mitochondrion rather than the chloroplast is involved in cytoplasmic male-sterility and circumstantial evidence suggests that variant mitochondrial translation products may be involved in the expression of male-sterility (Forde and Leaver, 1980). In sorghum evidence indicating mitochondrial rather than the chloroplast DNA is rather weaker. Dixon and Leaver (1982) found that mitochondria from cms line, Martin nucleus in 9E cytoplasm, synthesizes an altered form of cytochrome c oxidase, sub-unit I, COI, which is 14000 Daltons larger than the normal form (42000 Daltons by 9E while 38000 Daltons by normal) synthesized by mitochondria from other forms. Bailey-Serres *et al.* (1986) provided evidence to show that synthesis of variant COI by 9E cytoplasm mitochondria was due to mutation in coding region of mitochondrial gene and not due to alteration in RNA processing, translation or post-translation processing. These workers isolated two different mt genes involved in formation of 42 kd forms of COI.

References

Allard, R.W. (1960). *Principles of plant breeding.* John Wiley & Sons, New York.

Bailey-Serres, J.; D.K. Hanson; T.D. Fox and C.J. Leaver (1986). Mitochondrial genome rearrangement leads to extension and relocation of the Cytochrome c oxidase subunit I Gene in sorghum. *Cell*, 47: 567-576.

Banga, S.S. and K.S. Labana (1983). Male-sterility in Indian mustard (*Brassica juncea* (L.) Cess). Genetics and cytology of Ms1. In: Proc.6th Intern. Rapeseed Conf., Paris, pp. 349-353.

Bateman, A.J. (1955). Self-incompatibility systems in angiosperms. III. *Cruciferae. Heredity*, 9: 53-68.

Beadle, G.W. (1932). Genes in maize for pollen sterility. *Genetics*, 17: 413-443.

Bhojwani, S.S. and S.P. Bhatnagar (1988). *The embryology of Angiosperms.* Publ. Vikas Publishing House, Pvt.Ltd., New Delhi, pp. 119-141.

Boutry, M.; M. Briquet and A. Goffeau (1983). The alpha sub unit of a plant mitochondrial F_1 — ATPase is translated in mitochondria. *J.Biol.Chem.*, 258: 8524-8526.

Braun, C.J.; G.C. Brown and C.J., Levings (1992). Cytoplasmic male-sterility. In "Plant Gene Research, Cell Organelles". Ed. R.G. Herman. *Springer Verlag*, pp. 219-245.

Briggs, F.N. and P.F. Knowles (1967). *Introduction to plant breeding.* Reinhold Publishing Corporation, New York.

Casady, A.J.; E.G. Hayne and D.E. Weibel (1960). Inheritance of female-sterility in sorghum. *J. Heredity*, 51: 35-38.

Chatterjee, M. and T.A. More (1991). Techniques to overcome barrier of interspecific hybridization in Cucumis. *Cucurbit Genet. Coop.*, 14: 66-67.

Crane, M.B. and W.J.C. Lawrence (1934). *The genetics of garden plants.* Mac Millan & Co. Ltd., London.

Deshpande, R.B. (1940). A sterile mutant in safflower (*Carthemus tinctorius*, L.). *Curr. Sci.*, 9: 370-371.

Dixon, L.K. and C.J. Leaver (1982). Mitochondrial gene expression and cytoplasmic male-sterility. *Plant Mol.Biol.*, 1: 89-102.

Doggett, H. (1972). Recurrent selection in sorghum populations. *Heredity*, 28: 9-29.

Durbin, R.D. and T.F. Uchytil (1977). Cytoplasmic inheritance of chloroplast coupling factor 1 subunits. *Biochem.Genet.*, 15: 1143-1146.

Duvick, D.N. (1965). Cytoplasmic pollen sterility. *Advances in Genetics*, 13: 1-56.

Duvick, D.N. (1967). Influence of morphology and sterility on breeding methodology. In *Plant Breeding* Ed. K.J.Frey, Publ.Iowa State University Press. Ames, Iowa, pp. 85-138.

East, E.M. (1940). The distribution of self-sterility in the flowering plants. Proc*Amer.Phil.Soc.*, 82: 449-518.

East, E.M. and A.J. Mangelsdorf (1925). A new interpretation of hereditary behaviour of sterile plants. Proc. *Nat.Acad.Sci.*, II: 116-183.

East, E.M. and S.H. Yarnel (1929). Studies on self-fertility. *Genetics*, 14: 455-487.

Forde, B.G. and C.J. Leaver (1980). Nuclear and cytoplasmic genes controlling synthesis of variant mitochondrial polypeptides in male-sterile maize. Proc.*Nat.Acad.Sci.*USA, 77: 418-422.

Fox, T.D. and C.J. Leaver (1981). The *Zea mays* mitochondrial gene coding cytochrome oxidase subunit II has an intervening sequence and does not contain TGA codons. *Cell*, 26: 315-323.

Frank, S.A. (1989). The evolutionary dynamics of cytoplasmic male sterility. *An Nat.*, 133: 345-376.

Gableman, W.H. (1956)., Male-sterility in vegetable breeding. Brook haven symposia in Biology No. 9: 113-122.

Hanson, M.R. and M.F. Conde (1985). Functioning and variation of cytoplasmic genomes: Lessons from cytoplasmic nuclear interactions affecting male-fertility in plants. *Int.Rev.Cytol.*, 94: 213-267.

Heslop-Harrison, J. (1975). Incompatibility and the pollen-stigma interaction. *Ann.Rev.Plant Physiol.*, Vol. 26: 403-425.

Jones, H.A. and A.E. Clarke (1943). Inheritance of male-sterility in onion and the production of hybrid seeds. Proc*Amer.Soc.Hort.Sci.*, 43: 189-194.

Jones, H.A. and G.N. Davis (1944). Inbreeding and heterosis and their relation to development of new varieties of onions. *U.S. Deptt. Agric.Tech.Bull*, 874: 1-28.

Kalloo, G. (1988). *Vegetable breeding*, Vol. I. Publ.C.R.C. Press, Florida, USA.

Levings, C.S. III (1990). The Texas cytoplasm of maize: cytoplasmic male sterility and disease susceptibility. *Science*, 250: 942-947.

Levings, C.S. III and D.R. Pring (1976). Restriction endonucleases analysis of mitochondrial DNA from normal and Texas cytoplasm male sterile maize. *Science*, 193: 158-160.

Lewis, D. (1954). Comparative incompatibility in angiosperms and fungi. *Adv. In Gen.*, 6: 235-285.

Lewis, D. and K. Crowe (1956). The genetics and evolution of gynodioecy. *Evolution*, 10: 115-125.

Lim, S.M. and A.L. Hooker (1972). Disease determinant of *Helminthosporium maydis*, Race T. *Phytopathology*, 62: 968-971.

Meyers, V.G. (1973). A study of reciprocal hybrids between upland cotton (*G.hirsutum* L.) and experimental lines with cytoplasm from other species. *Crop Sci.*, 13: 439-444.

Meyers, V.G. (1975). Male-sterility from *Gossypium harknessii*. *J.Hered.*, 66: 23-27.

Mital, S.P. and T.A. Thomas (1961). A functional sterile mutant in tomato variety "Meeruti". *Indian J.Hort.*, 18: 150-151.

Murthi, A.N. and J.B. Weaver, Jr. (1974). Histological studies on five male-sterile strains of upland cotton. *Crop Sci.*, 14 (5): 658-663.

Nair, C.K.K. (1993). Mitochondrial genome organization and cytoplasmic male-sterility in plants. *J.Biosci.*, 18: 407-422.

Nair, R.V. and A. Rekha (1996). Incompatibility in cocoa germplasm. *Indian J.Genet.*, 56(2): 137-140.

Nuttal, V.M. (1963). The inheritance and possible usefulness of functional male-sterility in *Solanum melongena*. *Can.J.Cytol.*, 5: 197-199.

Pandey, K.K. (1973). Heat sensitivity of esterase isoenzymes in the styles of *Lilium* and *Nicotiana*. *Incompatibility Newsletter*, 3: 35-36.

Pearson, O.H. (1962). Bull. *Calif.Agri.Exp.Stn.*No.352, p. 22 (Quoted by Duvick, 1967).

Pring, D.R. and C.S. Levings III (1976). Proc. *Am.Phytopathol.Soc.*, 3: 232 (Abst.).

Pring, D.R.; M.F. Conde and K.F. Schertz (1982). Organelle genome diversity in sorghum male-sterile cytoplasms. *Crop Sci.*, 22: 414-421.

Quinby, J.R. (1980). Interactions of genes and cytoplasms in male-sterilty in sorghums. In Proc.35th *Corn Sorghum Res.Conf. Am.Seed Trade Assoc.Chicago.* III, pp. 5-8.

Ram, H. H. (1988). *Vegetable breeding. Principles and practices.* Kalyani Publishers, Ludhiana.

Reddy, B.V.S.; J.M. Green and S.S. Bisen (1978). Genetic male-sterility in pigeonpea. *Crop Sci.*, 18: 362-364.

Ross, W.M.; C.O. Gardner and P.T. Nordquist (1971). Population breeding in sorghum. In *Grain sorghum producers Association*, Grain Sorghum Research and Utilization Conference, 7th Biennial Program. March, 2-4, 1971. *Lubbock, Texas*, pp. 93-98.

Schertz, K.F. (1977). Registration of A_2T_x 2753 and BT_x 2753 sorghum germplasm. *Crop Sci.*, 17: 983.

Schertz, K.F. and D.R. Pring (1982). Cytoplasmic sterility systems in sorghum. In "*Sorghum in eighties*". Proc. *International Symposium on Sorghum ICRISAT*, Hyderabad, India, pp. 373-383.

Sears, E.R. (1937). Cytological phenomenon connected with self-sterility in the flowering plants. *Genetics,* 22: 130-181.

Sears, E.R. (1947). *Genetics and farming.* Yearbook Agri.USDA, pp.245-255.

Sharma, D.K. and R.N. Singh (1970). Studies on some pollination problems in mango. *Indian J.Hort.,* 27: 1-15.

Shi, M.S. (1981). Preliminary report of breeding and utilization of late *japonica* natural double purpose line. *J.Hubei. Agric.Sci.,* 7: 1-3.

Shivanna, R.R. and N.S. Rangaswamy (1969). Overcoming self-incompatibility in *Petunia axillaries.*1. Delayed pollination, pollination with stored pollen and bud pollination. *Phytopathology,* 19: 372-380.

Siddiq, E.A.; M.I. Ahmed; B.C. Viraktamath; J. Ali and N.T. Hoan (1995). Status of hybrid rice in India. In "*Hybrid Research and Development*" Proc. Seminar held 1993. Indian Society of Seed Technology, New Delhi, Ed. M. Rai and S. Mauria, pp. 139-156.

Singh, S.P. and H.H. Hadley (1961). Pollen abortion in cytoplasmic male-sterile sorghum. *Crop Sci.,* 1: 430.

Smith, R.L. and M.K.U. Chowdhury (1989). Mitochondrial DNA polymorphism in male-sterile and fertile cytoplasms of pearl millet. *Crop Sci.,* 29: 809-814.

Sneep, J.; B.R. Murthy and H.F. Utz (1979). Current breeding methods. In *Plant breeding perspectives.* Eds. J. Sneep and A.J.T. Hendrickson. Centre for Agricultural Publishing and Documentation. Wageningen, the Netherlands, pp. 104-233.

Stephens, J.C. and R.F. Holland (1954). Cytoplasmic male-sterility for hybrid sorghum seed production. *Agron.J.,* 46: 20-23.

Stevens, V.A.M. and B.G. Murray (1982). Studies on heteromorphic self incompatibility systems. Physiological aspects of the incompatibility systems of *Primula obconica. Theor.appl. Genet.* 61: 245-256.

Sujata, V.; S. Sivaramakrishnan; K.N. Rai and K. Seetha (1994). A new source of male-sterility in pearl millet: RFLP analysis of mitochondrial DNA. *Genome,* 37: 482-486.

Ullstrup, A.J. (1972). The impacts of Southern corn leaf blight epidemics of 1970-1971. *Ann.Rev.Phytopathol.* 10: 37-50.

Vranceanu, A.V. (1974). (Sunflower) Editura Academiei Republicii Socialiste, Bucharest, Romania (Original in Rumanian).

Webster, O.J. (1965). Genetic studies in *Sorghum vulgare* (Pers). *Crop Sci.,* 5: 207-210.

Wilson, J.A. and W.H. Ross (1962). Male sterility interactions of *Triticum aestivum* nucleus and *T. timopheevii* cytoplasm. *Wheat inf.Ser.* 14: 29-30.

Worstell, J.V.; H.J. Kidd and K.F. Schertz (1984). Relationship among male-sterility inducing cytoplasms of sorghum. *Crop Sci.,* 24: 186-189.

4

Plant Genetic Resources

Man has received a gift from nature in the form of plants. Because of their diverse nature, plants are useful as food, vegetables, fruits, oil, fibre, medicine, fuel, source of sugar, starch, etc. and fodder for cattle. Plants with beautiful flowers satisfy the aesthetic feelings of man, whether utilized to embellish a garden or as cut flowers in drawing rooms. Plants also furnish raw material for manufacturing many industrial products. Enormous diversity exists among plant species, not only among various plants of different families but also within a given species. On this earth about 2,65,000 plant species have been recorded to exist of which 1,85,000 occur in tropical regions and 80,000 in the temperate regions (Singh, 1995).

4.1 EVOLUTION OF CULTIVATED PLANTS

The modern man evolved from the pre-human form about 50,000 years ago. During the early Stone Age, i.e. Paleolithic period, the man did not domesticate either animals or plants. He sustained himself by hunting and gathering food. However during the Neolithic period (late stone age) man learned to cook, make pottery, weave wool and fibres from plants into cloth. It is during this period that the man domesticated plants. Domestication is a process that brings wild species under human management. It is estimated that man focused his interest in agriculture about 10,000 years ago. He selected some wild plants for cultivation, which he thought, were useful to him. Domestication of plants began at different times in different areas. All important crops were domesticated about 5000 years ago in the old world and about 3000 years ago in the new world (Briggs and Knowles, 1967). Probably cereals were the first plants to be domesticated. This may be so because they were easy to store from one harvest to another and simple to process. Legumes were subsequently domesticated followed by vegetables, fibre crops, root crops, oil seed crops, probably in that order. Domestication of fruit crops was not far behind these crops.

The wild plant progenitors of the presently cultivated forms had several undesirable traits. In one way or the other, the cultivated plants must have arisen from the wild species. Most cultivated plants are very old, others are of more recent origin and still others mainly remain at wild stage. Cases are also known which demonstrate that plants may return from cultivated stage to the wild (Muntzing, 1977). Species of plants those were important earlier, for example, certain dye and spice plants, are in our days unessential and have, therefore, been abandoned and may now have a troublesome existence as weeds.

The way in which our domestic animals and cultivated plants have arisen, and the differences between the properties of the wild original material and domesticated races were first thoroughly discussed by Charles Darwin. The cultivated plants have more or less abnormal characters, which are useful for man but which make them incapable of existing as wild forms.

The cultivated plants have very often larger cells and external dimensions than their wild relatives. Moreover the cultivated plants have a good yield of economic produce that is made possible by a large surface of leaves and an intensive metabolism (Schwanitz, 1960). This intensive metabolism is also correlated with a special ability of storing organic matter in some special parts of the plant. The cultivated plants are very efficient at utilizing the irrigation water and higher levels of nutrients supplied through fertilizers or manures as compared to their wild relatives. Only in such favourable, man made conditions, the cultivated plants exhibit great potential for growth and production. When the conditions are unfavourable, the cultivated plants are usually inferior in yield to more primitive but better adapted strains. The cultivated plants usually possess enhanced number or size of parts of plants, which determine yield of economic produce. However the relation between the two components of yield may be very different in different crops (Schwanitz, 1959). For example, increase in yield of cultivated *Seteria italica*, as compared to its wild ancestor *Seteria viridis*, appears to be due to an increase in the number of flowers in the inflorescence and hence more number of grains per panicle. But in case of *Capsium annum*, the fruits are few in number but are enormously enlarged. In maize, sorghum, pearl millet, number of seeds is much larger in varieties/hybrids presently cultivated than in the ancient, primitive representatives of these species.

The cultivated plants differ from their wild ancestors by the loss of certain means of seed dispersal, loss of bitter and poisonous substances which in the wild state, protect the seed from being eaten, loss of mechanical means which protect the fruit (e.g. spines or thorns) and loss of resting period or dormancy which is usually observed in wild species before their seeds are capable of germination (Schwanitz, 1957). The cultivated plants produce palatable foods and feeds, unusual colour of seeds and fruits, double flowers instead of single in ornamental species, self-pollination and annual habit (Briggs and Knowles, 1967). Mutation, genetic recombination between different forms, and polyploidy are the agencies which are responsible for these differences between the wild forms and cultivated forms. Recombination has strong effect only when it is accompanied with selection. This selection may be either natural or artificial.

Mutations are the building blocks of evolution. Genetic diversity arises predominantly due to mutations. Hybridization between two different mutants followed by recombination further enhances the genetic diversity.

The oldest cultivated forms of *Brassica oleracea* existed as early as Stone Age, but we do not know their appearance. In the fifth century B.C., the Greeks cultivated two varieties of leaf Kale and Romans knew a whole series of different Kale varieties, which were, however, not so clearly differentiated from each other as modern varieties, such as cauliflower and white cabbage (Schwanitz, 1957). The cabbage, cauliflower, broccoli, Brussels sprouts and kohlrabi have been derived directly from wild cabbage, which still grows in the coastal regions of Europe and North Africa (Allard, 1960). All these variants emerged as mutants from the wild cabbage.

What selection can accomplish can be seen from a beautiful example of sugar beet. The sugar beet was originally derived from wild form, *Beta maritima*. The oldest cultivated forms were vegetables, later followed by white and red fleshed salad beets, and during the 18th century it was possible to produce mengels. In 1747 Marggraf discovered that the beet root contained sucrose, and in the beginning of the 19th century Achard succeeded to produce the first real sugar beets after selection in certain mengel beet (Muntzing, 1966). Beets first used in the manufacture of sugar, early in the 19th century contained less than 7.5 per cent sugar on average. By continuous selection, the breeders could raise the sugar content from 7.5 per cent to more than 20 per cent, which required about 100 years and thus created a quite new and important cultivated form of plant.

Selection among the spontaneous mutant forms has also been responsible for the development of new cultivated plant species, the sweet lupines (Sengbusch and Zimmermann, 1947; Schwanitz, 1957). The starting material was three wild species, referred to as white, yellow and blue lupines. The first step towards improvement involved selection for large plant and seed size. This led to improvement in seed quality also. Although these plants had high protein content in the seed, these were undesirable because the seed contained high concentrations of bitter and poisonous alkaloids. Further selection resulted in isolation of 6 sweet lupin individuals, representing the three original species. Thus the selection led from wild types to sweet lupins. These forms, however, had an undesirable attribute, viz., the pods opened before harvest. V. Sengbusch and Zimmerman (1947) from a total of about 10 million plants succeeded in isolating a single individual in which ripe pods remained closed till harvest. Another step towards improvement was development of a plant in which the ripe pods did not break off from the axis of inflorescence. The pods remained attached. Six further steps of mutation represented improvement, which led to the transformation of wild state of plants to the cultivated ones (Muntzing, 1966).

Origin of many of our most important cultivated plants was through polyploidy. A polyploid plant has more than two sets of chromosomes (genomes) in somatic cells. There are two kinds of polyploids, viz., autopolyploid and allopolyploid. In the autopolyploid each set of chromosomes, is identical or at least very closely similar to each other. On the other hand, in allopolyploid the basic sets of chromosomes, making up the multiple sets of chromosomes are differentiated from each other. Allopolyploid is the plant derived from hybridization between two differentiated species followed by doubling of the genomes. In some cases cultivated varieties have arisen by autopolyploidy. This is true of timothy, *Phleum pratense*. The 42 chromosomes in this species are all derived from the 14 chromosomes occurring in the small semi-wild pasture timothy, *Phleum nodosum* (Nordenskiold, 1949). Similar spontaneous autopolyploid origin is known for a large number of ornamental plants, potato, banana, apples, etc.

A large proportion of cultivated plants are allopolyploids. Prominent among these are wheat, cotton, tobacco, and many forage grasses. Evolution of cultivated varieties of wheat is via allopolyploidy. In wheat and *Aegilops* only allopolyploid species have been found in nature, while no autopolyploids have been ever known (Kihara, 1966). According to Kihara (1966) the common wheat, *Triticum aestivum*, which is hexaploid (i.e. possesses six sets of chromosomes per somatic cell) has arisen from hybrids between the cultivated emmer (i.e. a tetraploid wheat, such as *Triticum dicoccum*) and *Aegilops squarrosa.* The evidence suggests that amphidiploid between the einkorn, *Triticum urartu* (possessing 14 chromosomes or two sets of chromosomes, *AA*) and a species presumably of sitopsis section of *Aegilops*, gave rise to wild emmer wheat, *T. dicoccoids* with genomic constitution *AABB*, which in turn gave rise to most cultivated hexaploid wheats (Jain, 1989).

The commercial tobacco, *Nicotiana tabacum* does not occur as a natural species. The somatic chromosome number of this tobacco i.e. 2n, is 48. On the other hand, the related tobacco species have 24 chromosomes in somatic cells. According to Goodspeed and Clausen (1928) *Nicotiana tabacum* arose following the doubling of chromosome number of the F_1 between *N. sylvestris* and *N. tomentosa.* This was followed by long periods of evolution, which probably happened in South America.

In a similar way it was possible to trace the origin of several other polyploid cultivated crop species.

Crossing of distinct taxonomic species with preservation of improved types from the segregating material also contributed to evolution of cultivated plants from the wild types. Gene differences are considerably larger in the inter-specific crosses. Moreover, structurally also the chromosomes of two species differ. As a result the F_2s from inter-specific crosses exhibit disharmonious recombinations, which do not appear useful to man. Nevertheless, where vegetative propagation is possible such types could be propagated.

Inter-specific crosses have been predominantly responsible for "nobalization" in sugarcane. There are three species of cultivated sugarcane in the genus *Saccharum* viz., *Saccharum officinarum*, *S. sinese* and *S. barberi* while *S. spontaneum* and *S. robustum* are wild types. *S. officinarum* includes the tropical "noble" canes indigenous to New Guinea region (Stevenson, 1965). They are characterized by thick stem, soft rind, low fibre, high sugar content and high cane yield. The Dutch research workers in Java applied the term "noble" to the tall, large barreled, colourful canes of this species. However, these canes exhibit low tolerance to biotic and abiotic stresses. The

S. sinense and *S. barberi* are indigenous to north India, east Pakistan and Burma-China region. The canes of these species are characterized by thin stalks, great vigour, early maturity and wide adaptability. These species can tolerate light frost and drought. However, they are poor in cane yield, intermediate to low in sugar content but resistant to red rot disease. It is believed that the north Indian sugarcanes viz., *S. sinense* and *S. barberi* originated by natural hybridization between the migrating forms of *S. officinarum* and *S. spontaneum* (Parthasarathy, 1946).

Saccharum officinarum was crossed at Coimbatore, with locally available wild species *S. spontaneum* with an object to combine hardiness to withstand rigorous climate of sub-tropical India and also disease resistance in *S. officinarum* (Sreenivasan, 1989). One of the seedlings of this inter-specific cross gave CO 205 which became commercially acceptable in sub-tropical India and later replaced the then existing cultivars belonging to *S. barberi* (Sreenivasan, 1989). In the early history of sugarcane breeding, *S. officinarum*, *S. barberi* and *S. spontaneum* were used for evolving improved varieties. Crossing of the sugarcane varieties with noble cane, *S. officinarum* to combine high yield of sugar of *S. officinarum* clones with hardiness and disease resistance of other species was termed by Javan breeders as "nobalization".

A form of inter-specific hybridization known as introgressive hybridization is probably responsible for the evolution of cultivated forms of some crop species. Introgressive hybridization is the hybridization that introduces some genes of one species into the genotype of another species (Anderson, 1949). This involves hybridization between two species followed by back-crossing of the recombination products for several generations to one of the parental species.

Cultivated forms of sorghum, according to Doggett (1970) were developed from wild sorghum of Africa by disruptive selection in the Ethiopian area. Disruptive selection consists in selecting of more than one level of a particular character in a population, against selection for only one level of a character in directional selection. Great diversity was created through the disruptive selection and by isolation and recombination, due to diverse ecological habitats and to movement of people through the African continent. The wild and cultivated sorghums became distinct at specific level and a separate intermediate population occupying abandoned cultivation and field margins became established. The wild and the cultivated sorghums have been affecting each other continuously by introgression and gene exchange. Hence, there are resemblances between the cultivated and the wild types in any area. After the sorghum reached South China it came in contact with a diploid wild type, viz., *Sorghum propinquum*. Introgression of this plant probably accounts for the distinctive characteristics of Kauliangs as a group (Doggett, 1965).

High yielding modern incipient races of maize, viz., Tuxpeno, Celaya and Chalqueno, have very complex origin (Wellhausen, 1966). Many different, more or less, primitive varieties including teosinte (*Zea mexicana*) have played their part in their genetic constitution. Introgressive hybridization of maize from South America, presumably with *Tripsacum* (wild gamma grass) has been responsible for distinctive features of North American maize varieties (Allard, 1960).

4.2 PLANT GENETIC RESOURCES

There are diverse types of plant species growing in various parts of the world. Natural evolution led to building up of genetic diversity in the natural populations of various plant species. Evolution is the change in hereditary characteristics of the successive generations of a population. Mendelian variation (arising from mutation, and further enhanced by recombination of genes following hybridization between different mutant forms), inter-specific hybridization (including introgressive hybridization), polyploidy and genetic drift are the agencies which have been instrumental for evolution of cultivated plant species. Domestication caused further differentiation of small part of diversity of wild plant species, which became adapted to human requirements. Plant breeders manipulated the genetic diversity of domesticated species and made it suitable for modern agricultural systems (Hintum, 1995). The sub-continent of India is floristically rich with about 50,000 plant species, of which 17,000 species are of flowering plants (Nayar, 1988). Plant genetic resources are the repositories of diverse genes, which exist, in the entire population of plant species. Plant genetic resources comprise of cultivars, land races, advanced or improved cultivars, wild relatives of crops, weedy species, genetic stocks, parental lines of hybrids, GMOs, etc. which possess the genes of economic value and are source of crop improvement. Swaminathan (2000) quoted an old saying "diversity is the spice of life". The plant diversity constitutes the basic raw material required for crop improvement programmes. The diversity also confers ability to adapt to the changing crop requirements. It should be noted that throughout the agricultural history, the crop plants have adapted to a range of environments and cropping/farming systems, and much of the existing crop diversity has been generated and maintained on farms through thousands of years of adaptation to subsistence agricultural systems, where stability and sustainability are more important than production. The natural selection as well as the selection exercised by the early farmers, in the choice of plants suitable for their requirements, contributed to the process of plant genetic diversity among the traditional populations of different crops. As a matter of fact diversity or genetic variability is essential among the varieties in agro-eco systems to minimize risk of crop failure due to fluctuations in the environmental conditions, including the emergence of new pathotypes/biotypes of pests. The evolutionary processes, such as Mendelian variation, inter-specific hybridization and polyploidy, took thousands of years, beginning long before the civilization itself and continuing today, to generate plant genetic diversity among the land races. The plant breeders have learned to exploit and utilize the products of these forces to accelerate the evolutionary process through genetic manipulation. However, in developing the high yielding varieties adapted to specific ecological pockets, breeders promoted uniformity. The high yielding and widely adapted varieties of various crop plants developed by the plant breeders, predominated large areas of production and became a potential hazard, because the very success of the high yielding varieties prompted the breeders to repeatedly use these varieties as one of the parents in plant improvement programmes. For example, during eighties in the Sorghum Improvement Programme, the plant breeders throughout the country largely utilized *SPV* 475, evolved at *ICRISAT*, Hyderabad. In other crops also similar situation is common. This attitude of breeders further narrows down the genetic base of future varieties. Moreover, because of the convincing superiority of currently developed high yielding varieties, the farmers have discarded the land races and other varieties. Presently cultivation is confined to a limited number of varieties. Khor (2000) surmises that by 2005, India is expected to produce 75 per cent of its rice with just 10 varieties as compared to 30,000 varieties traditionally cultivated. According to Swaminathan (2000) expansion of human settlements and increasing specializations in agriculture led to destruction of natural habitats rich in plant diversity and thus narrowing the composition of food basket.

The wild relatives of crop plants constitute the most valuable component of genetic resources required for incorporating desirable genes for biotic and abiotic stresses. The natural wealth occurs in diverse habitats occupying arid, semi-arid to humid tropical and temperate vegetation zones. Changing land use patterns, rapidly increasing pressure on land both for agricultural and forestry development programmes, residential and industrial requirements in urban areas have posed serious threat to the existing genetic diversity.

Genetic diversity existing in plant species inhabiting various ecological pockets in the world constitutes a common heritage of mankind and a treasure, which supports life. This diversity is being lost at an unprecedented rate due to anthropogenic causes, particularly in the tropical regions of the world, which house bulk of the plant genetic diversity. The centres of diversity are concentrated in tropics, encompassing largely the developing countries. The gene-rich countries are ironically often food deficit and poor nations and hence these countries are not capable to use modern biological technology for converting the genetic diversity into economic wealth. Three negative processes are associated with the loss of plant genetic resources, viz., genetic erosion, genetic vulnerability and genetic wipeout.

Genetic Erosion

Genetic erosion refers to gradual loss of plant genetic resources. In the centres of diversities (to be described in section 4.3) of various crops, the primitive cultivars, their wild and weedy relatives are getting replaced at very high rate (Singh, 1989). After 1961 several hybrids of crops, like maize, sorghum, pearl millet, cotton, sunflower, castor, tomato and several vegetable crops and high yielding varieties of wheat, rice, and other crops were evolved and given to farmers for cultivation in India. Consequently, these high yielding varieties replaced the land races on large scale. According to Harlan (1975) land races have a certain genetic integrity. These are recognizable morphologically. Different land races differ in their ability to adapt to soil types, time of sowing, days to maturity, plant height, nutritive value, reaction to diseases and insects, etc. Most important, they are genetically diverse. Moreover, they are balanced populations. Loss of land races due to genetic erosion is irreparable. Their loss means loss of all genes contained in them. In India genetic erosion is a serious problem in some vegetable crops as well as cereals. Release of okra variety "Pusa sawani" in 1959-60 swept off the native varieties out of cultivation. Similarly there is gradual erosion of germplasm diversity of watermelon due to release of 'sugar baby' and its large scale cultivation. Indigenous varieties of sorghum, wheat, and rice are facing genetic erosion due to the release of high yielding varieties.

Genetic Vulnerability

Genetic vulnerability is a risk for crop that finds itself incapable of defending itself against diseases and/or insects or changing environmental conditions. This damage has crept in due to monoculture, which has narrowed down the genetic base of crops. For example, during seventies a large acreage of pearl millet was under *HB*3. The HB series, based on male-sterile line Tift 23*A* (carrying A_1 cytoplasm) was highly susceptible to downy mildew. During 1970s this led to devastation of pearl millet fields. In Sri Lanka, where 2000 varieties of rice were being grown in 1959, today only 100 can be seen. Such type of narrowing down of genetic base among the varieties of crops grown could be seen in United States of America, Indonesia and few other countries. About 75 per cent of the crop of potato is confined to four varieties, to fifty per pent area under soybean to 6 varieties and to nine varieties in about fifty per pent of the entire acreage under wheat in U.S.A. Famines or crop failures could be responsible for narrowing down of genetic base. Singh (1989) has quoted several examples of such destructions. For instance the great Bengal Famine of India in 1943 was associated with brown spot disease of rice (*Cochliobolus miyabeanus*). Epidemic of southern corn blight in U.S.A. caused by *Cochliobolus heterostrophus* during 1970-1971 on maize hybrids possessing the T cytoplasm could be attributed to narrowing down of genetic base.

Genetic Wipeout

Rapid and wholesale destruction of genetic resources comprises of genetic wipeout. Total crop failures or famines may lead to such destruction of genetic resources. Discarding of the genetic collection by the breeder because the collection is not presently useful for the institution is also one of the causes of genetic wipeout. Bigger genetic wipeouts may result when the accessions preserved in gene banks are not successfully and timely regenerated with rejuvenated seed or propagating material.

India is the second largest producer of vegetables in the world (Vishnuswarup, 1998). The tropical Asian region, India and China possess maximum diversity (Arora, 1995). For example the Indian centre of diversity is the primary centre of diversity for brinjal, wax gourd, cucumber, ridge gourd, sponge gourd, bitter gourd, drumstick while it is secondary centre of diversity for watermelon, okra, bottle gourd, amaranth. Similarly Indian centre has rich diversity in genetic resources of various fruit plants of indigenous and exotic origin and hence serves as a reservoir of large number of indigenous minor fruits. About 50 different indigenous fruits and their wild relatives, totaling to about 400 species are concentrated in various agro-ecological regions of India. The indigenous fruits and their wild relatives include *Mangifera indica*, *Emblica officinalis*, *Aegle marmelos*, *Musa* sp. *Tamarindus indica*, *Sizygium* sp., *Artocarpus* sp., *Ziziphus* sp., *Citrus* sp., *Buchanania* sp. etc.

Wild relatives of crop plants constitute the most valuable component of genetic resources required for incorporation of genes responsible for resistance against biotic and abiotic stresses.

As mentioned earlier, due to change in the land use patterns, there is a severe threat to these wild species. Information on diversity and distribution of wild relatives of crop plants and related species occurring in different phyto-geographical regions has been synthesized by Arora and Nayar (1984).

Among the native flowering plant species, there are many, which are precious for use in Indian system of medicine. These are facing loss due to encroachment over forest lands for new human settlements and development of plantations through large scale afforestation of economic trees. Secondly in Ayurvedic and Unani sector, there are about 5000 licenced medicine manufacturing units whose demand for raw material has increased extensively. Entire mature plant, with its roots and fruits is collected which results in drastic loss of seed bearers in the ecological habitats, which are mostly forests. Domestication of all these medicinal plants, which are in regular bulk demand in the industry, is a means of conservation of this natural wealth (Gupta and Rana, 1995).

India exhibits extreme variation in altitude and climate. The eco-physiographical conditions of the Indian sub continent are highly diverse. These provide basis for occurrence of a wide range of plant species diversity. On the basis of distribution of flora, the Indian sub-continent has been divided into eight regions by Chatterjee (1939), viz., (1) Western Himalayas, (2) Eastern Himalayas, (3) Central Himalayas, (4) Assam, (5) Gangetic Plain, (6) Indus plain, (7) Deccan and (8) Malabar. The planning commission, nevertheless, demarcated the country into fifteen agro-climatic zones (Sehgal *et al.*, 1990).

4.3 CENTRES OF DIVERSITY

From the beginning of civilization value of plant diversity was realized. Plant collecting, also referred to as "gene prospecting", began very early.

The Indian Agricultural Research Institute (IARI) organized a scientific expedition to Afghanistan in 1939, recognizing the role of plant genetic resources in crop improvement. During the expedition a good collection of wheat was made (Joshi, 1988). Further in 1940's the Institute organized a trial of potato varieties and species collected in central and South America by the Empire Potato Collecting Expedition, mounted by the Great Britain. Potato material was grown at the Potato and Wheat Breeding Station set up by the Institute at Shimla in 1936. The Institute started a small section of Plant Introduction under the Division of Botany. Dr. B.P. Pal (then Head of the Division of Botany) and Dr. Harbhajan Singh, who joined the Institute in 1946, were the stalwarts and pioneers of plant genetic resource work in India. During the second five year plan the Plant Introduction Section was raised to that of Plant Introduction and Exploration Organization. The National Bureau of Plant Genetic Resources (NBPGR) was established at New Delhi in 1976. The Consultative Group on International Agricultural Research (CGIAR) founded the International Board of Plant Genetic Resources (IBPGR) in 1974 with its head quarters at Rome, Italy.

Plant species are spread throughout the world. Nevertheless, certain areas show tremendous variability. According to a Russian Botanist, Nikoli Ivanovich Vavilov, crop plants evolved from wild species inhabiting regions, which show diversity. Vavilov (1926, 1950) and his colleagues carried out systematic collection of plants and pioneering research and conservation of plant diversity. Vavilov established a gene bank in 1940 at St. Petersburg, which still exists as N.I. Vavilov Institute and contains some 3,80,000 plant specimens from more than 180 places in the world. Vavilov and his colleagues studied the agro-morphological patterns of plant accessions in relation to their geographical distribution, reaction to diseases and insects, etc. Based on observations Vavilov conceived the idea that vast genetic variation within a particular crop species is generally concentrated within a small geographical area. Certain areas of the world manifested much diversity of plants. Often one species or one genus was identified with one centre, but sometimes it was found in two or more centres. Vavilov considered the centres of diversity of various crop plant species as the centre of their origin. Those centres were called as 'Primary centres' where the cultivated species evolved from wild species. Vavilov concluded that a centre of origin was characterized by dominant alleles and that the frequency of recessive alleles increased and diversity decreased towards periphery. Vavilov recognized 'Secondary centres' where certain crop species showed considerable diversity of forms, although they did not originate there. Secondary centres are probably those where two or more species crossed together and natural as well as artificial selection pressure was exerted subsequently (Hays *et al.*, 1955). Alexander Van Humboldt was probably the first author to refer to the origin of crops. The next study was by Alphonse De Condolle (1855), and then came Charles Darwin (quoted by Zeven and deWet, 1982). The centres of diversity recognized by Vavilov were: (I) China, (II) India, (IIa) Indo-Malayan, (III), Central Asia (including now Pakistan, Punjab, Kashmir, Afghanistan and Turkestan), (IV) Near East, (V) Mediterranean coastal and adjacent regions, (VI) Ethiopia, (VII) South Mexico and Central America, (VIII) South America (Peru, Ecuador, Bolivia), and (VIIIa) Isle of Chiloe (Chile). According to him agriculture in these 8 regions developed independently, because of the differences in agricultural methods, implements and domestic animals.

During his visit to Turkey, Harlan (1951) was impressed to observe tremendous plant diversity in small areas to which he coined term "micro centres". According to him, in these centres evolution appeared to proceed at a rapid rate than in other areas, particularly the larger geographical areas. For wheat Harlan identified 3 micro centres in Turkey. Undoubtedly many more may exist elsewhere.

With the increasing knowledge of cultivated, weedy and wild plants it is becoming evident that some parts of Vavilov's theory needs to be changed. One point in Vavilov's theory proposes that primary centre is marked by high frequency of dominant alleles. It may not be possible to indicate such a centre for wheat, maize, *Vicia faba* (Zeven and de Wet, 1982). Besides this it has been pointed out that a great diversity may also arise from variation of environment.

Vavilov's theory that where the greatest diversity is found is also the centre of origin is no longer tenable. This was shown for crops such as *Triticum dicoccum* and *Hordeum vulgare* in Ethiopia. These crops exhibit great diversity there, but no wild relatives are present (Zeven and de Wet, 1982). Zhukovskij (1968) heralded the idea about 'mega gene centres'. Because so many crops originate outside one of the Vavilov's centre of origin it was necessary to enlarge the areas in which species were domesticated. These mega centres engulf much of the world's land surface and extend over vast areas.

According to Simmonds (1979) there is a need to modify Vavilov's views because:

(i) We can only rarely speak of a centre of origin in the sense of a small, well defined patch of country.
(ii) There now seems to be little connection between source of wild ancestors, area of domestication and area of evolutionary classification. Sometimes two or more of these coincide and sometimes they do not.
(iii) We now think of a geographical continuum of domestication and crop evolution wherever agriculture is practiced, whether new or old world. The idea of 12 or so areas of outstanding evolutionary activities disappears with the idea of localized origin.

According to Zhukovskij (1970) the centres of diversity and centres of origin do not coincide very often. In his opinion secondary centres are more important as a source of valuable breeding material. Zhukovskij (1968) proposed slightly modified version of Vavilov's original system. He proposed twelve mega gene centres, viz., Sino-Japanese centre, Indonesia and Indochina, Australia, Hindustan centre, Central Asia centre, Near East, Mediterranean, Africa, Europe and Siberia, Central America, South America and North America (Figure 4.1).

In the Chinese-Japanese centre, China was considered as primary centre while Japan as secondary. Plants with their primary centre in this area include various forms of *Oryza sativa*, several millets and sorghum, naked oats, *Hordeum sativum* ssp. *humile*, *Fagopyrum tataricum*, *Glycine max*, *Phaseolus angularis*, various species of *Brassica* and related genera, *Colocasia esculenta*, number of fruits, nuts, tea, cinnamon and various fibres and bamboos.

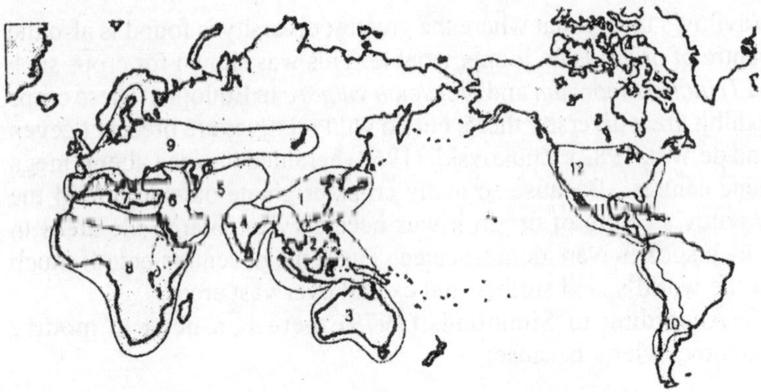

1. Sino-Japanese
2. Indonesia-Indo China
3. Australia
4. Hindustan
5. Central Asia
6. Near-east
7. Maditerranean
8. Africa
9. Europe and Siberia
10. South America
11. Central America
12. North America

Fig. 4.1: Centres of diversity (Zhukovskij, 1968)

The Indonesia and Indochina centre contains several species and sub-species of *Oryza, Citrus, Musa, Saccharum*, and several other tropical plants. The third primary centre is Australia, which contains species of *Gossypium*, *Eucalyptus*, *Nicotiana*, *Citrus*, *Oryza*, *Acacia*, *Macadamia* and is a secondary centre for *Trifolium subterraneum*. The fourth centre viz., Hindustan, is a primary centre of rice, sorghum and other cereals, *Saccharum*, *Phaseolus* and other pulses, *Cucumber*, sesame, species of citrus, cotton, jute and other fibres, coconut, black pepper and various bamboos.

The fifth centre, Central Asia embraces certain sub species of *Triticum aestivum* and *Secale cereale*, various legumes, *Brassica campestris*, linseed, various vegetables and number of fruits. A centre in Near Eastern contains various species of *Triticum*, *Secale* and other cereals, legumes, beet and a number of fruits and nuts. The Mediterranean centre is regarded as one of the oldest and contains species of *Triticum* and other cereals and grasses. It also embraces various legumes, flax, beet, *Brassica*, olive, various fruits and vegetables. Africa is a primary centre proposed by Zukovskij (1968) for sorghum, *Oryza glaberrima*, various legumes, *Cucurbits*, species of *Gossypium*, coffee, date and oil plam. It is regarded as secondary centre for species of *Triticum*. Europe and Siberia is another primary centre, which was not included in centres of origin/diversity by Vavilov (1926, 1950). Beet, many forms of *Brassica*, clover and other legumes and a range of fruits belong to this centre. Tenth centre is Central America, which embraces maize, species of *Phaseolus*, *Solanum*, various cucurbits, sweet potato, cotton, red pepper and cocoa. Edible potatoes and other tubers, certain species of *Gossypium*, *Arachis*, *Lycopersicon*, *Manihot*, *Frageria*, *Helianthus*, *Ananas*, *Hevea* and several other crops of lesser importance are contained in eleventh primary centre, South America. The last centre, i.e. twelth is North America, which embraces species of *Vitis*, many other fruits and nuts, *Helianthus*, *Nicotiana*, *Frageria*, *Hordeum*, *Lupins* and *Zizania*. He also illustrated importance of microcentres.

The origin of some of the crops, according to Vavilov (1926, 1950) is shown in Table 4.1.

SN	*Centre*	*Plant species*
1.	Chinese	*Panicum miliaceum, P. italicum, Andropogon sorghum, Avena nuda, Glycine hispida, Phaseolus vulgaris, Saccharum sinense, Dioscoria batatas, Raphanus sativus, Brassica rapa, Papaver somniferum, Cinnamomum camphora, Sesamum indicum*, **Vegetable crops:** *Brassica* sp., *Allium, Solanum melongena, Cucumis, Cucurbita moschata* **Fruits:** *Pyrus* sp., *Prunus* sp., *Citrus* sp. *(Oranges, Kumquats), Camelia sinensis.*
2.	Indian	*Oryza sativa, Sorghum bicolor, Cicer arietinum, Phaseolus aureus, Vigna sinensis, Elusine coracana* **Vegetable crops**: *Solanum melongena, Cucumis sativus, Lactuca indica, Raphanus indicus, Moringa pterigosperma, Luffa acutangula, Momordica charantia. Gossypium arboreum, Corochorus olitorus, Indigofera tinctoria, Cannabis indica, Piper nigrum.* **Fruits:** *Mangifera indica, Syzygium cumini, Zizyphus mauritiana, Citrus sinensis, C. nobilis, C. medica, C. aurantirum, Artocarpus heterophyllus, Musa* spp. (AB, AAB group). *Saccharum officinarum*
2. (a)	Indo-Malayan	*Dioscorea* spp. **Fruits:** *Citrus maxima, Musa* spp. *Cocos nucifera*

(Contd.)

(*Contd.*)

SN	Centre	Plant species
3.	Central Asia	*Triticum aestivum, T. compactum, Secale cereale, Lens esculenta, Phaseolus aureus, Cicer arietinum, Sesamum indicum, Linum usitatissimum, Carthemus tinctorius.* **Vegetables:** *Cucumis melo, Daucus carota, Raphanus sativus, Allium cepa, Allium sativum, Spinacea oleracea, Pisum sativum.* **Fruits:** *Pistacia vera, Prunus armeniaca, Pyrus communis, Amygdalus communis, Vitis vinifera, Juglans regia, Malus pumila.* **Oilseeds:** *Linum usitatisimum, Sesamum indicum, Carthamus tinctorius* **Fibre:** *Gossypium herbaceum*
4.	Near Eastern	*Triticum monococcum, T. durum, T. turgidum, T. timopheevi, T. aestivum, T. macha, Hordeum distichum, Secale cereale, Avena sativa, Lens esculenta, Sesamum indicum, Linum usitatissimum.* **Vegetables:** *Brassica campestris.* **Fruits:** *Ficus carica, Punica granatum, Malus pumila, Pyrus, Prunus, Vitis vinifera, Pistacia vera*
5.	Mediterranean	*Triticum durum, T. dicoccum, Avena strigosa, Lens esculenta, Pisum sativum, Linum usitatissimum, Sinapis alba, Brassica* sp. **Vegetables:** *Brassica* sp. (Both cabbage and turnip), *Allium* sp., *Beta vulgaris, B. maritima, Lectuca sativa,*
6.	Ethiopia	*Hordeum sativum, Triticum durum, T. dicoccum, Pennisetum spicatum, Cicer arietinum, Lens esculenta, Pisum sativum, Linum usitatissimum, Carthamus tinctorius, Sesamum indicum, Ricinus communis, Coffea arabica.*
7.	South Mexican and Central America	*Zea mays, Phaseolus vulgaris, Ipomea batatas, Capsicum annum, C. frutescens, Gossypium hirsutum, G. purpurascens, Theobroma cacao.* **Fruits:** *Carica papaya*

(*Contd.*)

(*Contd.*)

SN	*Centre*	*Plant species*
8.	South America (Peruvian, Ecuadorean, Bolvian)	*Solanum andigenum, Ipomea batatas, Zea mays, Gossypium barbadense* **Vegetables:** *Lycopersicum esculentum, L. peruvianum* **Fruits:** *Psidium guajava, Nicotiana tabacum*
8. (a)	Chili	*Solanum tuberosum*
8. (b)	Brazilian-Paraguayan	*Manihot utilissima, Arachis hypogea, Hevea brasiliensis,* **Fruits:** *Ananas comosa, Anacardium occidentale, Eugenia.*

*Based on Vavilov's centres of diversity. Only primary centres shown.

India is certainly a centre of crop diversity. In addition to the crops mentioned in Table 4.1, several other crops originated in this centre (Arora, 1988). Prominent among these are:

Vigna mungo (Urid), *V. aconitifolia* (Moth), *Cajanus cajan* (Pigeonpea), *Macrotyloma uniflorum* (Kulthi), *Carissa congesta* (Karonda), *Manilkara hexanxdra* (Khirni), *Grewia subinaequalis/ G. asiatica* (Phalsa), *Aegle marelos* (Bael), *Garcinia indica* (Kokum), *Emblica officinalis* (Aonla), *Citrullus lanatus* (Tinda), *Trichosanthes dioica* (Parwal/Pointed gourd), *Colocasia esculenta* (Taros), *Coccinia indica* (Kundri), *Curcuma domestica* (Turmeric), *Zingiber officinale* (Ginger), *Elettaria cardamom* (Cardamon), *Piper nigrum* (Black pepper), *Cinnamomum zeylanicum* (Cinnamon/ Dalchini), *Crotolaria juncea* (Sunhemp), *Sesabania sesban* (Dhaincha) and *Camellia sinensis* (Tea)

4.3.1 Law of Homologous Series

Vavilov noticed parallelism in variability for certain characters in related species on the basis of observations recorded on accessions. Characters found in one species also occur in other related species. Vavilov called this parallelism "law of homologous series." Several characters occasionally make their appearance in several varieties or races descending from same species and more rare in progeny of widely distinct species. Any character that is variable in one species in a given group shows tendency for variation also in another related species. There are, for example, similarities in fibre colour (white, brown and green), degree of hairiness, colour of seed, form of leaf, opening of bolls among different species of cotton, viz., *Gossypium arboreum, G. herbaceum, G. hirsutum and G. barbadense* (Briggs and Knowles, 1967).

In *Triticum*, with same chromosome number, say $2n = 42$, as in *T. aestivum*, variation is duplicated in other hexaploid *Triticums* such as *T. compactum, T. spelta, T. sphaerococcum* and *T. macha.* Similaraties are manifested in family *Cucurbitaceae*, e.g., Watermelon (*Citrullus vulgaris*), *Cucumis* sp. Cantaloups, squashes, where fruits are round or oblong, and simple or segmented, variation in colour of fruits etc. Similar variations could also be noticed in different genera of *Cruciferae* e.g. *Brassica, Eruca, Raphanus, Sinapis, Capsella* with respect to form of leaves, arrangement of blossoms, hairiness of fruits, etc.

4.4 CONSERVATION OF PLANT GENETIC RESOURCES

After the domestication of useful plants, the early man was harvesting seeds and fruits from a much larger number of plant species than those which find place in the modern agriculture (Sturtvant, 1985). Of the thousands of plant species known to us the modern man derives his food from just 25 to 30 plant species, including wheat, rice, maize, sorghum, oats, barley, pigeonpea, gram, groundnut, banana, tomato, citrus, potato etc. (Harlan, 1975). The tendency in more recent years has been to restrict resources of food. For example, more people in India now consume predominantly rice and wheat. To quote an example, sorghum was cultivated in the state of Madhya Pradesh (a state of India) on about 20 to 22 lakh hectares during sixties. A large population inhabiting rural sector consumed sorghum in the form of unleavened bread. Presently sorghum is being grown on just 9.7 lakh hectares (1996). Wheat is predominantly consumed in these sorghum areas. Such a tendency leads to loss of plant genetic diversity in the form of earliest cultivated forms and the land races evolved from the wild populations. Thousands of land races of most of the crop species grown in India before the arrival of "Green Revolution" have been replaced by very few high yielding varieties (including hybrids). Moreover, breeding efforts to evolve high yielding varieties aimed at purification of genetically mixed land varieties, led to uniformity and reduced genetic variability within a variety.

It is needless to say that the plant genetic diversity generated by the evolutionary forces and subsequently during the domestication, should be conserved because the genes contained in them may be required any time in future. To quote an example, earlier tall varieties of wheat and rice were considered as desirable in India. However, it was realized during the sixties that dwarf plants respond better to higher levels of nitrogen and irrigations. Consequently, Norin genes for shorter internodes of wheat from Japan and Dee-Geo-Woo-Gen (Dee-Geo means short leg in Chinese) gene for semi-dwarf plant stature of rice from China became precious. The cytoplasmic-genetic male-sterility gene WA of rice from Hainan Island, helped to produce hybrid rice on large scale. Wild and weedy relatives could be the source of resistance against diseases and insect pests.

In any biologically evolving system, when evolution sets in motion, extinction of unfit organisms is a biological necessity in the process of natural selection. Before extinction takes place, a species becomes rare. Usually the endemic species become rare due to their narrow habitat (Nayar, 1989). There is now a general agreement that the current rates of species loss exceed any thing found in the past 65 million years. The International Union for Conservation of Nature and Natural Resources (IUCN) has estimated that if the current trends continue, we may lose upto one quarter of the world's species by the year 2050 (Swaminathan, 1995). Considering all these facts a need has been urgently felt to conserve the plant genetic resources.

Conservation of plant genetic resources includes the sustainable utilization of germplasm as well as its maintenance (preservation and protection) in a particular state. It is important to recognize that conservation of plant genetic resources is more than just storage. Plant genetic resource activity involves broadly four steps, viz., (*i*) exploration and collection, (*ii*) evaluation and classification, (*iii*) conservation and (*iv*) utilization. Nevertheless, after collection of plant genetic resources from various places and their transportation to the gene banks, the material should pass through quarantine and phytosanitary stages.

Either introduction or exchange of germplasm from other countries may be responsible for inadvertent introduction of diseases and/or insects, which did not exist earlier in the country where such germplasm has been received. As pointed out by Leppik (1970) the centres of origin of host plants may also be the centres of origin of their pathogens and/or insect pests. In other words, by introducing germplasm from such areas there is always a potential danger of introducing genetically diverse pathogens and/or insect pests or even their new races/biotypes. There are several such examples. The well known Irish potato famine of 1845 is believed to be due to the introduction of late blight pathogen (*Phytophthora infestans*) with potatoes introduced from North America to Europe. The Mediterranean fruit fly (*Ceratitis capitata*) introduced from Mexico to Florida in 1929 spread to 28 countries and was eradicated in U.S.A. at the cost of 10 million dollars (Rajak, 1988).

Plant quarantine is designed to legally preclude the introduction of exotic pests and diseases and their spread into the country.

The plant genetic resources are broadly classified into following three categories (Singh, 1989):

(a) Basic genetic resources, (b) Derived genetic resources, and (c) Genetic resources at molecular level.

The basic genetic resources, according to Singh (1989), comprise of wild species, weedy species and the land races. While the derived resources comprise of obsolete varieties, advanced cultivars, parents of released hybrids, mutants and breeding lines. DNA libraries *in-vitro* represent the genetic resources conserved at molecular level.

For operational purposes the germplasm collections could be classified into two categories viz., the base collection and the active collection (Gautam *et al.*, 1998a).

Base collection comprises of the genetic resources preserved for future use in long term storage conditions. Normally, the seeds from base collection are not distributed to the users. It acts only as a reserve stock.

The active collection is made up of accessions maintained for a short to medium term availability for multiplication and distribution to users. These accessions are stored generally upto 10 to 15 years. The active collections are kept under conditions, which ensure that the viability of accessions shall remain above 65 per cent for a period under storage. These are generally held at temperatures between 0 to 10°C depending upon the species being conserved, prevailing ambient environment and cost factors. On the other hand, the base collections are stored at sub zero temperature (less than 0°C) with 3 to 7 per cent seed moisture content (Gautam *et al.*, 1998a).

The methods of conservation of plant genetic resources are

classified into two categories viz., *in-situ* conservation and *ex-situ* conservation. The *in-situ* conservation has been defined as the conservation of ecosystems and natural habitats of the plant species and maintenance and recovery of viable populations of species in their natural surroundings and in case of cultivated species in the surroundings where they have developed their distinctive properties.

In-situ conservation is thus continued maintenance of a plant population within its ecosystem to which it is adapted. This system allows for continued adaptation of plants to the environment.

4.4.1 In-Situ Conservation

In-situ conservation in contrast to *ex-situ* conservation approach allows for continued dynamic adaptation of plants to the environment. The distribution pattern of wild plants in different geographical regions and the areas of their concentration, where rich diversity of wild species still continues to grow, are especially valuable from the point of view of *in-situ* conservation programmes. *In-situ* conservation approaches have to be different for cultivated and non-cultivated gene resources. These include conservation of land races or traditional crop cultivars 'on farm' and conservation of wild relatives and wild food plant species in protected areas.

In-situ conservation of plant genetic resources could be executed through establishment of natural or biosphere reserves, national parks, or by enacting special legislation to protect the endangered or threatened species. In this system wild species and their associated natural ecosystem are preserved together to maintain the genetic integrity of populations and allow natural evolution.

The 'on farm' conservation of crop genetic resources can be defined as the continued cultivation and management of a diverse set of crop population by the farmers in agro-ecosystem where a crop has evolved. It is a dynamic process and is aimed at evolutionary processes that continue to shape the diversity. It is very well taken into consideration that the farmers have developed and nurtured the crop genetic diversity and this process still continues among many farmers. The major challenge of implementing 'on farm' conservation of crop genetic resources is to encourage the farmers in the areas of diversity to continue same thing that many already do, i.e. growing diverse mixtures of land races for both subsistence and commercial purposes.

In-situ conservation of plant genetic resources has several advantages as compared to the *ex-situ* approaches, viz.,

(i) It usually allows increased probabilities of conserving a large range of potentially useful alleles.

(ii) It is especially useful for those plant species, which cannot be established or regenerated outside their natural habitats.

(iii) It allows evolution to continue. It is valuable option in the light of the probability that the parasites and crop species co-evolve. This provides a rich source of resistance to the breeders.

(iv) *In-situ* approach assures protection of associated species. When the plant species of present economic importance are conserved *in-situ* within a natural ecosystem, conservation of many subsidiary species, which are presently not economically valuable but are part of nature's heritage, can be achieved as a bonus.

(v) It facilitates research on species in their natural habitats.

The plant genetic resources, which could be conserved *in-situ*, are categorized in four groups, viz., wild crop relatives, semi domesticated (weedy) crop relatives, perennial species and land races of crop species.

Conservation of plant biodiversity is a need of time. Our ancestors, in India, realized this problem much before the advent of modern science and devised unique ways for conservation of plant biodiversity. They tapped the weak point of human psychology, i.e. their fear for unknown, future, god and linked conservation practices with religion. This is a cost free approach, which is followed religiously and faithfully. The task of maintaining the plants from one generation to the next is regarded as a pleasant duty and performed with religious fervour (Ahuja *et al.*, 2001).

According to Ahuja *et al.* (2001) our ancestors were environmentalists who studied each plant carefully to understand its economic utility. They attached different plants to different gods. The most important plants were included in every day rituals while the relatively less important ones were included in annual or biannual and more specific worships. In these rites not only important food crops, such as *Oryza sativa* and sesame, but weeds and wild races were also used. However, most of the conservation of plant biodiversity, *in-situ*, was confined to trees. For example, Ahuja *et al.* (2001) mentioned a list of plants considered as sacred throughout the India, some of which are *Aegle marmelos* (Bel), *Ficus religiosa* (pipal), *Ficus benghalensis* (Banyan or vata), *Musa paradisica* (Kela), *Curcuma domestica* (Haldi), *Cyanadon dactylon* (Doob), *Mangifera indica* (Mango), *Oryza sativa* (Rice), *Sesamum indicum* (Til), *Cocos nucifera* (Narial, coconut), *Saraca asoca* (Ashok), *Santatum album* (Chandan). Among other plants worshiped by great majority of Indians is tulsi (*Osimum sanctum*, scared basil). This plant could be found in almost every house of Hindus. Further the authors have listed plants scared to specific tribes/religions in India, plants associated with various gods in India, plants associated with various deities (such as serpent worship, Varuni worship, Moon worship, Sitala worship, village worship, etc), plants used in various ceremonies and vratas etc.

Because of the importance of various trees mentioned above, practically in every village either of these are planted and looked after.

Further conservation of plant biodiversity was recognized as an important duty of man because many plants are venerated. Mehra (1996) commenting on Surpala's Vrikshayurveda says that the text was written with possibly an eye on conservation of plant biodiversity.

4.4.2 Ex-situ Conservation

Maintenance of plant genetic resources outside their original habitats involves *ex-situ* conservation. This does not give scope for natural evolution. Plant genetic resources could be conserved *ex-situ* by preserving the germplasm resources in (*a*) Seed gene banks (*b*) Field gene banks and (*c*) *In-vitro* techniques, including cryopreservation. Conservation could be achieved even through botanical garden, arborata, 'on farm' and like means.

(a) Seed Gene Banks

The effective and relatively cheaper means of conservation of plant genetic resources is through long term cold storage of seeds. Seeds are grouped into two categories based on response to dehydration (Roberts, 1972). Majority of the plant species possess seeds which tolerate desiccation to a considerable extent and are known as "orthodox". These seeds could be stored at low temperatures (–20°C), after drying to low levels of moisture (5 to 8 per cent) for long durations. A second group of plant species, called recalcitrant seeds, on subjecting to drying of seeds suffer injury and, therefore, can not be stored at sub-zero temperatures without freezing damage. The recalcitrant seeds are completely intolerant of drying to moisture content that could enable them to withstand lower storage temperatures (Chin and Pritchard, 1988). Most, but not all the recalcitrant seeds producing species, are tropical and include for example coconut, mango, walnut, avocado, cocoa, jackfruit and oak. The *ex-situ* conservation of orthodox seeds in the gene banks has been the most effective strategy for plant genetic resources conservation.

The germplasm collection could be classified into three categories, depending upon their use and the type of storage conditions to be maintained. These are base collections, active collections and working collections.

Base collection is a collection that is preserved for further use in long term storage conditions. It acts only as a backup/reserve collection. The requirement for long term storage of base collection is met with by storing the seeds at temperatures – 18 to – 20°C in sealed containers and seed moisture between 4 to 6%. The seed moisture contents may be raised in exceptional crops where there is a strong evidence of problem like seed breakage during handling at very low moisture content. The active collections comprise of accessions stored for a period of 10 to 15 years. The active collections are used for regeneration, multiplication, distribution, characterization and documentation. Active collections are kept under conditions, which ensure that the viability of seeds of accessions shall remain above at least 65 per cent for the period of storage. These collections are generally stored at temperatures around 0°C and seeds moisture around 8 per cent. The working collection is the third category of plant genetic resources, which are also conserved. These comprise of genetic resources, which are regularly used by the plant breeders, entomologists, plant pathologists etc. who are not responsible for their perpetuation (Singh, 1989). Such accessions are stored for about three to five years at 5 to 10°C and seed moisture between 8 to 10 per cent.

Among the species with orthodox seeds, there is a genetic variability for storability. Some species can be stored for a long period while others do not respond favourably when stored for long. For example, seeds of plants belonging to *Leguminosae*, e.g. *Cassia, Trifolium* and *Lupins* sp., can be stored for over 100 years. However, seeds of groundnut have short cold storagability.

(b) Field Gene Banks

The plant germplasm, which is difficult to manage through seeds and may be perennial in nature, can be conveniently maintained through field gene banks. Many crops producing recalcitrant seeds and some which can not be multiplied even vegetatively are to be conserved as living field collections. In vegetable crops the germplasm of bulbous and rhizomatous crops are being maintained in field banks at NBPGR (Gautam, *et al.* 1998b). According to Singh (1989) the field gene banks do not truly represent the range of genetic variability that exists in a crop species.

(c) In-Vitro Techniques

Interest in *in-vitro* conservation of plant genetic resources has emanated mainly from the developments in the field of plant tissue culture and biotechnology. *In-vitro* conservation is complementary to the conventional conservation methods, especially for those plant species which can not be conserved through orthodox seeds and which are poor seed producers. Recalcitrant seed producing plant species, we know, can not be stored in gene banks due to their inherent weakness. Storage/conservation of such plant species could be done through various *in-vitro* methods.

Another problematic case for conventional storage is of those species that are propagated clonally. This may be for reasons such as sterility or poor seed production or because of high degree of heterozygosity that can only be maintained via vegetative propagation. This heterozygous group is represented by plant species such as potato, cassava, yam, sweet potato, taro, banana and plantain, apples, pears, various *Allium* sp., hops and a number of species (Langhe, 1984). Conventionally these problem crops are conserved in field gene banks. According to Withers and Engels (1990) this is not very satisfactory because:

(i) The field gene bank can not be considered to be a long term conservation mode and in that sense only an active and not a base collection and

(ii) Plants growing in the field gene bank are open to risks of damage by severe weather, disease, insects etc.

Moreover, field gene bank is costly to maintain. Hence a safer and effective alternative mode of conserving such plant genetic resources is *in-vitro* conservation including the cryopreservation (Withers, 1995).

There are two alternative options for *in-vitro* conservation of plant genetic resources (Chandel *et al.*, 1988): slow growth for short and medium term conservation and cryopreservation for long term conservation. Cryopreservation can also be used for short and medium term conservation. Slow growth method refers to any tissue culture procedure, which tends to reduce the growth rate of plant tissue *in-vitro* (Withers, 1991). The objective of this method is to attain longest period of subculture without detrimental effects on the plant tissue. Slow growth of cultures could be achieved by using one or combination of following techniques.

Slow Growth Storage

As mentioned in preceeding paragraph there are two options for *in-vitro* conservation, viz., for base collection, which has to be conserved for long periods, cryopreservation is desirable while for active collection *in-vitro* alternative provides for slow growth storage. Some cultures grow sufficiently slowly so that it is convenient to maintain them under normal growth conditions and sub-culture them at the intervals of 6 months or more. Nevertheless, in most cases, it is necessary to apply slow growth conditions (Ng and Ng, 1991; Withers, 1991; Engelmann, 1991). The culture medium in which desired regeneration of plant is achieved is very important for *in-vitro* conservation.

The explant material used for slow growth storage should be able to maintain high degree of genetic stability. Generally organized tissues, such as meristem, apical and axillary bud from actively growing plant parts are preferred than unorganized cultures (cell, callus, protoplast). Among the explants, meristem and axillary buds are more desirable. The nutrients in the culture medium could be increased or reduced in order to slow down the growth rate of culture. Usually six to twelve cultures of each accession are maintained. The culture vessels are sealed, labeled properly and incubated at defined temperatures and light conditions. Reduction of the culture temperature to slow down the growth is also commonly used. Cultures which are maintained at temperatures around 25°C may be transferred to 10 to 15°C to slow down growth rate. Tropical species are generally cold sensitive. The physiological damages induced by low temperature are known as 'chilling injury'. Another alternative to slow the growth rate is addition of inhibitory levels of growth retardants (maleic hydrazide, abscicic acid, cycocel, phosphone *D*) to the growth medium. Mannitol and sorbitol act as osmotic inhibitor, and may be added to fresh culture medium. Minimal growth can be achieved by lowering the available oxygen level to the tissues. The simplest method to achieve this is to cover the tissue with mineral oil layer. Although little work has been done, judicious use of combination of treatments mentioned above may be useful for slow growth rate *in-vitro* conservation.

The slow growth storage has limitations (Withers, 1995) viz.,

(i) It involves application of stress which could lead to selection of stress tolerant genotypes from within a heterogeneous population, and
(ii) *In-vitro* cultivars are generally regarded as clones. Somaclonal variations are known in tissue culture (Scowcroft, 1984) which could be considered as desirable for developing new varieties, but is detrimental during conservation.

Slow growth storage has been applied successfully to more than fifty species. For example, it is routine for potato and sweet potato. For more details on slow growth storage reader is referred to Ng and Ng (1991), Withers (1991).

Cryopreservation

Cryopreservation is a very different phenomenon from slow growth storage. For long-term storage, cryopreservation (freeze preservation in liquid nitrogen) using liquid nitrogen (liquid phase at–196°C) is currently in practice. At this ultra low temperature all the metabolic activities of the cells cease, and theoretically plant material can be stored for an indefinite period. Evidence from other biological systems indicates that the only significant threat to stability at cryogenic temperatures is free radicle damage. Most of the cryogenic procedures include application of chemicals to protect against the action of free radicles generated, for example, by back ground radiation (Withers, 1995). The real challenge in developing a cryopreservation method is the achievement of a safe transition from and to the storage temperature. Research has accordingly concentrated on measures to protect tissues against stress induced during these transitions, by optimizing treatments at the stages of pregrowth, cryoprotection, cooling, warming and recovery.

Cryopreservation of Pollen

Using *in-vitro* techniques pollen can be preserved for a considerable period. Storage of pollen is important in germplasm conservation with regard to affecting fertilization outside, seasonal, geographical barriers, besides ensuring supply of pollen (Withers, 1980). Using *in-vitro* technique pollen can be employed to produce haploid plant and subsequently homozygous diploid (dihaploid). Pollen conservation in liquid nitrogen is feasible at both low (–20°C) and ultra low (i.e. – 196°C) temperatures (Omura and Akihama, 1980). But this method does not conserve whole genotypes of plant and has, therefore, a complementary role in long term conservation. Cryopreservation of pollen of tomato and brinjal varieties has been successfully accomplished (Seshadri, 1988).

A routine method was published by Withers and King (1980) for cryopreservation. This has proved applicable either unchanged or with limited modifications. The basic technique is as follows:

Selection of Material

Generally rapidly growing meristems, which are small, relatively resistant to freezing, possessing fewer or smaller vacuoled cells and dense cytoplasm are preferred (Engelmann, 1991). The preservation of meristem and shoot tips, is important in the context of germplasm conservation because they have greater genetic stability than callus and suspension cultures, due to their organized growth (*D*'Amato,

1975). These systems are, however likely to generate somaclonal variation (Larkin and Scowcroft, 1981). Cryopreservation has achieved highest degree of refinement with suspension cultures (Withers, 1985).

1. *Pregrow*: This refers to period in culture medium before cryopreservation begins. Pregrow for 3 to 7 days in normal medium or medium with 3 to 6 per cent mannitol. This is to induce freeze tolerance (Mannitol, sorbitol, sucrose act as ostmoticum).
2. *Cryoprotection*: Just before freezing, compounds such as cryoprotectants, are applied which include a mixture of 0.5 M DMSO (dimethyl sulphoxide), 0.5 M glycerol and 1 M sucrose in culture medium at 4°C for 1 hour. Dispense into ampoules.
3. Cool at –1°C per minute to –35°C, hold for 40 minutes and then plunge into the liquid nitrogen.
4. Thaw in water bath at + 40°C.
5. Transfer the cell suspension without washing to a plate of semi solid medium.
6. Once the cells resume normal growth, transfer them to liquid medium.

Similar approaches for cryopreservation of callus culture have been successful. The *in-vitro* conservation method has following advantages:

(i) Multiplication rate is higher.
(ii) Plants are grown in asceptic condition and hence can be kept free from disease.
(iii) Very small space is required.
(iv) Genetic erosion is reduced.
(v) Ready availability of material for distribution.
(vi) Easy transportation with limited quarantine restrictions.
(vii) Storage through pollen and anther culture.

Vitrification, encapsulation/dehydration are other approaches for conservation of plant genetic resources. An important opinion of Withers (1995) needs consideration. "A further massage to convey in connection with *in-vitro* conservation is that it is a complementary approach to the traditional conservation system and not a replacement for them. In certain situations and locations field gene banks may be logically more appropriate as primary method of choice".

References

Ahuja, U.; R. Thakrar; P. Nathwani and S.C. Ahuja (2001). Role of religion in conservation of plant biodiversity. *Asian Agri.History*, 5: 225-248.

Allard, R.W. (1960). *Principles of plant breeding*. John Wiley and Sons, Inc., New York.

Anderson, E. (1949). *Introgressive hybridization*. John Wiley and Sons, Inc., New York.

Arora, R.K. (1988). Indian gene centre — priorities and prospects for collection In *Plant genetic resources — Indian Perspective*, Eds. R.S. Paroda *et al.*, NBPGR, New Delhi, pp. 66-75.

Arora, R.K. (1995). Genetic resources of vegetable crops in India — Their diversity and conservation. In *Genetic resources of vegetable crops* NBPGR, New Delhi, pp. 29-39.

Arora, R.K. and E.R. Nayar (1984). *Wild relatives of crop plants in India*. NBPGR, India.

Briggs, F.N. and P.F. Knowles (1967). *Introduction to plant breeding*. Reinhold Publishing Corp., New York.

Chatterjee, D. (1939). Studies on endemic flora of India and Burma. *J.Royal Asiat. Soc.*Bengal, N.S. *Science*, 5: 19-57.

Chandel, K.P.S.; R. Pandey; R. Chaudhary; S.M. Balachandran and N. Sharma (1988). Biotechnology — its role in conservation of plant genetic resources. In *Plant Genetic Resources*, Eds. R.S. Paroda *et al.*, NBPGR, New Delhi, pp. 499-516.

Chin, H.F. and H.W. Pritchard (1988). Recalcitrant seeds, a status report. International Board of Plant Genetic Resources, Rome, Italy.

D'Amato, F. (1975). The problem of genetic stability in plant tissue and cell cultures. In *Crop genetic resources for today and tomorrow* Eds. O.Frankel and J.G. Hawkers. Cambridge Uni. Press, London, p. 333.

Doggett, H. (1965). Disruptive selection in crop improvement. *Nature*, London, 206: 279-280.

Doggett, H. (1970). *Sorghum*. Tropial agricultural series, The Longmans.

Ellis, R.H.; T.D. Hong and E.D. Roberts (1985). *Hand book of seed technology for gene banks*. Vol.I. Principles and Methodology, Hand books for gene banks No.2 IBPGR, Rome, Italy.

Engelmann, F. (1991). *In vitro* conservation of horticultural species. *Acta Hort.*, 298: 327-334.

Gautam, P.L.; S. Kochhar; S.K. Pareek and Ramnath (1998a). Plant genetic resources conservation and exchange in the national agricultural research system. Paper presented in "National Consultation on Access to Biological Diversity and Benefit Sharing, Innovations, Incentives and Institutions. I.I.M. Ahmedabad, April, 10-12, 1998.

Gautam, P.L.; A. Kumar and B.M. Singh (1998b). Germplasm collection, maintenance and utilization in improvement of vegetable crops. Paper presented in Summer School on "Advanced Technologies in the improvement of vegetable crops including cole crops." I.A.R.I., New Delhi, May 4-24, 1998.

Goodspeed, T.H. and R.E. Clausen (1928). Interspecific hybridization in *Nicotiana.* VIII. The *Sylvestris-Tomentosa-Tabacum* hybrid triangle and its bearing on the origin of tabacum. University of California Publications in Botany, 11: 245-256.

Gupta, R. and R.S. Rana (1995). Conservation and sustainable use of medicinal plant genetic resources in India. In *Genetic Research and Education: Current Trends and the Next fifty years*, Eds. B. Sharma *et al.*, Indian Society of Genetics and Plant Breeding, New Delhi, pp. 30-39.

Harlan, J.R. (1951). Anatomy of gene centres. *Amer.Nat.*, 85: 97-103.

Harlan, J.R. (1956). Distribution and utilization of natural variability in cultivated plants. *Brookhaven Symposia in Biology*, 9: 191-208.

Harlan, J.R. (1975). Crops and man. American Society of Agronomy/Crop Science Society of America, Madison, Wisconsin.

Hays, H.K.; F.R. Immer and D.C. Smith (1955). *Methods of plant breeding.* McGraw Hill Book Company, Inc., New York.

Hintum, Van T.J.L. (1995). Hierarchical approaches to the analysis of genetic diversity in crop plants. In *Core collections of Plant Genetic Resources*, Eds. T. Hodgkin, A.H.D. Brown, Th.J.L. van Hintum and E.A.V. Morales. John Wiley and Sons.

Jain, K.B.L. (1989). Cereals — Wheat and Barley. In *Plant breeding – theory and practice*, Ed. V.L. Chopra. Oxford and IBH Publishing Co., New Delhi, pp. 77-99.

Joshi, A.B. (1988). India's role in global genetic resources activities. In *Plant genetic resources*, Ed. Paroda, R.S.; R.K. Arora and K.P.S. Chandel, N.B.P.G.R., New Delhi, pp. 3-6.

Khor, Martin (2000). Development. Time for a paradigm shift. In *The Hindu survey of the Environment, 2000.* M/s Kasturi and Sons, Chennai 600 002, India.

Kihara, H. (1966). Factors affecting the evolution of common wheat. *Indian J.Genet.*, 26A: 14-28.

Langhe, E.A.L.De (1984). The role of *in-vitro* techniques in germplasm conservation. In *Crop genetic resources: Conservation and education.* Eds. J.H.W. Holden and J.T. Williams. Allen and Unwin, London, U.K., 131-137.

Larkin, P.J. and W.R. Scowcroft (1981). Somaclonal variation — a novel source of variability from cells of plant improvement. *Theor. Appl.Genet.* 60: 197-214.

Leppik, E.E. (1970). Relative resistance of *Cucumis* spp. to diseases and insects. Plant Introduction Investigations, 43-54.

Mehra, K.L. (1996). Biodiversity perspective. In "Surpala's Vriksha-yurveda (Nalini Sadhale tr). *Agri-History Bulletin No.1.* Asian Agri. History Foundation, Secunderabad, India, pp. 71-74.

Muntzing, A. (1966). On the evolution and breeding of cultivated plants. *Indian J.Genet.*, 26A: 3-13.

Nayar, M.P. (1988). Prospects of *in situ* conservation of native endangered species. In *Plant genetic resources*, Eds. Paroda *et al.* National Bureau of Plant Genetic Resources, New Delhi, pp. 44-48.

Nayar, M.P. (1989). *In situ* conservation of wild flora resources. In "National symposium on conservation and sustainable management of India's Genetic Estate" *WWF*, New Delhi, November, 3-4, 1989.

Ng, S.Y.C. and N.Q.Ng (1991). Reduced growth storage of germplasm. In *In vitro methods for conservation of plant genetic resources*. Ed. J.H. Dodds, Chapman & Hall, London, U.K.11-39.

Nordenskiold, H. (1949). Synthesis of *Phleum pratense*, L. from *P.nodosum* L. *Herditas*, 35: 190-202.

Omua, M. and T. Akihama (1980). Pollen preservation of fruit trees for gene banks in Japan. FAO/IBPGR *Plant Genet. Resources Newsletter*, 43: 28.

Parthasarthy, N. (1946). The probable origin of North Indian sugarcanes. *Journal of Indian Botanical Society*, 133-150. (M.P.O. Iyengar Commemoration Volume).

Poehlman, J.M. (1987). Breeding field crops. 3rd Edition. AVI Publishing Co., Inc., West Port, Connecticut.

Pritchard, H.W. (1991). Preservation of tree seeds and embryos for genetic conservation. Proc. 28th Annual Meeting, Society of Cryobiology, 7 to 12 July, 1991. University of Leuven, Belgium, 42.

Rajak, R.L. (1988). Plant quarantine regulations in relation to exchange of plant genetic resources. In *Plant Genetic Resources*, Eds. Paroda *et al.*, NBPGR, New Delhi, pp. 49-58.

Roberts, E.H. (1972). *Viability of seeds*. Publ. Chapman and Hall, London, p. 253.

Schwanitz, F. (1957). *Die Entstehung der Kultur pflanzen.* Berlin — Gottin Heidelberg. Springer-Verlag.

Schwanitz, F. (1959) Selection and race formation in cultivated plants. *Cold Spr.Harb.Symp.Quant.Biol.*, 24: 107-114.

Schwanitz, F. (1960). Das Ertragsproblem in enwicklungsphysiologischer Sicht. *Zuchter*, 30: 45-56.

Schgal, J.C.; D.K. Mondal; C. Mondal and S. Vadi (1990). Agro-ecological regions of India. NBSS Tech.Bull. No. 24, NBSS and UP, Nagpur, p. 73.

Seshadri, V.S. (1988). Genetic resources and their utilization in vegetable crops. In "Plant Genetic Resources". Eds. Paroda *et al.*, NBPGR, New Delhi, pp. 335-343.

Simmonds, N.W. (1979). *Principles of crop improvement*. Longman, London and New York.

Singh, R.B. (1989). Crop genetic resources. In *Plant breeding — Theory and practice*. Ed. V.L. Chopra, Oxford & IBH Publishing Co., New Delhi, pp. 433-452.

Singh, R.B. (1995). New dimensions of conservation and utilization of biodiversity and FAO's role. In "Genetic Research and Education: Current trends and the next fifty years". Ed. B. Sharma *et al.* Indian Society of Genetics and Plant Breeding, New Delhi, pp. 121-128.

Sreenivasan, T. (1989). Sugarcane. In *Plant Breeding — Theory and Practice*, Ed. V.L. Chopra. Oxford and IBH Publishing Co. Pvt. Lt d., New Delhi, pp. 253-274.

Stevenson, G.C. (1965). *Genetics and breeding of sugarcane*. Longman, London.

Sturtvant, A.H. (1985). Edible plants of the world. In *Flora Dietica* Ed. U.P. Hedrick. Dover Publisher, New York.

Swaminathan, M.S. (1995). Genetic resources for sustainable advances in crop productivity. In *Genetic Research and Education: Current Trends and the Next Fifty Years*. Eds. B. Sharma *et al.*, Indian Society of Genetics and Plant Breeding, New Delhi, pp. 1-11.

Swaminathan, M.S. (2000). Biodiversity. Equity in benefit sharing. In *The Hindu survey of Indian environment, 2000*. Kasturi and Sons, Chennai, 600-002, India.

v.Sengbusch, R. and K. Zimmermann (1947). 20 Jahre susslupinenforschung und zuchtung in Deutschland. *Forsch und Fortschr*, Jahrg, 21/23, p. 20.

Vishnuswarup (1998). Utilization of hybrid vigour in meeting the vegetable demands by the turn of century. In *Summer School on advance technology in crop improvement of vegetable crops, including çole crops*. May, 4-24, 1998. Eds. Chakravarti, A. and N. Singh, IARI, New Delhi, p. 24.

Vavilov, N.I. (1926). Studies on the origin of cultivated plants. *Bull.Appl. Bot.*, 26(2): 248.

Vavilov, N.I. (1950). The origin, variation immunity and breeding of cultivated crops. *Chron.Bot.*, 13: 1-34.

Welhausen, E.J. (1966). The origin and breeding of maize. *Indian Journal of Genetics and Plant Breeding*, 26A: 45-59.

Withers, L.A. (1980). Tissue culture storage for genetic conservation. IBPGR Publication AGP: IBPGR/80/8 International Bureau of Plant Genetic Resources, Rome.

Withers, L.A. (1985). The preservation of plant cell, tissue and organ cultures and seed for genetic conservation and improve agriculture practices. In *The effects of low temperature on biological systems*, B.W.W. Grout and J.G. Morris Arnold, London.

Withers, L.A. (1991). *In vitro* conservation. *Biol.J.Linnean Soc.*, 43: 31-42.

Withers, L.A. (1995). *In vitro* conservation of plant germplasm. In *Genetic research and education: Current trends and the next fifty years*, Eds. B. Sharma *et al*. Indian Society of Genetics and Plant Breeding, New Delhi, pp. 93-113.

Withers, L.A. and P.J. King (1980). A simple freezing unit and cryopreservation method for plant cell suspènsions. *Cryoletters*, 1: 213-220.

Withers, L.A. and J.M.M. Engels (1990). The test tube gene bank — a safe alternative to field conservation. *IBPGR New Asia Pacific*, 3: 1-2.

Zeven, A.C. and P.M. Zukovsky (1975). *Dictionary of cultivated plants and their centres of diversity*, PUDOC, Wageningen, p. 219.

Zeven, A.C. and J.M.J.de Wet (1982). *Dictionary of cultivated plants and their regions of diversity*. Centre for Agricultural Publishing and Documentation. Wageningen, Netherlands, p.259.

Zhukovskij, P.M. (1968). New centres of origin and new gene centres of cultivated plants including specifically endemic micro centres of species closely allied to cultivated species. *Bot.J.*, 53: 430-460. (Russian Bot Z.).

Zukovskij, P.M. (1970). The world gene pool of plants for breeding. Mega-gene centres and endemic micro-gene centres. *Nauka*, Leningrad RO42, p. 87 (Russian).

5

The Genetic Composition of Crop Plants

Genotype is the particular assemblage of genes possessed by an individual. The genotypic composition of plants in the population is influenced by several agencies, such as the mode of reproduction, the kind of pollination mechanism prevailing, selection pressure being exercised (both natural and artificial), migration (both immigration and emigration), mutation and the size of population. As will be shown later, when the population is small, the genes are subject to random sampling variation consequently the gene frequencies in small populations change and hence the genetic properties.

By virtue of the mating system predominantly prevailing in the self pollinated crop plants, all the individuals in the population are homozygous at most, if not all the loci. However, as will be shown in section 5.1, individuals in population of self-pollinated crop species may not all be genotypically alike i.e., the population may be heterogeneous. Contrary to this, individuals in the populations of cross-fertilized crop plants are often heterozygous at most of the loci. The population of these crop species is mostly heterogeneous. Only in some exceptional cases, such as the single-cross hybrids between the completely inbred lines of maize, the individuals are heterozygous at almost all the loci, but the population is expected to be homogenous. But the populations of double-cross, double- top-cross maize hybrids, synthetics, and composites are heterogeneous.

Asexually propagated crop plants reproduce by vegetative means. The progeny of an individual is a clone. A clone is a group of individuals descended by mitosis from a common ancestor. Hence, the clones are genotypically similar to their parental stocks, in the absence of mutation. The clonally propagated crops are virtually all cross-pollinated and hence tend to maintain variability in heterozygotes (Simmonds, 1962). Selection of a clone probably implies selection for high heterozygosity. Asexually propogated crop plants often carry recessive mutations in heterozygous state. A dominant lethal mutation may be eliminated promptly in these populations. But the individuals carrying deleterious recessive mutations in heterozygous state continue indefinitely in such populations, unless self-fertilization takes place or another recessive mutation at the locus on homologous chromosome takes place and becomes homozygous. Such populations, therefore, carry a considerable "genetic load".

The extent to which population departs from perfect genetic constitution is known as genetic load. According to Wright (1951) "The best adapted form in a species is usually one which is close to average in all the quantitatively varying characters". Fisher (1930) states that natural selection working on whole complex of the characters tends to favour organisms clustering about the mean values for all the characters. This suggests that there will be selection against extreme phenotypes in the population and if the reproductive wastage is to be avoided, to achieve phenotypic uniformity among individuals, population should carry least number of genes with extreme phenotypes. Nevertheless, population should maintain genetic variability and remain flexible to meet the long term evolutionary challenges. According to Mather (1943) successful species of plants have developed a mechanism to overcome the conflict between demands of immediate fitness and long range flexibility via building a linkage system. Even when two genes exist at a locus, variability is in free form, according to Mather (1943), when genes are in homozygous state, i.e. *AA* and *aa*. Selection can disfavour either *AA* or *aa*. But the variability shall be in the bound form in Aa. Heterozygote can resemble one or the other homozygote but not both of them. This variability is known as potential variability since it shall become homozygous in certain circumstances. Potential variability also exists wherever two or more gene differences influence the same character. Suppose genes *a* and *b* produce undesirable phenotypes, then potential variability will be in bound form in *AAbb* and *aaBB*. Release of the bound variability will be slow, because firstly heterozygote should form (i.e., *AaBb*) and then *AABB*. A population shall be in advantageous state when many genes are scattered along the chromosome, in heterozygous state.

Segregation and recombination are the mechanisms which contribute largely to the genetic variability, in addition to mutation, in sexually reproducing crop species. Asexually propagated crops are some what deprived of these mechanisms, except the mutation.

5.1 SELF-POLLINATED CROP PLANTS

Populations of self-pollinated crop plants are characterized by individuals homozygous at all the loci. For example, self-fertilization of individuals *AA* and aa shall lead to progenies with genotypes *AA* and *aa*, respectively. However, if chance mutation of *A* to *a* takes place or *AA* hybridizes with *aa*, heterozygous individuals may be produced. Self-fertilization of heterozygous individual and also the homozygotes, for about 8 to 9 generations results in a population comprising of completely homozygous individuals, *AA* and *aa*, in equal proportions (Table 5.1). An individual heterozygous at *n* loci, in the initial population of self-pollinated crop, ends up in 2^n homozygotes, at the end of complete inbreeding. With *n* gene pairs in heterozygous state, the proportion of homozygous individuals after *m* generations of self-fertilization is given by:

$$\left[\frac{2^m - 1}{2^m}\right]^n$$

where *n* represents the number of heterozygous loci in the initial population and *m* the number of generations of self fertilization. For instance, about 88 per cent population shall be homozygous at the 4 loci after 5 rounds of self-fertilization. In this case $n = 4$ and $m = 5$. A reader should confirm, the percentage homozygous loci.

Table 5.1: Attainment of homozygosity by a self-pollinated population*

Generation of self fertilization	*Percentage of genotypes in progeny*			*Per cent heterozygosity*
	AA	*Aa*	*aa*	
0	–	100.00	–	100.00
1	25.00	50.00	25.00	50.00
2	37.50	25.00	37.50	25.00
3	43.75	12.50	43.75	12.50
4	46.87	6.25	46.87	6.25
5	48.44	3.12	48.44	3.12
6	49.22	1.56	49.22	1.56
7	49.61	0.78	49.61	0.78
8	49.80	0.39	49.80	0.39
9	49.90	0.19	49.90	0.19

* This assumes absence of differential fertility among adult individuals and differential viability among the individuals in the progeny. This also assumes absence of linkage.

It should be clear to the reader that the final result of selfing an heterozygous individual leads to a population which comprises of homozygous individuals at all the loci, nevertheless, population shall be heterogeneous. For example, a population comprising of 100 per cent individuals with genotype *AaBb*, will end up (assuming no linkage) in *AABB*, *AAbb*, *aaBB* and *aabb* in equal proportions, after complete inbreeding. Whenever, genes are linked the approach to complete homozygosity will be delayed. It can be confirmed easily that with *n* gene pairs in heterozygous state, the F_1 will form 2^n kinds of gametes (Table 5.2) and the F_2 generation will comprise of 3^n kinds of genotypes when these gametes unite at random. When one gene pair is under consideration smallest perfect population in the F_2 would be made up of 4 plants, 1*AA*, 2*Aa* and 1*aa*. With 2 gene pairs 16 plants make up a smallest perfect population. With *n* gene pairs, to generalize, 4^n plants comprise of the perfect smallest population.

Table 5.2: Showing influence of number of heterozygous loci on kinds of gametes and genotypes in F_1 and F_2 of self-pollinated plants

No. of loci at which F_1 is heterozygous	*Kinds of gametes formed by F_1 (expected)*	*Kinds of genotypes in F_2 (expected)*	*Smallest perfect population in F_2*
1	2	3	4
2	4	9	16
3	8	27	64
10	1024	59,049	10,48,576
20	10,48,576	3,48,67,84,401	3,48,67,84,401
n	2^n	3^n	4^n

There is a form of mating known as inbreeding. Inbreeding means mating together of individuals that are related to each other by ancestry. For example, let us consider a population comprising of individuals *AA* and *aa*. Progeny resulting from mating of these individuals shall be entirely *Aa*. This is F_1 generation. If the plants are hermaphrodite, like sorghum or wheat, self-fertilization results in F_2 generation. But in case of maize, pollen grains from one of the F_1 plants can be used to pollinate the silk of another plant of same strain. This is known as sib-mating, i.e. brother-sister mating. The progeny so resulting shall be designated by S_2 rather than F_2. Once again selecting some plants as pollen parents from S_2 to fertilize other plants from this very population leads to formation of seeds of next generation. The progeny, produced by planting ear to row seeds, is again full-sib progeny (within the row all plants result from mating of same two parents) and designated as S_3 progeny rather than F_3. The genotypic composition of S_2, S_3 ..... progenies differs, obviously, from F_2, F_3 ..... progenies.

The degree of relationship between individuals in a population depends upon the size of population. Thus pairs of individuals mating at random are more closely related to each other in the small population as compared to the larger population. Essential consequence of two individuals having a common ancestor is that both of them may carry replicate of one of the gene in their common ancestor and if they mate they pass on these genes to their progeny. Thus the inbred individuals, i.e. individuals produced by inbreeding, may carry two genes at a locus that are replicates of one and the same gene in previous generation. These two genes are known as identical by descent. Thus mating between closely related individuals also leads to homozygosity. The speed at which homozygosity is attained may be slower with this kind of inbreeding as compared to self-fertilization. As will be explained in chapter 11, since intense inbreeding leads to loss of vigour in maize, sib-mating is adopted to produce inbred lines.

Inbreeding does not have any harmful effect on the individuals of self-pollinated crops. According to Mather (1943) homozygosity is the normal state for self-pollinated crops. Self-pollinated crops during the course of evolution developed a genetic organization called "homozygous balance". One of the attributes of this genetic organization is that inbreeding does not lead to loss of vigour. Moreover, F_1 hybrids between the lines are fully vigorous and may even exhibit heterosis. Mather (1943) explained heterosis displayed by the self-fertilized crops on the grounds that inbreeders have descended from outbreeders. Stebbins (1950) provided evidence to show that inbreeding species often descended from outbreeding ancestors. In wheat and tomatoes, cross-fertilization has been supplanted by self-fertilization (Allard, 1960). This is true for peas also. Though evidence strongly suggests that inbreeders arise from outbreeders, reverse is rarely true.

During the outbreeding portion of their evolutionary history, the self-fertilizing species were endowed with the genetic organization that characterizes heterozygous populations of cross-pollinated crops. Inbreeding fixes some of the heterozygous balance, characteristic of outbreeding populations and at the same time leads to rise of homozygous balance. Heterosis manifested by the inbreeders, is according to Mather (1943) and Darlington and Mather (1949), a relic of outbreeding past.

There is a general opinion that genotypic variability in the population is related to breeding system. Accordingly it is believed that inbreeders are composed of homozygous or little variable lines while the outbreeders are composed of a mass of different heterozygous individuals (section 5.2). This conventional view needs modification in case of inbreeders (Simmonds, 1962). While working with one of the barley composite crosses, Jain and Allard (1960) found evidence favoring heterozygous advantage at five out of eight loci considered. Barley is regarded as strongly self-pollinated crop, but the barley used in their experiment showed 1 to 2 per cent cross-pollination. According to Jain and Allard, a low degree of cross-pollination, coupled with some heterozygote advantage, could maintain hybridity indefinitely in self-pollinated crops. There is much more variability in heterozygous phase than we think in self-pollinated crop species.

Regardless of the form in which variability occurs in crop species, its survival depends upon selection pressures exerted in the system of agriculture (Simmonds, 1962).

Many rice varieties, some old cotton varieties, old land races of wheat, sorghum before the green revolution in India were populations of inbred lines showing considerable genetic variability among lines but negligible heterozygosity within the lines. Nevertheless, according to Jain and Allard (1960), this may be an oversimplification. There may be much more hybridity in land races than is usually supposed. After the green revolution era in India, hybrids of sorghum, cotton and rice, high yielding varieties of wheat, hybrids of tomato, brinjal, chilli, etc. were released for general cultivation. The hybrids were highly heterozygous but exhibited low variability.

5.2 CROSS-POLLINATED CROP PLANTS

Because of the predominantly cross-pollinating mechanism prevailing in these crop species individuals within a population are heterozygous at most of the loci and the population is heterogeneous. Some mechanisms such as monoecy, protandry and protogyny encourage, but do not enforce, cross-pollination in these species. However, dioecy, self-incompatibility and male-sterility are the devices, which enforce complete or near complete cross-fertilization. Populations of cross-pollinated crop species are freely inter-interbreeding groups. Pollen grains produced on a plant can pollinate and ultimately fertilize females of any other plant. Like the strictly self-fertilized plants (e.g. cleistogamous) a male gamete is not forced to fertilize female gamete within the same flower, which produced the pollen grains. Such populations are, therefore, known as random mating. Strictly speaking the term random mating should be applied to breeding systems that meet two requirements (Allard, 1960) viz., each individual in the population has an equal chance to produce offspring and any female gamete is equally likely to be fertilized by any male gamete. This theoretical form of random mating is seldom met in nature. Due to artificial selection by the breeder or natural selection, acting via differential fertility and/or differential viability some individuals in the population may be deprived to produce progeny and contribute to next generation. Further, because of maturity differences among plants in the population, the second requirement may not be exactly satisfied. A reproductive community of sexual and cross-fertilizing organisms which shares in a common gene pool was designated as "Mendelian population" by Dobzhansky (1951). This assumes that each individual in this population produces gametes and contributes to a common gene pool from which any two random gametes unite to produce next generation. The term 'panmixia' is sometimes used as a synonym of random mating and hence the population is called panmictic. To describe the behaviour of genes and genotypes and the properties of large random mating populations Hardy and Weinberg independently described a postulate known as Hardy-Weinberg law.

5.2.1 Hardy-Weinberg Law

A British mathematician, G.H. Hardy and a German physician Dr. W. Weinberg, independently proposed an hypothesis in 1908 about the gene frequency and genotype frequency in the large random mating population. According to them in a large random mating population both the gene and genotype frequencies remain constant from generation to generation, in the absence of migration, mutation and selection. This is known as Hardy-Weinberg law. Because the gene frequencies and genotype frequencies do not change from generation to generation, the population is said to be in Hardy-Weinberg equilibrium. This law is of fundamental importance in population genetics.

In what follows we shall assume a single locus with two alleles, viz., A and a. Let us suppose that there are N diploid individuals in the population. Among these N individuals D are dominant homozygotes (AA), H are heterozygotes (Aa) and R are recessive homozygotes (aa), such that $D+H+R=N$. Proportion or percentage of a specific genotype in the entire population is its genotype frequency. Because the genotype frequencies are expressed in per cent or proportions sum of the frequencies of all the genotypes in the population shall be 100 or 1. Because we are considering a population of diploid individuals there will be $2N$ genes in the population. The proportion or percentage of a specific gene in the population is its gene frequency. Thus homozygous dominants (D) shall contain $2D$ genes of A kind while the heterozygotes (H) carry $2H$ genes out of which half, i.e., H, are A, and rest a. Thus the gene frequency of gene A denoted by $\hat{p}$ is given by

$$\hat{p} = \frac{2D+H}{2N} = \frac{D+½H}{N} \qquad \text{...(5.1)}$$

Whenever genotype frequencies are expressed in proportions, sum of genotype frequencies in the entire population is unity, i.e., $N = 1$, and (5.1) can be simply written as $D + ½H$. In an analogous manner the gene frequency of a can be derived. The gene frequency of gene a is denoted by $\hat{q}$, which shall be:

$$\hat{q} = \frac{2R+H}{2N} = \frac{R+½H}{N} \qquad \text{...(5.2)}$$

And again, when genotype frequencies are expressed in proportions, (5.2) can be written as $R + ½H$. In the expressions (5.1) and (5.2) hat (^) has been used on p, and q, which symbolizes expected values of.

Because we are considering only two genes at the locus, sum of p and q is unity. Hence, following relationship holds true:

$$p + q = 1$$

$$\therefore \; p = 1 - q,$$

$$\text{or } \; q = 1 - p,$$

For better understanding of the reader let us consider an hypothetical population of 100 individuals (this is a finite population, but this figure is used for explanation purpose and simplicity for understanding of the reader). Among these, let there be 25 individuals with genotype *AA*, 10 with *Aa* and 65 with genotype *aa*. Genotype frequencies are therefore:

AA 25/100 = 0.25 or 25%, *Aa* 10/100 = 0.10 or 10% and *aa* 65/100 = 0.65 or 65%

A reader can verify that sum of genotype frequencies is 1 or 100.

There are two genes *A* and *a*. The gene frequencies using relationships (5.1) and (5.2) would be:

$$\hat{p} = D + ½ H = 0.25 + ½ (0.10) = 0.30$$

or 30 per cent, and

$$\hat{q} = R + ½ H = 0.65 + ½ (0.10) = 0.70$$

or 70 per cent.

And $p + q = 0.30 + 0.70 = 1$ or 30% + 70% = 100%.

Random mating of individuals in the population could be shown in two ways, viz., as random union among the gametes produced in the population and actual random mating among the individuals of different genotypes. Let us initially elucidate random mating through random union among the gametes. Genotype *AA* will produce only one kind of gametes, i.e. *A*, whereas *Aa* will produce two kinds of gametes in equal proportions, i.e., *A* and *a*. Therefore, the proportion of gametes *A* produced by the hypothetical population is given by following relationship:

$$\frac{2D+H}{2N} \; or \; \frac{D+½H}{N} \text{ or simply } D+½H$$

$$= 0.30$$

This is same as the gene frequency of *A*. By similar reasoning the proportion of gametes *a* produced in this population would be:

$$\frac{2R+H}{2N} \; or \; \frac{R+½H}{N} \; or \; R+½H$$

$$= 0.70$$

We, therefore, see that the gene frequency for *A* is same as gametic frequency of *A* and same relationship is true for gene *a* and gametes *a*. Assuming that the frequencies of gametes *A* and *a* are same in both the male and female individuals in the population, random union will result in genotypes *AA*, *Aa* and *aa* with frequencies 0.09, 0.42 and 0.49, respectively as shown below:

Male Gametes

		A 0.30	*a* 0.70	Genotype Frequency
Female Gametes	*A* 0.30	*AA* 0.09	*Aa* 0.21	Genotypes of progeny Frequencies of progeny
	a 0.70	*aA* 0.21	*aa* 0.49	

We see that the frequencies of genotypes *AA*, *Aa* and *aa*, are not same in progeny as in the parental generation. Let us find out the gene frequencies in progeny generation. Thus using (5.1) and (5.2) we get:

$$p = D + ½ H = 0.09 + ½ (0.42) = 0.30, \text{ and}$$

$$q = R + ½ H = 0.49 + ½ (0.42) = 0.70$$

Once the frequency of gene *A* is known the gene frequency of *a* can be obtained using following relationship.

Because $p + q = 1$

$$q = 1 - p$$

$$= 1 - 0.30$$

$$= 0.70$$

Therefore, one thing is clear. After one generation of random mating in the initial population, the genotype frequencies have changed but not the gene frequencies. The original population was, therefore, not in Hardy-Weinberg equilibrium. Let us examine what happens when random mating takes place in progeny generation *I*, considered above. The gametic frequencies in progeny generation *I* are $A = \hat{p} = 0.30$ and $a = \hat{q} = 0.70$. Random union among the gametes shall lead to three genotypes *AA*, *Aa* and *aa* with genotype frequencies 0.09, 0.42 and 0.49, respectively. Thus the gene and genotype frequencies of progeny generation *I* and *II* are same. Earlier we had concluded that progeny generation *I* was not in Hardy-Weinberg equilibrium. It should be clear to the reader that one generation of random mating leads to Hardy-Weinberg equilibrium with respect to a single locus with two alleles.

Now the actual random mating among the individuals in the hypothetical population will be used to demonstrate the attainment of equilibrium. This is a longer method. Let us assume that the frequencies of three genotypes are same among the male and female individuals, i.e. 0.25, 0.10 and 0.65, respectively. Random mating among the three kinds of genotypes will result in nine kinds of matings as shown below:

		Genotypes of males			
		AA 0.25	*Aa* 0.10	*aa* 0.65	*Genotype frequency*
Genotypes of females	*AA* 0.25	*AA* × *AA* 0.0625 1	*AA* × *Aa* 0.0250 2	*AA* × *aa* 0.1625 3	Frequencies of matings Kinds of matings
	Aa 0.10	*Aa* × *AA* 0.0250 4	*Aa* × *Aa* 0.0100 5	*Aa* × *aa* 0.0650 6	
	aa 0.65	*aa* × *AA* 0.1625 7	*aa* × *Aa* 0.0650 8	*aa* × *aa* 0.4225 9	

Assuming absence of maternal inheritance mating types 2 and 4, 3 and 7, and 6 and 8 shown in the table above are similar and could be pooled. Therefore, the total mating types are six and not nine. The mating types, their frequencies, the kind of progeny resulting from the matings and their genotype frequencies are summarized in Table 5.3.

Table 5.3: Result of random mating in the hypothetical population

SN	*Types of mating*	*Frequency of mating*	*Genotypes among progeny and their frequencies*			*Frequencies of matings in generalized terms*
			AA	*Aa*	*aa*	
1	*AA* × *AA*	0.0625	0.0625	–	–	D^2
2	*AA* × *Aa*	0.0500	0.0250	0.0250	–	$2DH$
3	*AA* × *aa*	0.3250	–	0.3250	–	$2DR$
4	*Aa* × *Aa*	0.0100	0.0025	0.0050	0.0025	H^2
5	*Aa* × *aa*	0.1300	–	0.0650	0.0650	$2RH$
6	*aa* × *aa*	0.4225	–	–	0.4225	R^2
	Total	1.0000	0.0900	0.4200	0.4900	*1*

From the table 5.3 it is clear that in the progeny frequencies of genotypes *AA*, *Aa* and *aa* are different from the genotype frequencies in the original hypothetical population. After one generation of random mating among the individuals, the progeny generation I shall attain equilibrium. These results are in agreement with those shown for random mating among the gametes.

In the last column of table 5.3, frequencies of matings among the individuals of 3 genotypes are shown in generalized symbols. From the reasoning of Mendelian genetics it is easy to find out the frequencies of progenies with different genotypes resulting from each of the mating type. For example, mating *AA* × *AA* will produce all the progeny with genotypes *AA*. The second type of mating in the table 5.3, i.e., *AA* × *Aa*, will produce half the progeny with genotype *AA* and rest *Aa*. Twenty five per cent progeny resulting from mating *Aa* × *Aa* shall be *AA*. Hence, the total number of *AA* progeny resulting from the matings shall be:

$$D^2 + \frac{1}{2}(DH) + \frac{1}{4}(H^2) \qquad \ldots (5.3)$$

It can be verified that this is equal to $(D + \frac{1}{2} H)^2$. And since $p = D + \frac{1}{2} H$, the frequency of genotype *AA* in the population, can be expressed as p^2. Because p in the present example equals 0.30, p^2, the genotype frequency of *AA* in progeny shall be 0.09.

We can now summarize the various properties of the random mating population with respect to the Hardy-Weinberg law.

(i) A large random mating population is stable with respect to its gene frequency and genotype frequency from generation to generation, in the absence of migration (both immigration and. emigration), mutation (no differential mutation rates, i.e

$$A \underset{n}{\overset{m}{\rightleftarrows}} a \Big)$$

and selection (both artificial and natural). That is genetic properties remain constant from one generation to the next.

(ii) The genotype frequencies in the progeny generation, produced by random mating among the parents, are determined entirely by the gene frequencies of parents, and lastly,

(iii) Population attains Hardy-Weinberg equilibrium after one generation of random mating, if the earlier population was not in equilibrium (with respect to a single locus with two alleles).

It is, in fact, interesting for a reader to confirm that all the populations with same gene frequencies will attain same equilibrium frequencies after random mating. For example, populations with genotypes *AA*, *Aa* and *aa* with genotype frequencies (0.25, 0.10, 0.65), (0.30, 0.00, 0.70), and (0.20, 0.20, 0.60) shall attain equilibrium values of 0.09, 0.42, 0.49, respectively, after one generation for random mating.

Populations also attain Hardy-Weinberg equilibrium following random mating for sex-linked genes (we have considered autosomal genes in hypothetical example). We have considered only two alleles at a locus but population also attains an equilibrium when more than two alleles are present at a locus. Lastly, population also reaches an equilibrium with respect to more than one locus, nevertheless, it takes some time depending on the number of loci being considered and the strength of linkage between genes at various loci. A reader may refer to Li (1955) for more details.

A basic difference between the genotypic make up of cross-fertilizing and self-fertilizing crop plants should be apparent to the reader from foregoing discussion about Hardy-Weinberg law. Due to random mating, all the genotypes existing in the population are maintained in subsequent generations, in the absences of forces that tend to alter the gene frequencies. However, in the self-pollinated crops only homozygotes are maintained as such from generation to generation, in the absence of selection, mutation and migration. Suppose a population has three genotypes viz., *AA*, *Aa* and *aa*, in certain proportions. In a self-pollinated crop, the population will be predominantly made up of *AA* and *aa* at the end of about eight to nine generations (see table 5.1). On the other hand, all the three genotypes will remain in the population after random mating. There may be some change in the genotype frequencies after one generation of random mating, but not in subsequent generations. Gene frequencies may also change, due to random drift, if the population size is reduced. This requires some elucidation. If the size of population is small the genes passed through the gametes to their progeny are subject to random sampling variation. As a result, the gene frequency in subsequent generation changes. For example, let us assume that there are two alleles *A* and *a* at a locus and further assume that there are two populations one comprising of 10,000 individuals and another of ten. 20,000 gametes shall be produced by a population with 10,000 individuals, 10,000 *A* and 10,000 *a* and they will be subject to sampling variation of 70.7 (the standard deviation of gamcte number shall be 0.353 per cent) as shown below:

$$\text{Sampling variance of gametes produced} = \sqrt{\frac{p \times q}{2N}}$$

where *p* and *q* are the number of gametes and *N* is the total number of gametes produced.

$$\text{Thus} = \sqrt{\frac{10{,}000 \times 10{,}000}{20{,}000}} = 70.7$$

And this is 0.353% of 20,000 (the total gamete number). On the other hand, population made of 10 individuals shall produce 20 gametes, 10 of which shall be *A* and rest *a*. The sampling variation to which they will be subject to shall be $\sqrt{\frac{10 \times 10}{20}}$ = 2.23, which is 11.2% of the total number of gametes produced. Hence, the frequency of genes will fluctuate more in random direction in the smaller population.

5.3 GENOTYPIC VARIABILITY AND BREEDING SYSTEM

A great majority of wild plants are cross-pollinated. By contrast with the wild plants, close inbreeding is comparatively common among crop plants. For example, rice, wheat, barley, sorghum, peas, mung, urid, vicia, groundnut, soybean, tomato, tobacco, cotton, are predominantly self-pollinatcd. Nevertheless some cross-pollination has been noted in sorghum, vicia, tomato and cotton. On the other hand, maize, rye, brassicas are cross-pollinated annuals. Close inbreeding among the wild organisms is rare, and this rarity may be attributed to the long term evolutionary disadvantage of inbreeding. Inbreeders are less flexible because of paucity of genotypic variability.

Of course the classification of various crops as self-pollinated and cross-pollinated is crude and no rigorous distinction can be made between these categories. Diploid coffees are self-incompatible and hence outbred, but the tetraploid coffee, *Coffea arabica*, is often successfully self-pollinated to be classed as inbreeder (Sybenga, 1961). The diploid potatoes have self-incompatibility mechanism and are rigorously cross-pollinated, however, the tetraploid potatoes are basically self-fertile but do not tolerate enough selfing (Simmonds, 1962).

The correlation between breeding system and habit is interesting. The perennials are predominantly cross-pollinated and annuals are self-pollinated. It is known since the times of Darwin that self-fertility, (which is prerequisite for inbreeding) is common among the annuals, than among the perennials, irrespective of the fact whether they are cultivated perennials or wild (Williams, 1951).

In case of cross-pollinated crops also there appears a relationship between variability and agriculture (Simmonds, 1962). The maize varieties before the release of four maize hybrids in India, exhibited high genotypic variability and were highly heterozygous. However, after the release of hybrids open-pollinated varieties have been largely replaced. The advanced agriculture has, therefore, influenced both the variability and heterozygosity of maize. There is not as much genetic variability among maize hybrids as existed among land races.

The same thing appears to be true for perennial cross-pollinated crops. The primitive clonal crops of banana of Uganda and South-East Asia, comprised of mixtures of many distinct clones. Consequently, variability was high and heterozygosity was very high. However, in the advanced agriculture, there is much reduced variability because most of the commercial banana fields are planted under one of the two clones or one of their somatic mutants. Broad conclusion could, therefore, be drawn, that advanced agriculture reduces variability, as is evidenced by replacement of open-pollinated varieties of maize by hybrids. Even in case of wheat, rice, cotton, sorghum, tomatoes only a few land races are in cultivation.

5.4 ADAPTATION AND GENOTYPIC CONSTITUTION

It is very well known that phenotype is a joint product of genotype and environment. When a number of varieties are grown at several locations the phenotype comprises of interaction of genotype with environment also. So genotype × environment component also contributes to the phenotypic variation. This suggests that the effects of genotype and environment need not always be independent of each other. A specific change in environment may not have similar effect on various genotypes. Genotype x environment interaction is a cause of concern to the plant breeder. Relative rankings of a number of varieties, when grown over locations, are not same at each location. Which variety should be chosen for the entire region for which he is working? Should he choose a specific variety for a particular location, even though it is not superior at other locations? This aspect of varietal performance concerns adaptation. Adaptation is the property of a genotype, which permits its survival under selection (Simmonds, 1962). A variety having wide or good adaptability is one, which is consistently superior over a wide range of environments (Frey, 1964).

Various methods were developed to characterize phenotypically stable varieties. By phenotypically stable variety is meant a variety which gives considerably higher yield when grown over locations and years, exhibits least genotype × environment interaction for the character under consideration. A technique of regression analysis was proposed by Finlay and Wilkinson (1963) which was somewhat modified by Eberhart and Russell (1966) to characterize varieties suitable for specific circumstances or wide adaptation. According to Eberhart and Russell (1966) a variety is designated as widely adapted, which gives significantly higher yield than average of all varieties over several locations, the regression coefficient of variety on environments is approaching or equal to one and deviation expected on the basis of linear regression is least at several locations. This is also known as generally adapted variety. When regression coefficient (b_i) is greater than one, i.e., variety is more sensitive to favourable change in the environment, mean performance superior than average of all the varieties over locations and deviation from the expected performance is least, such a variety is known to be specifically adapted to favourable environments. The ability to adapt to varying environments could be explained on the basis of homeostasis. Homeostasis refers to the property of an organism to adjust itself to varying conditions or to the self-regulatory mechanism of an organism, which permits it to stabilize itself in the fluctuating inner and outer environments. The term homeostasis was actually proposed by Cannon (1932) for the ability of human body to develop and maintain itself in a steady state in spite of the exposure to a wide range of environments. This term is widely applied in biology. A variety is said to be well buffered in relation to the environment when it is not too sensitive to changes or adversities and is able to continue or complete the development over a wide range of differences in temperatures, humidity and other components of environment (Briggs and Knowles, 1967). Stability in performance of a variety could be attained according to one of the two approaches suggested by Allard and Hansche (1964), viz., populational buffering and individual buffering. A variety utilizes populational buffering when it is made up of a number of genotypes, each of which is well adapted to somewhat different environments. On the other hand, individual buffering is exploited for achieving stability in performance when a variety is made up of individuals which are themselves well buffered such that each member of a population is well adapted to a range of environments. Single-cross hybrids of maize, sorghum, pearl millet and pure line varieties of wheat, gram, soybean etc. depend upon individual buffering. Multiline varieties, double-cross hybrids, synthetics, composites or other genotypically heterogeneous populations are based on populational buffering.

A reader is referred to a classic paper by Simmonds (1962) dealing with various aspects of variability in crop plants and the breeding systems. He proposed four kinds of adaptations by various populations of crop plants. Specific genotypic adaptation is characteristically a property of varieties of self-pollinated annual crop species, bred for high performance in a limited specific environment. Many vegetable crops bred for cultivation in a small area or to meet specific demands such as development of pea varieties for frozen trade, tomatoes for glass house cultivation, etc. are examples of genotypes, which are specifically adapted. A variety, which is widely adapted, is a genotype with general adaptation. Maize hybrids developed during sixties were evaluated over several locations in the country and were generally adapted, e.g. Ranjit, Deccan, Ganga 1. Likewise, sorghum hybrids CSH 1, CSH 6, CSH 5 and CSH 9 exhibited general genotypic adaptation.

Such varieties are widely grown for many years and though they fail in trials several times they still remain favourite with farmers. Although the attitude of farmers to conserve old varieties may be responsible for cultivation of such varieties, stability in performance is another important aspect for their continued use. A potato variety "Majestic" released in Britain in 1911, still occupies about fifty per cent of the main crop acreage of the country. Although it has no quality of yield, resistance to blight, virus and scab disease, it survives because farmers regard it as more reliable (Simmonds, 1962). It appears that some genotypes are well buffered. According to Lerner (1954) homeostasis is a property of heterozygotes. Shank and Adams (1960) compared the performance of inbred lines and their hybrids in respect of six attributes of maize. They observed much greater coefficient of variability for inbred lines than the hybrids for all the six characters, which suggested that hybrids were more stable than inbred lines.

Specific population adaptation is similar to specific genotypic adaptation and is the aspect of specific adaptation of a heterogeneous population, in which plants of different genotypes are mixed. Thus specific adaptation of population occurs when performance of a mixture is superior than the weighted mean of the component lines. This type of adaptation implies some kind of productive interaction between the components. Borlaug (1959) compared the performance of various multilineal mixtures. Over the 21 comparisons, the mixtures exceeded the component mean by 4 to 21 per cent in eight cases, lay within ± 3 per cent of the component in 9 cases and were inferior in 4 cases. The fourth kind of adaptation was designated by Simmonds (1962) as general population adaptation. It is analogous to general phenotypic adaptation and is the capacity of a heterogeneous population to adapt to a variety of environments. When a population is adapted to a wide range of environments, rather than a specific location, general population adaptation is said to be operative. Jones (1958) showed on the basis of very extensive yield trials that double-cross hybrids of maize were less variable than single-cross hybrids. Although the mean yields were same, the percentage standard deviations for double-crosses and single-crosses were 12.3 and 21.4 per cent, respectively.

References

Allard, R.W. (1960). *Principles of plant breeding*. John Wiley & Sons, New York.

Allard, R.W. and P.E. Hansche (1964). Some parameters of population variability and their implication in plant breeding. *Advances in Agronomy*, 16: 281-324.

Briggs, F.N. and P.F. Knowles (1967). *Introduction to plant breeding*. Reinhold Publishing Corporation, New York.

Cannon, W.B. (1932). *The wisdom of the body*. W.W. Norton & Company, Inc. New York.

Darlington, C.D. and K. Mather (1949). *The elements of genetics*. The Macmillan Co., New York.

Dobzhansky, Th. (1951). Mendelian populations and their evolution. In *Genetics in the 20th century*, The Macmillan Co., New York, pp. 573-589.

Eberhart, S.A. and W.A. Russell (1966). Stability parameters for comparing varieties. *Crop Sci.*, 6: 26-40.

Finlay, K.W. and G.N. Wilkinson (1963). Analysis of adaptation in plant breeding programme. *Aust.J.Agric.Res.*, 14: 742-754.

Fisher, R.A. (1930). *The genetical theory of natural selection*. Clarendon Press, Oxford.

Frey, K.J. (1964). Adaptation reaction of oat strains selected under stress and non-stress environmental conditions. *Crop Sci.*, 4: 55-58.

Hutchinson, J.B. (1958). *Genetics and improvement of tropical crops*. Cambridge.

Jain, S.K. and R.W. Allard (1960). Population studies in predominantly self-pollinated species. I. Evidence for heterozygote advantage in a closed population of barley. *Proc. nat. Acad. Sci.*, Wash., 46: 1371-1377.

Jones, D.F. (1958). Heterosis and homeostasis in evolution and applied genetics. *Amer.Nat.*, 92: 321-328.

Lerner, I.M. (1954). *Genetic homeostasis*. Edinburgh.

Li, C.C. (1955). *Population genetics*. The University of Chicago Press, Chicago.

Mather, K. (1943). Polygenic inheritance and natural selection. *Biol. Rev.*, 18: 32-64.

Shank, D.B. and M.W. Adams (1960). Environmental variability within inbred lines and single-crosses of maize. *J. Genet.*, 57: 119-126.

Simmonds, N.W. (1962). Variability in crop plants, its use and conservation. *Biol.Rev.*, 37: 422-465.

Stebbins, G.L. (1950). *Variation and evolution in plants*. Columbia University Press, New York.

Sybenga, J. (1961). Genetics and cytology of coffee. *Bibliogr.genet.*, 19: 217-316.

Williams, W. (1951). Genetics of incompatibility in Alsike clover, *Trifolium hybridum*. *Heredity*, 5: 51-73.

Wright, S. (1951). The genetical structure of populations. *Ann.Eug.*, 15: 323-354.

6

Quantitative Characters

Plant breeder encounters, broadly speaking, two categories of characters for improvement. Red versus white flower colour, round versus wrinkled seeds, determinate plant growth versus indeterminate plant growth are some examples of trait differences, which fall under the category called qualitative characters. The different classes in which individuals for these characters fall are clear cut. There is no overlapping of characters and hence no confusion during classification. Such trait differences are usually governed by one or two gene pairs, each with large effect which can be measured even in the presence of segregation at other loci and even in the presence of effects of environmental agencies. These characters are very less influenced by the environmental agencies. For instance, the colour of flowers of soybean variety JS-335 is purple. It will continue to be purple whether it is grown in extreme south or extreme north of India or it is grown in kharif or rabi season.

On the other hand, most of the characters of economic importance, like yield of fruits of vegetable crops, yield of vegetative parts in case fodder crops, yield of lint in case of cotton, yield of grains in cereals, pulses, oil seeds, etc. plant height, days required for maturity etc. are called quantitative characters. Variation observed for such characters is continuous, ranging from one extreme to another with middle expression being common. Further, there is overlapping of character differences and one cannot assign individuals to a specific class unambiguously. Because observations for such trait differences are recorded in terms of weight, length and time, these are also called metrical characters. These characters are influenced by environmental agencies, such that phenotypes are not same when grown over locations and/or seasons. For example, a sorghum variety, which matures in 90 to 95 days at Coimbatore, takes about 105 to 110 days at Indore. Yield of a tomato variety, say Avinash, is not same even at a given location in successive years, despite receiving same dose of nutrients and other inputs.

6.1 MULTIPLE FACTOR HYPOTHESIS

Mendel studied the inheritance of seven pairs of trait differences of pea. He concluded from his investigations that a single gene pair, with two alleles at a locus, governed the inheritance of these characters. For instance, round seed is completely dominant (*R*) to wrinkled (*r*) seed.

He further concluded that tall plants (5 to 6 feet) were completely dominant (*T*) to short plants (9 to 12 inches), and one gene pair was responsible for the inheritance. It can be regarded as a qualitative character because the difference between two categories is very large. But a close look at the tall group of plants, if grown by you in field, would reveal that there is a continuous range of plants from "tall" talls to "short" talls, with intermediates being more frequent in tall category of plants. Similar graduated series from "tall" shorts to "short" shorts with intermediate types more frequent could be noticed in the short category. Mendel himself and rediscoverers of Mendelism avoided the study of continuous variation because they thought it would only confuse their analyses.

However, a few scientists were concerned, about continuous variation, in England. Francis Galton, a cousin of Charles Darwin and Karl Pearson, had clearly shown such variation to be at least partly hereditary, even though they failed to discover the mode of inheritance. Failure of Galton and his associates to discover the genetic basis of continuous variation, for instance for a trait like stature of man, could be attributed to uncertain ancestries and small family size. This group of scientists, "Biometricians", regarded that continuous somatic variation is due to continuous genetic variation. They were unable to understand as to how genes with discontinuous effects, proposed by Mendel for trait differences studied by him, could explain the inheritance of continuously varying characters. Galton used statistical methods to understand the resemblance between relatives. But he was not able to delineate the genetic basis of continuous variation.

Two important steps had to be taken before the biometrical methods and genetical methods could be brought together to understand the mode of inheritance of quantitative characters. Johansen (1903) had clearly shown that phenotype of a quantitative character e.g., seed weight of beans, was governed jointly by the genotype (particular assemblage of genes possessed by the plant) and environment surrounding it. Moreover, the magnitude of effect of these two agencies on phenotype was same (chapter 8).

Yule in 1906 proposed that there need be no conflict between the particulate inheritance and the continuous inheritance if it is assumed that many genes having small, similar and additive effects govern the inheritance of quantitative characters. A Swedish plant breeder, Nilsson Ehle (1911) found a naturally occurring model for the type of inheritance proposed by Yule. He observed that in wheat and oats the kernel colour is governed by three gene pairs. When genes at any of the locus segregated (say *A, a*) alone, gave a phenotypic ratio in F_2 of 3 red: 1 white kernel coloured plants. It did not matter whether a single gene pair alone was segregating at first, second or third locus (say *A, a, B, b* and *C, c*). A gene for red colour at any of the three loci was dominant to its recessive counterpart. When genes segregated at any of the two loci together, gave a ratio of 15 red: 1 white in F_2. Plants had red kernels whether they carried one dominant gene at one locus or two genes at same locus or one dominant gene at each of the two loci (*AaBb*). Further, when three genes segregated simultaneously, the F_2 had a phenotypic ratio of 63 red: 1 white. However, there was difference in degree of redness when genes at more than one locus were segregating. The intensity of redness was associated with the factor dosage. For example, most intense red coloured kernels were *AABBCC*. They bred true because when sown the progeny did not segregate.

The next degree of redness was associated with five dominant genes (e.g. *AABBCc, AaBBCC, AABbCC*). While the least red kernels were genetically with only one dominant gene at any of the three loci, viz., *Aabbcc, aaBbcc, aabbCc*. This shows that the effect of genes at the three loci was to add a degree of redness to the kernels. The white kernel was genetically *aabbcc* and its progeny did not segregate. The factors in wheat and oat had large measurable effect on the kernel colour, which made Mendelian analysis of various phenotypic classes possible. This was realized by Nilsson Ehle and also by E.M. East, and hence suggested that continuous variation could be expected for quantitative traits if large number of genes with small, similar and cumulative effects segregated. With phenotypic variation proportional to factor dosage, variation would follow Galton's frequency curve and would be nearly continuous. Continuity would be completed by the blurring effects of environmental agencies, whose variation is truly continuous. Consequently, phenotypic ranges of various genotypes would overlap.

This showed that different genes could have similar and cumulative effects and Mendelian genes have the properties necessary to explain the mode of inheritance of quantitative characters. Conclusive evidence that several segregating genes, with similar and additive effects explain the inheritance of metrical characters was provided by E.M. East in 1916.

Flower size (corolla length) in tobacco is not much influenced by the environmental agencies. East, therefore, selected two true breeding varieties of *Nicotiana longiflora*. Corolla length is a quantitative character. Tobacco is a self-pollinated crop and hence the two varieties of *N. longiflora* selected by East should have been homozygous at each locus and the only source of variation expected was non-genetic. The mean length of corolla of smaller parent was about 43.0 mm. It showed very little variation (s = 1.74 mm). Similarly the larger parent had mean corolla length of about 93.0 mm and the standard deviation was comparable with smaller parent (2.37). The F_1 between the two parents was relatively intermediate with regard to corolla length. The F_1 between two true breeding lines is homogeneous, and hence should be comparable with parents for its variability. Standard deviation of F_1, which is a measure of variability of F_1, was comparable to parents (2.92). This is as per the Mendelian expectations. The individuals in F_2 generation should be more variable due to segregation of genes, contributed by the 2 parents and recombinations between the linked genes. This was found in case of F_2. The standard error was 6.50 mm. Although non-heritable agencies contribute to phenotypic variation, F_2 cannot be expected to be oversensitive to this cause of variation. The greater variability observed among the plants in F_2 generation was attributed to genetic causes. The quantitative character, corolla length behaved up to this point as per the Mendelian expectations. It means that the transmission of factors concerned with quantitative characters follow the Mendelian laws of inheritance. However, East wanted confirmation to his concepts. He anticipated that if quantitative character followed the Mendelian mode of inheritance then plants taken from various parts of F_2 generation should produce F_3 families with different means and also different variabilities. For example, if a cross is made between *AAbb* and *aaBB*, the F_1 will be *AaBb*. If we assume that contribution of $A = B = 2$ units to a character and $a = b = 1$ unit, then individuals in F_2 generation would be:

Genotype	*Frequency*		*Value*
AABB	1/16		8
AABb, *AaBB*	4/16		7
AaBb	4/16		
AAbb	1/16	6/16	6
aaBB	1/16		
Aabb	2/16	4/16	5
aaBb	2/16		
aabb	1/16		4

From the table it will be clear that if individual *AABB* is selected from F_2, the mean of F_3 progeny shall be 8 units and exhibit least variability. On the other hand, progeny of *aabb* shall have very low mean and least variability. Individuals *AABb*, *AaBB* when chosen shall produce F_3 progeny with different mean than F_3 progeny of *AABB* and *aabb*, respectively and shall be more variable. East, observed means and variances of F_3 progenies as per the expectation. In self-pollinated crops, homozygosity increases at the expense of heterozygotes. Consequently, heritable variability within progeny families, subsequent to F_3 generation is expected to decline or at least not increase as compared to their ancestral F_3 selections. Some what expected results were obtained in F_4 and F_5 generations by East.

All this evidence was enough to show that quantitative traits or characters showing continuous variation are not only heritable to a large extent but the genes, which determine their inheritance, follow Mendelian pattern.

The next question to be answered was regarding number of genes responsible for inheritance. Now suppose the character is governed by a single pair of gene with two alleles, viz., *A*, *a*. Suppose for simplicity that the parent with larger corolla length measured 90 mm and parent with shorter corolla length 40 mm. Because one gene pair governs the inheritance, the larger parent would be genotypically *AA* and smaller parent *aa*. Contribution of gene *A* to offspring would be 45 mm and that of smaller parent 20 mm. Hence the F_1 would have corolla length measuring 65 mm, assuming absence of dominance. In the F_2 generation 25 per cent individuals should have corolla length equal to the larger parent, 50 per cent would have corolla equal to F_1 in length and rest i.e. 25 per cent, would have corolla length equal to smaller parent. The three categories should be clearly visible in F_2 generation. This kind of distribution was not noticed, however. Hence, more than one gene pair was involved. If two gene pairs, with additive gene effects (no dominance) were involved, there would be five phenotypic classes, as shown below.

Genotype	*Frequency*	*Corolla length (mm)*
AABB	$\frac{1}{16}$	90.0
AABb *AaBB*	$\frac{4}{16}$	77.5
AaBb *AAbb* *aaBB*	$\frac{6}{16}$	65.0
Aabb *aaBb*	$\frac{4}{16}$	52.5
aabb	$\frac{1}{16}$	40.0

AABB has four genes governing corolla length, and each gene contributes 22.5 mm to corolla length. Since *AABb*/*AaBB* have three genes with increasing effect and one gene with 10 mm contribution, these individuals have corolla length which measures 77.5 mm. But even such a simple situation did not exist. Five genotypes could have been identified in F_2 generation, if they existed. Mendelian expectation proposes that if *n* gene pairs govern the inheritance of a character, smallest perfect population would be 4^n in F_2 generation. With one gene pair 4 individuals would be the perfect smallest population and it should be possible to get at least one plant similar to parent with larger phenotypic manifestation and 1 with smallest manifestation. With 2 gene pairs $4^2 = 16$ individuals would be smallest perfect population. Assuming involvement of 4 gene pairs, $4^4 = 256$ would be smallest perfect population. East had raised 444 plants in F_2 generation and could not locate a single plant measuring corolla length equal to either parent. Hence it can be concluded that more than four gene pairs with small, similar and additive effects were involved in the inheritance of corolla length. This explanation for inheritance of quantitative characters was called as multiple factor hypothesis. The multiple factor hypothesis, however, assumed that genotype of a plant simply adds or subtracts to a character by a definite number of genes but does not visualize a coordinated genetic system of the type now regarded as basic to quantitative inheritance.

The multiple factor hypothesis was applied to data from a number of organisms by East and his collaborators and also by Fisher. East showed that inheritance of a number of quantitative characters in tobacco and maize could be fully accounted for by this hypothesis. Fisher carried out the integration of biometrical sciences and genetics still further.

The essential features of multiple factor hypothesis are two, viz., that the genes concerned with continuous variation are inherited in Mendelian fashion; and that they have effects on the character under consideration similar to one another, supplementing each other and sufficiently small in relation to the non-heritable variation or at least in relation to the total variation. In this way smooth, continuous variation of the type observed could arise from discontinuous effects of the genes concerned. The distinction between genes concerned with qualitative characters and those concerned with metrical attributes lies in the magnitude of their effects on phenotype relative to effects of other sources of variation. A gene with the effect large enough to cause recognizable discontinuity, even in the presence of segregation at other loci or the non-heritable agencies, can be studied by Mendelian techniques. Such genes are regarded as major genes. However, genes whose effects are not large enough and are influenced by genes at other loci as well as by non-heritable agencies may be called as minor genes. But by virtue of pleiotropic effects, the same gene may operate as major gene for one character and function as minor gene for a quantitative character. The term "polygene" was introduced by Mather (1943) to refer to the genes with minor effects. Mather (1943) in contrast to the multiple factor hypothesis believed that each polygene is dependent on and interacts with genes at other loci. Variation caused by simultaneous segregation of polygenes was called as polygenic variation.

Because the effect of each gene governing inheritance of quantitative character is very small, which is difficult to be measured and effect is mimicked by non-heritable agencies, effects of these genes can not be measured individually. Such genes cannot be followed individually by simple Mendelian techniques. Hence, how can we be sure that these genes are in truth borne on chromosomes residing in the nucleus of a cell? Although there are differences in the manifestations of reciprocal crosses, with respect to continuously varying characters, these differences are not much greater than those observed for qualitative characters. It means that contribution by both the parents to the heredity of metrical traits is equal. The properties of nuclear borne genes are two, viz., segregation and linkage.

Members of a pair of alleles separate cleanly from each other when an individual forms germ cells. Thus an F_1 between *AA* and *aa* receives genes *A* and *a* from its parents. When F_1 forms gametes the genes *A* and *a* separate cleanly (without being contaminated by each other because they are present on homologous chromosomes within the same nucleus) and go to different gametes. Random union of gametes results in production of 3 genotypes, viz., *AA*, *Aa* and *aa*. When *n* gene pairs are in heterozygous condition in F_1 they will form 2^n gametes and form 3^n different genotypes in F_2. In this way F_2 generation shows more variability due to segregation. The estimates of standard deviations for P_1, P_2, F_1 and F_2 generations, in the example of *N. longiflora* are in agreement with what is said. Parents were long inbred so they exhibited less variability because of environmental variation in the experimental garden. The F_1 similarly showed relatively less variability but the F_2 generation was more variable. This was attributed to segregation of gene pairs responsible for corolla length. Relative variation of different degrees in various generations is the test of segregation.

When the two genes are located very closely on same chromosome they exhibit linkage. Consequently, the frequency of new combinations is less as compared to the parental combinations. The genes responsible for quantitative characters are many in number. And as shown by Breese and Mather (1960) several of the genes may be distributed on same chromosome. The genes for different quantitative characters may be intermingled on same chromosome. The test of linkage, to show that polygenes are nuclear borne (on chromosomes) may be of two kinds. Linkage between a major gene and polygenic system can be traced or else linkage may be sought among the polygenes affecting two quantitative characters. First kind of evidence is relatively easy. Sax (1923) showed that when two strains of *Phaseolus vulgaris*, one having large coloured seeds and the other with small and white seeds, were crossed yielded in F_2 generation plants with coloured seeds whose average seed weight was 30 centigrams. White seeds had average weight of 26 centigrams. The heterozygous plants with coloured seed, identified by progeny test, had seed weight of 28 centigrams. The colour *vs* white was governed by a single gene pair, viz., *P* and *p*. The coloured seeds were always heavier. This is one test of linkage between a major gene and genes for quantitative character. Here size of seed was measured in terms of seed weight.

Polygenes affecting different characters are distributed along the length of chromosomes. The F_2 generation, resulting from a cross of two lines differing for one quantitative character, will exhibit variability for that character due to recombination between linked genes. Since genes for different quantitative characters are intermingled on same chromosome (Breese and Mather, 1960) selection will pick out recombinant chromosomes. And because genes for second quantitative character are intermingled with first character on same chromosome, recombination for first character will mean recombination for second character as well. Selection favouring one quantitative character will drag genes for another quantitative character alongwith selections for first character although it was not selected for. Second character can thus show a correlated response to selection, which a breeder did not aim at altering it. Mather and Harrison (1949) showed linkage between polygenes for chaeta number and viability in case of *Drosophila melanogaster*.

6.2 GENE ACTION

The kind of gene action prevailing for a qualitative trait enables the choice of breeding procedure appropriate for improvement. For example, the details of breeding procedure may differ for trait which is determined by a dominant gene and that governed by a recessive gene. The choice of breeding programme depends upon the nature of gene action responsible for quantitative characters also. Gardner (1963) is of the opinion that in order to produce strains or hybrids that are superior than those already in commercial production, a breeder must devise an appropriate breeding procedure to accomplish this objective. But according to him information on following aspects is a prerequisite for planning details of breeding program:

(i) Additive genetic variance (s^2_A) resulting from additive effects of genes at all segregating loci.

(ii) Dominance variance (s^2_D) resulting from intra-allelic interaction of genes at all the segregating loci.

(iii) Epistatic variance (s^2_I), which results from inter-allelic interaction of genes between two or more segregating loci. It may be of three kinds:

(a) Additive × additive (σ^2_{AA}) resulting from genes showing additive gene effects at each of the two loci.

(b) Additive × dominance (σ^2_{AD}) which results from interaction of genes showing additive gene effect at one locus and showing dominance gene effect at other locus.

(c) Dominance × dominance variance (σ^2_{DD}) resulting from interaction of genes at two loci, each showing intra-allelic interaction.

(iv) Average degree of dominance.

(v) Genotype × environment interaction, and

(vi) Genotypic correlations among different quantitative characters of importance.

Because it is not possible to measure the nature of gene action at each locus in case of polygenic inheritance, gene action is estimated *en masse*. The average gene action is measured using gene models suggested by Hayman (1958), Gamble (1962) and Mather and Jinks (1971). Assuming simple situation of two loci, each with two alleles, viz., *A*, *a* and *B*, *b*, let us understand various kinds of gene effects, involved in quantitative inheritance. Let us assume that genes *A* and *B* tend to increase the expression of the character. Capital letter does not imply that these genes are dominant.

6.2.1 Additive Gene Effects

The simplest conception of the way genes act assumes that the substitution of one allele for another produces a plus or minus shift on the scale of measurement, and this shift is same regardless of genes present at other loci. This assumes that the genes involved are independent in their action in producing their effects and are not influenced by gene present at the same locus or genes present at other loci. The total effect of all the genes affecting the character is simple sum of their individual effects.

Let us assume an hypothetical situation to explain the gene action. Suppose there are two loci each with two alleles, viz., *A*, *a* and *B*, *b*.

Assume that measurements are made for some quantitative character on individual with genotype *aabb* and the measurement is 2 units. If gene *A* is substituted for *a*, in *aabb*, there is a shift of 2 units in positive direction and *Aabb* measures 4 units (Table 6.1). Further substitution of gene *A* for *a* in *Aabb* shifts the value of character by 2 units and consequently *AAbb* measures 6 units. Each substitution of gene *A* for *a* enhances the measurement of character by 2 units, irrespective of whether substitution is made in *aa* or *Aa*. Now if the value of *aa* is measured when genes at other loci are *Bb*, the character measures 5 units. This implies that substitution of gene *B* for *b* in *aabb* has shifted the value by 3 units in positive direction. If gene *A* is substituted for *a* in *aaBb*, the shift in value is 2 units and *AaBb* measures 7 units. Once again the substitution of gene *A* for *a* in *AaBb* is 2 units and *AABb* measures 9 units. This shows two facts:

Table 6.1: Additive gene action

Genotypes		First locus			
		aa	*Aa*	*AA*	*Mean*
Second locus	*bb*	2	4	6	4
	Bb	5	7	9	7
	BB	8	10	12	10
	Mean	5	7	9	

(i) Effect of substituting gene *a* for *A* in *aa* and *Aa* is same, i.e., 2 units. This implies absence of dominance. In other words, there is additivity in gene action at this locus. This can be shown in other way. Gene *a* contributes 1 unit to the heterozygotes (offspring) and *AA* contributes 3 units. Consequently, the F_1 measures 4 units, which is mid-way, between *aa* and *AA*. If *Aa* deviates from simple sum of effects contributed by genes *a* and *A*, dominance or intra-allelic interaction is suggested.

(ii) Effect of substitution of gene *a* for *A* is same, i.e., 2 units, irrespective of the presence of genetic constitution at *B*, *b* locus. When genes *bb* were present at *B*, *b* locus effect of substitution of *A* for *a* was 2 units and it remained so when phases were *Bb* and *BB*, at second locus.

This implies that genes at 2nd locus do not influence the effect of substitution at 1st locus. It points to absence of inter-allelic (or non-allelic) interaction.

Similar situation exists for substitution of gene *B* for *b*. In *aabb*, effect of substitution of *B* for *b* is 3 units. It is 3 units irrespective of whether genes at other loci are *aa*, *Aa* or *AA*. It is same whether gene at same locus is *b* or *B*. This is a completely additive model.

6.2.2 Dominance Gene Effects

Dominance is implied in gene action when the 2 genes at a locus interact. In other words, intra-allelic interaction is said to exist when effects of substitution at same locus are not identical. Once, again we take hypothetical example to explain this type of gene action. Let us assume that *aabb* measures 2 units for some quantitative trait. If gene *A* is substituted for *a* in *aabb*, the shift in value is 4 units. But when gene *A* is substituted for *a* in *Aabb*, there neither increment not reduction in the value of character. The value of *AAbb* and *Aabb* is same, which in classical genetics implies dominance. Dominance or intra-allelic interaction is suggested when heterozygote deviates, in plus or minus direction, from the mid-point. In the present example, mid point is 6 + 2/2 = 4. The F_1, *Aabb* deviates from this mid point and hence indicates presence of dominance (Table 6.2).

Shift in the value, for substitution of gene *a* by *A* is 4 units even when phases at 2nd locus are *Bb* and *BB*. Thus non-allelic genes do not influence the effect of gene substitution at *A*, *a* locus. However, there is evidence of complete dominance at *A*, *a* locus, because there is no change in value due to substitution of gene *A* for *a* in *Aa*.

Table 6.2: Dominance gene effects

Genotypes		First locus			
		aa	*Aa*	*AA*	*Mean*
Second locus	*bb*	2	6	6	5.00
	Bb	7	11	11	10.00
	BB	7	11	11	10.00
	Mean	5.75	9.75	9.75	

This model is thus evidence of dominance (complete) at *A*, *a* locus. Similarly at *B*, *b* locus there is complete dominance of gene *B* and effect of substitution of gene *B* for *b* is 5 units in *bb* (whether *aabb* or *Aabb* or *AAbb*) but no effect of substitution in *Bb*. Further, there is no evidence of digenic non-allelic interaction since the effect of substitution of gene *A* for *a* is not influenced by genic phases at *B*, *b* locus and vice versa.

The degree of dominance may be partial when either the mean measuement of phenotype of *Aa* lies between mid-parental value and *AA* or mid-parental value and *aa*. Degree of dominance is in overdominance range when *Aa* is greater than *AA* or lesser than *aa*. This phenomenon holds true for all the loci. When genes exhibit dominance and several loci govern the inheritance of a character fixation of a genotype is not practically possible because with complete dominance it is difficult to distinguish heterozygote from homozygote. The former will segregate in subsequent generation. Moreover a series of precisely placed crossovers must take place so that all dominant genes are carried on same homologue and are included in same gamete. Union of such gametes will result in an homozygote dominant at all loci influencing the inheritance of a character.

6.2.3 Digenic Non-Allelic Interaction

In the previous two sections we tacitly assumed absence of non-allelic interaction. This is, however, not a real situation. Genes located on same chromosome and some times even on the other chromosome within the same nucleus, influence each other's effects. Consequently, it is unrealistic to assume that the effect of gene substitution at one locus is not influenced by phases of genes at other locus. One of the simplest non-allelic interaction (digenic) is presented in hypothetical model (Table 6.3).

Table 6.3: Digenic Non-allelic interaction

Genotypes		First locus			
		aa	*Aa*	*AA*	*Mean*
Second locus	*bb*	2	4	6	4
	Bb	5	8	11	8
	BB	8	12	16	12
	Mean	5	8	11	

In this simple hypothetical model, effect of substituting gene *A* for *a* is 2 units, both in *aa* as well as *Aa*, when the phase at 2nd locus is *bb*.

Thus there is no dominance (i.e., there is additive gene action at *A*, *a* locus). Nevertheless, the effect of substitution of *A* for *a* is 3 units, when phase at 2nd locus is *Bb* and 4 units when phase is *BB*. There is no evidence of dominance because in both these situations heterozygote (i.e. *Aa*) fall on mid-point. There is thus additive gene action at *A*, *a* locus (i.e. no dominance) but digenic non-allelic interaction exists, since locus *B*, *b* does influence the effect of substitution at *A*, *a* locus.

Effect of substitution of *B* for *b* is 3 units, when genes *aa* are present at first locus. The effect of gene substitution at *B*, *b* locus is 4 units when phase at first locus is *Aa*. This suggests the effect of digenic non-allelic interaction. In complex situation there is partial dominance at first locus no dominance at second locus, but digenic non-allelic interaction is operating (Table 6.4).

Table 6.4: Complex gene model

Genotypes		First locus			
		aa	*Aa*	*AA*	*Mean*
Second locus	*bb*	2	4	7	4.25
	Bb	4	7	11	7.25
	BB	6	10	15	10.25
	Mean	4	7	11	

6.2.4 Geometric Gene Action

In section 6.2.1 it was assumed that effects of genes at the same locus and even the effects of genes at different loci add together to produce a given phenotypic effect. For example, x_a and x_b are the effects of genes at a^{th} and b^{th} locus. The phenotype would be $x_a + x_b$ if gene action is additive. However, genes may act in multiplicative fashion such that their joint effect is not the sum of their individual effects but the product. Thus it would be $x_a x_b$ instead of $x_a + x_b$. We usually measure a character on the scale that is convenient to us and not suitable for a character. We use a linear scale of measurement whereas the character is a product. For example, area of leaf of a capsicum plant is equal to length of leaf × breadth of leaf, i.e. a product. Instead of using a geometric scale to measure the area we use an arithmetic scale.

When the genes act in geometric fashion, the phenotypes form a geometric series such as 2, 4, 8, 16 ...... being the contribution of 1, 2, 3, 4..... active alleles, respectively. Variation observed for characters governed by genes with geometric action tends to be skewed into an asymmetric curve. The mean measurements of F_1 and F_2 generations tend to be closer to the mean measurement of one of the parents. Geometric mean of two numbers is given by square root of the product. For example if the two strains measure 4 and 9 units, then the geometric mean would be $\sqrt{4 \times 9} = 6$. Transformation of the raw data to an appropriate scale tends to induce normalcy to the population. For example, when data is recorded in percentages (or expressed as proportions) the distribution of population is not normal. Bliss (1937) suggested an angular transformation. Tables are available for such transformations. Additive gene action, on the other hand forms an arithmetic series, such as 2, 4, 6, 8, 10.... being the contribution of 1, 2, 3, 4, 5 ...... active genes, respectively. The population shows a normal curve, when the data are presented graphically. The statistical analyses assume that population is normally distributed.

6.3 VARIANCE COMPONENTS

Because the phenotypic value of a character, whether qualitative or quantitative, is governed by the genotypic value and environmental deviations, the variation in phenotypic expression of a trait is determined by genotypic variation and environmental variation (which comprises of all the non-genetic causes).

The primary object of measuring phenotypic variation is to partition it into different components, which determine phenotype. Relative magnitude of the components determines the genetic properties of the character of individuals in population, in particular the degree of resemblance between relatives. When a plant breeder selects a plants from segregating material he desires that the progeny of selected plant should resemble (or perform) as good as the selected one (i.e. parent). The degree of resemblance is chiefly determined by additive genetic variance. Hence, let us see how the phenotypic variance of individuals in selfing series (individuals derived by self-pollination of plants in F_2, F_3 and subsequent generations) or cross-pollinated/random mating population is made up of different components.

6.3.1 Components of Phenotypic Variance in Selfing Series

The theory for derivation of components of variances was developed by Fisher, Immer and Tedin (1932) and the ideas were further elaborated, modified, by Mather (1949), Mather and Jinks (1971).

Let us describe the components of phenotypic variation of individuals in selfing series by simplifying the genetic situation at the beginning. Individuals differ at a single locus with two alleles, and hence there are two kinds of individuals, viz., *AA* and *aa*. Let us assume that the mid-point, as origin, between two parents is *m*, which depends not only on genes in question but the other genotypes and the effect of environment, and hence mid-point reflects general circumstances of the observation. Let the mid-point (*m*) be origin from which *AA* deviates in positive direction by a value *d*, while *aa* deviates in the negative direction by a value – *d*. When *AA* and *aa* are hybridized, the heterozygote may fall on mid-point (i.e. $h = 0$). For example, let us suppose that arbitrarily assigned genotypic value of *AA* is +6 and that of *aa* – 6. The mid-point is naturally zero. If heterozygote is equal to zero, we say that there is no dominance (in other words there is additivity of gene effects). Whenever heterozygote deviates from the mid-point, the departure is denoted by *h*. *h* may fall on mid-point when there is no dominance, it may assume values equal to +*d* or – *d* when gene *A* is completely dominant or *a* is completely dominant. *h* is in between mid-point and + *d* when there is partial dominance of *A* and between mid-point and – *d* when *a* is partially dominant. The capital letter *A* here does not denote dominant gene but gene with increasing effect on trait. The mean measurements of *AA*, and *aa* shall be $m + d$ and $m - d$, respectively. The F_1, *Aa*, will measure $m + h$. The contribution of this locus to mean measurement of F_2 generation shall be:

$$\tfrac{1}{4}\,AA + \tfrac{1}{2}\,Aa + \tfrac{1}{4}\,aa$$

$$= \tfrac{1}{4}\,(m + d) + \tfrac{1}{2}\,(m + h) + \tfrac{1}{4}\,(m - d)$$

$$= m + \tfrac{1}{2}\,h$$

Contribution of this locus to crude mean square shall be:

$$\tfrac{1}{4}\,(d)^2 + \tfrac{1}{2}\,(h)^2 + \tfrac{1}{4}\,(-d)^2$$

$$= \tfrac{1}{2}\,(d^2) + \tfrac{1}{2}\,(h^2)$$

We did not include mid-parental value (*m*) because it is common to all individuals. From this crude mean square correction for mean,

$(\frac{1}{2}h)^2$, is subtracted to get mean square or heritable variation of F_2, i.e.

$$\tfrac{1}{2}(d^2) + \tfrac{1}{2}(h^2) - \tfrac{1}{4}(h^2)$$
$$= \tfrac{1}{2}d^2 - \tfrac{1}{4}h^2$$

If we assume that contribution to d and h, by genes at several loci which influence the inheritance of a given quantitative trait, is additive then in the absence of non-allelic interaction, heritable portion of variation of F_2 shall be:

$$\tfrac{1}{2}S(d^2) + \tfrac{1}{4}S(h^2)$$

If we write $S(d^2)$ as D and $S(h^2)$ as H, then heritable variation of F_2, in the absence of non-allelic interaction shall be

$$\tfrac{1}{2}D + \tfrac{1}{4}H,$$

and if we add to it E, the environmental variation, then variation of F_2 can be expressed as

$$\tfrac{1}{2}D + \tfrac{1}{4}H + E \qquad \text{...(6.1)}$$

In this expression D depends upon difference between individuals distinguished by being homozygous for two allelomorphs of various genes. In other words, it measures variation among homozygous individuals. It is known as additive genetic variation and is fixable portion of variation. H depends upon departure of heterozygote from mean of two corresponding homozygotes. It is the dominance variance and not fixable. Reason is obvious. The heterozygotes will segregate in next generation.

The mean measurements of backcrosses of Aa to AA (i.e. P_1) and to aa (i.e. P_2) shall be:

$$B_1 = AA \times Aa$$
$$= \tfrac{1}{2}AA + \tfrac{1}{2}Aa$$

and hence

$$\bar{B}_1 = \tfrac{1}{2}(d) + \tfrac{1}{2}(h)$$

Similarly $B_2 = Aa \times aa$

$$= \tfrac{1}{2}Aa + \tfrac{1}{2}aa$$

and therefore,

$$\bar{B}_2 = \tfrac{1}{2}(h) - \tfrac{1}{2}(d)$$

It can be verified that heritable variation of B_1 ($Aa \times AA$) and B_2 ($Aa \times aa$) shall be ¼ $(d-h)^2$ and ¼ $(d+h)^2$, respectively. If several loci are considered, which influence the inheritance of a quantitative character, and non-allelic interaction is assumed to be absent then heritable variation of B_1 can be written as

$$\tfrac{1}{4} D + \tfrac{1}{4} H - \tfrac{1}{2} DH$$

and heritable variation of B_2 as

$$\tfrac{1}{4} D + \tfrac{1}{4} H + \tfrac{1}{2} DH$$

And sum of heritable variances of B_1 and B_2 becomes $(V_{B_1} + V_{B_2})$

$$\tfrac{1}{2} D + \tfrac{1}{2} H$$

If we add environmental variation to the variation of B_1 and B_2, then total variation of $B_1 + B_2$ can be expressed as

$$\tfrac{1}{2} D + \tfrac{1}{2} H + 2E \qquad \text{...(6.2)}$$

Because additive genetic variation determines the degree of resemblance between relatives, breeders are interested to know how much observed variation for a continuously varying trait is attributable to additive variance. The ratio of genetic variation to total phenotypic variation is known as broad sense heritability (h^2_{bs}) and is expressed as percentage. Thus broad sense heritability (h^2_{bs})

$$h^2_{bs} = \frac{V_G}{V_G + V_E} \times 100 \qquad \text{...(6.3)}$$

Where V_G, V_E denote genotypic variance and environmental variation, respectively and $V_G + V_E$ equals phenotypic variation (V_P). But as is mentioned earlier it is the additive genetic variation, which determines the degree to which progeny of selected plants resemble their parents and we calculate narrow sense heritability as the ratio of additive genetic variance to total phenotypic variance (expressed in per cent). Thus

$$h^2_{rs} = \frac{V_A}{V_A + V_D + V_I + V_E} \times 100 \qquad \text{...(6.4)}$$

where V_A, V_D, V_I denote additive, dominance and non-allelic interaction variance components, respectively. Warner (1952) suggested use of heritable variation of F_2 generation and sum of backcrosses (B_1 and

B_2) for estimation of narrow sense heritabilitiy. Estimate of additive variance is got as:

$$2VF_2 = D + \frac{1}{2}H + 2E$$

$$\frac{-(V_{B_1} + V_{B_2}) = \frac{1}{2}D + \frac{1}{2}H + 2E}{\frac{1}{2}D}$$

And narrow sense heritabilitiy is estimated as

$$\frac{\frac{1}{2}D}{V_{F_2}} \times 100$$

$$\frac{\frac{1}{2}D}{\frac{1}{2}D + \frac{1}{4}H + E} \times 100$$

This can be delineated with an example. Observations were recorded on plant height (cm) of sorghum in F_2, B_1 and B_2 generations. The estimates of variances of 3 generations were:

$$VF_2 = \frac{1}{2}D + \frac{1}{4}H + E \qquad = 80.2036 \text{ cm}$$

$$VB_1 = \frac{1}{4}D + \frac{1}{4}H - \frac{1}{2}DH + E \qquad = 33.6440 \text{ cm}$$

$$VB_2 = \frac{1}{4}D + \frac{1}{4}H + \frac{1}{2}DH + E \qquad = 93.3680 \text{ cm}$$

We get estimate of ½ D as

$$2VF_2 = 160.4072$$

$$-(V_{B_1} + V_{B_2}) = \frac{127.0120}{33.3952}$$

And the narrow sense heritabilitiy is obtained as:

$$\frac{33.3952}{80.2036} \times 100$$

$$= 41.63 \text{ per cent}$$

This shows that additive variation is responsible for only 42%, phenotypic variation. Heritability is a property not only of the character being studied but also the population being sampled as well as the environment in which experimental material is grown. Because heritabilitiy is a measure of additive variance (V_A), it changes with the change of gene frequency in population. Gene frequency tends to change with selection. Because gene frequencies of two populations are different their estimates of additive variances will be different and consequently estimates of heritabilitiy are not same. A character in the population subjected to intense selection shall have lower heritabilitiy estimate than the population not subjected to selection. Heritabilitiy of a character may be usually larger in magnitude in earlier segregating generations than the later generations of a self-pollinated crop. Characters, which are predominantly controlled by additive variance, exhibit larger estimates of heritabilitiy. Environmental variance also influences the estimates of heritabilitiy. Better management can reduce environmental variance and consequently heritabilitiy can be enhanced. More variable environmental conditions reduce magnitude of heritabilitiy and more uniform conditions enhance it.

The portion of total genetic variability that can be utilised differs depending upon the crop. In asexually propagated crops, broad sense heritability may be useful, because entire genetic variability can be used.

According to Robinson (1966) heritability estimates in cultivated crops can be placed in various categories as:

(a) Low heritability (5 to 10 per cent).
(b) Medium heritability (10 to 30 per cent). For example, components of yield, components of plant height, etc.
(c) High heritability (30 to 60 per cent). For example, maturity characters, chemical composition of fruits, etc.

This classification is approximate and represents averages of heritability estimates over several crops, procedures of determination, several environments. Several mating designs could be employed to obtain estimates of additive variance and ultimately heritability (Refer to Dabholkar, 1999).

6.3.2 Components of Genotypic Variance in Random Mating Population

In the progeny generated by crossing of two true breeding lines and then selfing subsequently permits prediction of various genotypes that will comprise of the segregating generations and also their genotype frequencies. In random mating population, which is heterogeneous, the gene and genotype frequencies may not follow proportions expected for selfing series.

For deriving the heritable variation expected for a random mating population we assume a simple situation of a single locus with 2 alleles, *A* and *a*. We further assume that the frequencies of the 3 genotypes that are expected for this locus, viz., *AA*, *Aa* and *aa*, are p^2, $2pq$ and q^2, respectively, such that $p^2 + 2pq + q^2 = 1$. And hence $p + q = 1$. Let us assume that the genotypic values associated with *AA*, *Aa* and *aa* are $+d$, $+h$ and $-d$, respectively. These values are expressed as deviation from mid-point, i.e. $AA + aa/2$, which is assumed to be zero in the present case. Symbol h denotes departure of heterozygote from mid-point. It will be zero if there is no dominance (i.e. gene action is additive), will be in between 0 and $+d$ or 0 and $-d$ if there is partial dominance of gene *A* or *a*, respectively, and equal to $\pm d$, if *A* or *a* is completely dominant. h exceeds either $+d$ or is smaller than $-d$ when there is overdominance. The genetic components of genotypic variance of a random mating population will be as shown below in the absence of non-allelic interaction.

$$2pq\,[d + h\,(q - p)]^2 + 4\,p^2q^2h^2$$

The component $2pq\,[d + h\,(q - p)]^2$ is known as additive genetic variance and $4p^2q^2h^2$ is dominance variance. The total phenotypic variance will also include environmental variance in addition to these two components of genotypic variance. Let us derive these variance components. The genotypes, genotype frequencies and their genotypic values are shown below.

Genotypes	*AA*	*Aa*	*aa*
Frequencies	p^2	2pq	q^2
Genotypic values	$+d$	$+h$	$-d$

The total of this population (and also the mean because sum of genotype frequencies is 1) would be:

$$p^2d - q^2d + 2pqh \text{ or } d(p - q) + 2pqh; \text{ because } p + q = 1$$

Contribution of this locus to the heritable variance would be equal to sum of squared genotypic value multiplied by corresponding genotype frequency over the 3 genotypes from which correction factor is subtracted. Correction factor is square of total divided by the total of frequencies. Thus

$$p^2d^2 + 2pqh^2 + q^2d^2 - [d(p - q) + 2pqh]^2$$

$$= p^2d^2 + 2pqh^2 + q^2d^2 - p^2d^2 + 2pqd^2 - 4p^2qdh - q^2d^2 + 4pq^2dh - 4p^2q^2h^2$$

$$= 2pq[d^2 + h^2 - 2pdh + 2qdh - 2pqh^2]$$

$$= 2pq\ [d^2 + 2dh\ (q - p) + (1 - 2pq)\ h^2]$$

And because $p^2 + 2pq + q^2 = 1$, we substitute this value for 1 and recast the above expression as:

$$= 2pq[d^2 + 2dh\ (q - p) + (p^2 + 2\ pq + q^2 - 2pq)h^2]$$

$$= 2pq[d^2 + 2dh\ (q - p) + (q - p)^2\ h^2 + 2pqh^2]$$

$$= 2pq[d + h\ (q - p)]^2 + 4p^2q^2h^2 \qquad \ldots(6.5)$$

It should be clear to the reader that additive genetic variance does not suggest that gene action is additive. There can be any degree of dominance. Only when genotypic variance comprises of additive variance, the gene action could be additive. Additive variance shown in 6.5 would be $2pqd^2$ only when $p = q = ½$ or when there is no dominance, i.e. $h = 0$.

It will be apparent to the reader that both the components of genotypic variance are influenced by gene frequencies and degree of dominance.

Additive genetic variance is mainly responsible for degree of resemblance between relatives and the response to selection. This component can be estimated from the degree of resemblance observed among certain relatives developed by certain mating designs, such as diallel mating, North Carolina mating design and triple test cross (Dabholkar, 1999).

References

Allard, R.W. (1960). *Principles of plant breeding*. John Wiley & Sons, New York.

Breese, E.L. and K. Mather (1960). The organization of polygenic activity within a chromosome in *Drosophila*: II Viability, *Heredity*, 14: 375-440.

Briggs, F.N. and P.F. Knowles (1967). *Introduction to plant breeding*. Reinhold Publishing Corporation, New York.

Dabholkar, A.R. (1999). *Elements of biometrical genetics*. Revised edition Publ. Concept Publishing Co., New Delhi.

East, E.M. (1916). Studies on size inheritance in *Nicotiana*. *Genetics*, 1: 164-176.

Falconer, D.S. (1960). *Introduction to quantitative genetics.* Publ. Oliver and Boyd. Edinburgh.

Fisher, R.A.; F.R. Immer and O. Tedin (1932). The genetical interpretation of statistics of the third degree in the study of quantitative inheritance. *Genetics*, 17: 107-124.

Gamble, E.E. (1962). Gene effects in corn (*Zea mays* L.). I Separation and relative importance of gene effects for yield. *Can.J.Plant Sci.*, 42: 339-348.

Gardner, C.O. (1963). Estimates of genetic parameters in cross-fertilizing plants and their implications in plant breeding. In "Statistical genetics and plant breeding". NAS-NRC, 982.

Hayman, B.I. (1958). The separation of epistatic from additive and dominance variation in generation means. *Heredity*, 12: 371-390.

Johannsen, W.L. (1903). Ueber Erblichkeit in populationen und in reinen. Leinen. Gustav Fischer, Jena.

Mather, K. (1943). Polygenic inheritance and natural selection. *Biol. Rev.*, 18: 32-64.

Mather, K. (1949). *Biometrical genetics: The study of continuous variation.* Methuen, London.

Mather, K. and B.I. Harrison (1949). The manifold effects of selection. *Heredity*, 3: 1-52, 131-162.

Mather, K. and J.L. Jinks (1971). *Biometrical genetics. The study of continuous variation.* Chapman and Hall Ltd., London.

Nilson-Ehle, H. (1911). Lunds Univ.Arsskr. N.F.Afd.,2, Bd.5 Nr.2, pp. 1-122.

Robinson. H.F. (1966). Quantitative genetics in relation to breeding on the centennial of Mendelism. *Indian Journalof Genetics and Plant Breeding*, 26A: 171-187.

Sax, K. (1923). The association of size differences with seed coat pattern and pigmentation in *Phaseolus vulgaris*, *Genetics*, 8:552-560.

Warner, J.N. (1952). A method of estimating heritability. *Agron.*J., 44: 427-430.

Yule, G.V. (1906). On the theory of inheritance of quantitative compound characters on the basis of Mendel's laws — a preliminary note. Rept. Third Intern.Conf. Gen., 140-142.

7

Plant Introduction

Richness observed in the diversity of plant species existing in several countries can also be attributed to plant introduction. And hence plant introduction should also be regarded as a potent method of plant breeding. Undoubtedly selection in segregating material and mutations provide for extended variability and enhance the opportunities for plant improvement. Both these agencies, however, can not create new plant species altogether. Several thousand plant species exist on this earth, nevertheless, all of them are not found in each of the countries. Different species of plants were moved from one geographical region to another by several agencies. Plants were either exchanged or transported and introduced in another area, where they were not growing before. Plant introduction is defined as the transposition of a genetic entity from an environment, to which it is attuned, to one in which it is untried (Frankel, 1957). Plant introduction has been responsible for a variety of crops being grown in India. This does not undermine the importance of diversity of plants found in the Indian centre of diversity of plants. But many crops presently being grown in our country are not indigenous. For example, coffee, potato, tomato, okra, tapioca, tobacco, rubber, cocoa, guava, pineapple, cashewnut, cauliflower, cabbage, etc. were introduced from other countries. The invaders, explorers, traders, travelers, pilgrims, amateurs, etc. have been responsible for such plant introductions.

Alexander the Great, the Greek Commander, came upto North West India (the present day Pakistan) around 400 B.C. and through him or his men some exchange of vegetable crops must have taken place (Seshadri and Chatterjee, 2000). Colonizing traders, who came to India from 14th to 17th century from Portugal, France, Great Britain, Netherlands, were responsible for introductions of several crops in India, such as hot pepper or chillis, potato, cabbage, cauliflower, etc. Tomato was comparatively a late comer. Britishers are believed to have introduced tomato in India probably in the early nineteenth century (1828).

Coffee is endemic to Ethiopia and was introduced in India about 250 years ago by Baba Budan, a Mohammedan pilgrim, who probably brought few seeds from Mecca (Kumar, 1957). Cinchona (*Cinchona calsaya, C. succirubra*), grows wild in South America and found its way into East round about 1860, when Clement Markham, an Indian official, introduced seeds and seedlings of Cinchona from Peru to Nilgiris and later in Darjeeling district. Pararubber (*Hevea brasiliensis*), a commercially important plant, was also introduced by Clement Markham from Brazil in 1873. Portuguese introduced tapioca.

Several ornamental plants came to India as introductions, e.g., *Bougenvillea, Jacaranda mimosifolia* and *Spathodea campanulata.* Ornamental garden annuals like phlox; pink, aster, snapdragons and many others are foreign introductions.

The performance of introduced plant in another habitat may not be similar to its performance in its natural habitat. In many cases the introduced plant species or even a new variety may appear poorly adapted in the introduced habitat. However, after a few seasons many introductions appear promising. The introductions take some time to get acclimatized. The ability of a genotype to get accustomed or adapted to a new climate is referred to as acclimatization (Poehlman, 1987). Genetic variability for "fitness" is an essential prerequisite for ability of plant species to get acclimatized to a new environment. Natural selection exerts selection pressure for isolating well adapted genotypes within an introduced plant species. With acclimatization plants with genotypes well adapted to new environment tend to get enhanced in magnitude as compared to unadapted genotypes. Because the probabilities of recombination are more in cross-pollinated crop species, chances of these species getting acclimatized are more than the self-pollinated crop species. Mode of pollination, longevity of species, nature and intensity of environmental stress etc., influence the ability of plant species to get acclimatized to a new environment. Because of paucity of genetic variability within inbred lines, there will be slight change after their transportation, and in the absence of recombination and mutation, may be less capable to get acclimatized. As compared to the perennials, more generations of annuals grow in a given time span and hence frequency of recombinations will be comparatively more in annuals. Acclimatization of annuals will, therefore, be faster. Fitness is influenced by many types of environmental stresses and, therefore, these influence the extent of acclimatization of a given plant species. Different genotypes of a plant species interact with environments, which is reflected in fluctuations in performance of various genotypes and also unstability.

Plant introduction probably began when our ancestors uprooted plants, which they considered useful to them, from their natural habitats and transported them to the new surroundings near their dwellings. Subsequently, the invaders, pilgrims, visitors, traders, etc. whereever they went took back either the seeds or plant propagules, which satisfied their fancy or they thought would be useful to their society, to their own countries. These agencies did not only introduce new plant species in their own homelands, but also took plants growing in their country to the places they visited. Earliest record of plant exploration and introduction in the history is that of expedition organized by Queen Hatshepsut of Egypt to East Africa in 1500 B.C. to procure incense tree. Similarly citrus plants were brought back to Japan from China in 61 A.D. (Allard, 1960).

Systematic attempts at plant introduction began when organized expeditions were sent for plant collection to enrich the herbaria, botanic gardens, arboreta and museums. Outstanding example of the role played by plant introduction is the carrying of seeds of food crops such as cereals, pulses, millets, vegetables, plants of fruits and ornamental types by the colonizers and emigrants from Europe to America, Australia, New Zealand and Canada. According to an estimate of the United States Department of Agriculture, almost all of the cultivated crops and plants grown in United States are introductions from abroad, while only a small percentage are indigenous types which largely comprise of edible berries, millets and some fruit trees (Kumar, 1957). Australia is also concerned with plant introduction. When European colonizers settled in Australia, the continent was devoid of economic flora, which would satisfy the needs and aspirations of civilized man. Many plants, now growing in Australia are from other continents. The only Australian plants which have been contributed to the other parts of world are *Eucalyptus* spp., *Casurina* spp. and *Macadamia ternifolia* (Frankel, 1957).

7.1 PURPOSE OF PLANT INTRODUCTION

The purpose is obvious, what is not available in the country but exists in other parts of the world should be in possession, either for fancy, for direct use, or for future use. Kumar (1957) cited the reasons for plant introduction as:

(i) To serve the interest or fancy,
(ii) To serve some special purpose,
(iii) To enhance the living collection of plants in botanical gardens or arboreta, and
(iv) To serve the basic plant material for genetic improvement of economic crops.

Plant introduction may involve transposition of new crops, new varieties of established crops and new characteristics for established crops (Briggs and Knowles, 1967).

Introduction of new crop means any crop brought into an area where it was never grown before. For plant introduction to be successful it is necessary that varieties should fit into the prevailing photoperiodic conditions in the country where the plant is introduced. For example long day onion will not perform well in tropical conditions (Seshadri, 1989). Ability of new introduction to grow normally and produce adequate seed is also desirable. Of course, acclimatization may lead to development of desirable types in due course.

A number of ornamental plants, fruit plants were introduced. It may not be possible to describe introduction of all the plant species in India. A few examples will be cited so that the reader gets a glimpse of plant introductions, which enriched India's floriculture. In the beginning of this chapter it was mentioned that *Bougenvillea, Jacaranda mimosifolia, Spathodea*, phlox, pink, asters etc. were introduced in India. Gladiolus was grown in parts of Asia Minor 2000 years ago. Mr. Charles Gray of Coonoor was the first person to grow gladiolus in India during the nineteenth century (Mukhopadhyay, 2002). *Chrysanthemum* grows wild in China and Japan. There is evidence that cultivation of *Chrysanthemum* (According to Dutta, 1996, *Chrysanthemum morifolium* has changed to *Dendranthema grandiflora*) started in China about 500 B.C. It was introduced in Holland in about 1668, in France in about 1789 and in England in 1795 (Dutta, 1996). From the historical evidences it seems that *chrysanthemum* was introduced in India earlier to these introductions. Abul Fazal, a courtier of the emperor Akbar (1555-1605), mentioned the names of several flowering plants growing in 'Naseem Bag' in Srinagar, including *chrysanthemum* (Shevanti) in his book 'Ain-e-Akbari' (Rahudkar, 1991). Emperor Jehangir (1605-1627) introduced cherry (Shahalu) in Kashmir from Kabul. Mexico is the centre of diversity for dahlia, which was introduced in India in 1857 and grown in the garden of Agri-Horticultural Society, Calcutta (Gopal Sut, 1991). Gerbera, Statis (*Limonium latifolium*), and so many bcautiful flowering plants were introductions.

Introduction for agricultural and horticultural purposes includes plants of cereals, oilseeds, fibres, ornamental plants, fruit trees, etc. Again a few examples will be mentioned of such introductions. Maize has South Mexico and Central America as centre of origin. Portuguese traders brought it to India in 16th century. Centre of origin of pearl millet is Africa and was introduced in India in very early days (Krishnaswamy, 1951). Date of arrival of sorghum is not certain. Sir George Watt (1893) mentions that there was no specific Sanskrit name for sorghum. The most usual term being 'Yavana' meaning a stranger. Ethiopian plateau in African continent played major role in the development of African plants, including sorghum (Murdock, 1959). From Ethiopia cultivated sorghum moved to West Africa and then to Sudan. From Africa movement of sorghum to East was most probably brought about through trading routes between East Africa and India via Arabia. According to Krapovickas (1968) the Peruvian type of groundnut was transported to Mexico and then across the Pacific to Philippines from where it spread to China and India. Niger (*Guzotia abyssinica*) is native to tropical Africa. Evidence suggests that it originated in high lands of Ethiopia. According to a legend Ethiopian Queen, in the remote past occupied vast territory of Gujarat in India and made Ethiopians to migrate to India, who might have brought along with them niger to India (Getinet and Sharma, 1996).

Potato is indigenous to the Central Andean regions of South America. It was carried to Europe by the early Spanish explorers in 16th century. From Europe it was introduced in India (Pushkarnath, 1964). Cuzin (1976) suggested the origin of soybean in North and Central China, where its domestication was started (Hymowitz, 1970). From China it spread to Southern China, as well as Japan, India, Nepal, Russia around first century (Gai, 1996). Soybean appears to have been introduced first in Northern India either from China through Himalayan mountain ranges or from Java through Burma. The exact time of introduction is not definitely known (Tedia *et al.*, 2001). According to these authors, Andolle in 1884 believed that soybean was a recent introduction in India. Soybean (small black seeded) was grown since 1882 at Nagpur, and it appears that soybean was introduced there much earlier.

Grape (*Vitis vinifera*) is indigenous to the region stretching from Caucasus to Pakistan. Hence most of our grapes must have been introduced from these regions, although in India grapes are found growing wild in sub Himalayan tracts (Hays, 1957). *Annona squamosa, A. reticulata* and *A. muricata* are cultivated in India. Most of the species of genus *Annona* are indigenous to tropical America and few species are believed to be native of Africa. This fruit species was perhaps introduced in India by Portuguese (Mukherjee and Majumdar, 1966). Guava (*Psidium guajava*) was probably brought to India by Spanish settlers in 16th century.

Large scale cultivation of sunflower began in India in 1972, with the introduction of open-pollinated varieties of Soviet Union origin viz., Peredovik (EC 68414), Armavirskii (EC 68415). Subsequently one early maturing accession, viz., EC 101495 (Cernianka 66) was identified in the germplasm accession and released as Morden in 1979 for cultivation in Karnataka State (Virupakshappa, 1995).

It is interesting to know about the introduction of cauliflower in India. For detailed information reader may refer to Vishnuswarup and Chatterjee (1999) and Seshadri and Chatterjee (2000). As mentioned elsewhere in this book, the cole crops, including the cauliflower and cabbage have descended from a common kale-like ancestor, the wild cabbage (*Brassica oleracea* var. Sylvestris), still found growing in Western and Southern Europe and North Africa (Allard, 1960; Giles, 1941; Horne, 1954). Systematic and extensive cultivation of cauliflower was first started in Italy where 'originals' were developed. These 'original' Italian types were taken to France, England, Germany and Netherlands, where some local types were developed from them, such as 'Northerns' and 'Cornish' in Yorkshire and Cornwall (England), respectively, 'Angers' and 'Roscoff' in France (Horne, 1954) and Erfurt or its allied 'Snowball' in Germany (Giles, 1941) and Netherlands (Nieuwhof, 1970). These types were mainly for winter cultivation, except 'Erfurt', which was suitable for summer. Cauliflower was introduced into India in 1824. The Royal Agri-Horticultural Society, Calcutta also introduced seeds of English vegetables including cauliflower in 1824 from South Africa (Anonymous, 1970). For about a century (1822 to 1929) cauliflower underwent selection by local growers. Selections were made for early maturity and for adaptability to hot humid weather.

Indian cauliflowers, according to Vishnuswarup and Chatterjee (1999) are characteristically different from the types grown in Europe.

They can tolerate high temperatures and humid conditions. Indian cauliflowers are perhaps earliest maturing types. Giles (1941) is of the view that the Indian cauliflowers are dwarf selections of 'Erfurts' or snowball types. This view is also supported by Nieuwhof (1969). Studies carried out by Vishnuswarup and Chatterjee (1999) suggested that Indian cauliflowers are a separate group from European types. It appears that they have originated by simple selection and later perhaps by recombination as a result of natural crossing between different types. The Indian cauliflowers did not develop only from Cornish but also from some other types like Roscoff, Italians and Northerns, as evidenced by the characteristics of leaves and curd. The observations on self-incompatibility support the view that Indian types are different from Erfurt, because the Indian types are predominantly self-incompatible while the Erfurts are commonly self-compatible (Vishnuswarup and Chatterjee, 1999). Carrot, radish, spinach are other vegetables introduced in India.

Varieties of several crop species have also been introduced in India. Singh (1983) classified the introduction of varieties into two categories, viz., primary introduction and secondary introduction. When the introduced variety performs very well and is straight away released as variety for general cultivation, the introduction is termed 'primary'. On the other hand, when selections are made from the introduced variety or crossed with indigenous one to evolve a new improved variety, the introduction is known as 'secondary'.

Varieties Arkel and Bonneville of garden pea were introduced from U.K. and United States of America, respectively and released straight away for cultivation. Arkel is still the popular variety. Sugarbaby of watermelon has been introduced from U.S.A. and is so widely grown that it is responsible for genetic erosion of other varieties grown in India. Among vegetable crops other primary introductions are Nantes of carrot (from France), Golden acre of cabbage (from Denmark) and California wonder of capsicum from U.S.A. Blood Red Malta, Pineapple, Valencia, Washington Navel are direct introductions of *Citrus sinensis* (sweet orange). Mahsuri, a *japonica* × *Indica* hybrid of rice developed in Malaysia, has been introduced in Andhra Pradesh for general cultivation. It is being grown in some other states also. Variety Peredovik of sunflower could also be regarded as primary introduction.

There are several examples of secondary introductions. Advanced generation segregating material from cross (Penjamosib × Gabo 55) of wheat was received in India in 1963 from CIMMYT, Mexico. After evaluation of the segregating material at several locations, viz., I.A.R.I., Pusa, New Delhi, Bhowali and Wellington; Punjab Agricultural University, Ludhiana and G.B. Pant University of Science and Technology, Pantnagar, for adaptation to various Indian environmental conditions, reaction to prevailing races and biotypes of all the three rusts, selections such as S 227, S 308, S 307 and S 331 were made. Kalyansona, a superior variety of wheat, was developed from selection S 227 (Jain, 1989). Pusa Ruby of tomato was developed from a cross Sioux × Improved Meeruti. Punjab Padmini and Parbhani Kranti of okra were isolated from segregates of cross (Pusa Sawani × Ghana) selections. Pusa Sunehari of Muskmelon resulted from Haramadhu × Edisto. In these varieties Sioux of tomato, Ghana selection of okra, Edisto of muskmelon were introduced varieties. Potato varieties like up-to-date, Magnum Bonum, were introduced for developing hybrids and varieties. Kufri Kisan, for example, involves in its parentage introduction up-to-date from Scotland, Ekishrazu from Japan, Katahdin from United States of America. Phulwa was also one of the indigenous cultivar in its parentage.

Jawaharlal Nehru Krishi Vishwa Vidyalaya, Jabalpur and Gobind Ballabh Pant University of Science and Technology, Pant Nagar obtained about 3500 lines of soybean from U.S.D.A. Some introductions like Bragg and Monetta were released for general cultivation. This represents primary introduction. But Bragg was one of the parents used in evolution of varieties PK 1042 and MACS 450. Variety JS 71-05 was selected from an introduced variety of soybean, Lee. These are some examples of secondary introductions.

Male-sterile lines, either cytoplasmic-genetic or genetic, were introduced from other countries and used as seed parents to evolve hybrids. Burton (1958) developed a cytoplasmic-genetic male-sterile line of pearl millet, Tift 23A. This was made available to Indian breeders in 1962. Consequently, hybrids were developed and released for cultivation using Tift 23A as seed parent, viz., HB1, HB2, HB3, HB4 and HB5 (Gill and Virk, 1995). Tift 23A was also used to evolve new male-sterile lines, e.g. 5071A, 5141A and 5054A. Cytoplasmic-genetic male-sterile lines, CMS 2, CMS 124 and CMS 234 and restorer lines of sunflower (RHA 266, RHA 274) were introduced from United States of America (Virupakshappa, 1995). Hybrid based on CMS 234A and RHA 274 was developed and released as BSH-1 in 1980.

A hundred per cent pistillate line of castor released in USA in 1962, viz., TSP 10R (Texas S-pistillate 10), was obtained from Texas in 1965. Gujarat developed a hybrid, Gujarat castor hybrid 3 (GCH 3), in 1968 (Ramachandram and Prasad, 1995). Two male-sterile lines of rice, developed at the International Rice Research Institute, Philippines, viz., IR 58025A and IR 62829A have been utilised by the Indian plant breeders to develop hybrids of rice. A few have been released for cultivation, e.g. APHR 1, DRRH 1. A male-sterile line, Combine Kafir 60A, was introduced in India and utilized to develop the first hybrid of sorghum, CSH 1, released in 1964. Even the restorer parent of this hybrid, viz., IS 84 (yellow endosperm Feterita), was exotic in origin.

A number of exotic introductions, either varieties, wild related species, have been utilized to incorporate resistance against diseases and/or insect pests. For example, Sandhu *et al.* (1974) reported that one of the accession, EC 31830 (Asumtemkoko) from Ghana introduced by the division of Plant Introduction, IARI, New Delhi, and identified as *Abelmoschus manihot* spp. manihot, possessed high degree of symptomless carrier type of resistance to yellow vein mosaic virus. Crossing Pusa Sawani with accession EC 31830, as male, resulted in development of field tolerant variety of okra, Punjab Padmini. A variety of potato, Kufri Jeevan (M 109-3 × 698-D = SLB/E 427) had complex parental seedling origin (Anonymous, 1984). The seedling derived resistance to late blight from *Solanum demissum*. Segregating population of cross 3069d × 2814a (I) = SLB/Z 389 (b) was supplied by Dr. W. Black of Scottish Plant Breedeing Station, Edinburgh, U.K. The Central Potato Research Institute, Shimla, released a selection from this material, Kufri Jyoti as variety in 1968. The late blight resistance of this variety ows to *Solanum demissum*. Brown plant hopper is a serious pest of rice in India and it has several races/biotypes. Several lines from International Rice Research Institute, Philippines were screened for their reaction to BPH (Rao *et al.*, 1995). Leb Muey Nahng, Ptb 21 were identified as resistance sources. A few cultures from crosses RP 825 (Vijaya × Ptb 21), RP 1045 (RP 31-49-2 × Leb Muey Nahng) and RP 1015 (Sona × Manoharsali) showed high frequency of resistance against BPH. Some resistant varieties released are Sansali, Mansarovar and Suraksha. Leb Muey Nahng, Ptb 21 are also found to be resistant to gall midge of rice. Varieties Suraksha, Vibhaya are resistant to this pest.

Introductions were also made to utilize some desirable attributes. For example, 'Norin' dwarfing genes of wheat from Japan were utilized to develop short internoded, stiff strawed variety of wheat 'Gaines' by Dr. O.A. Vogel of Washington, U.S.A. Dr. N.E. Borlaug of CIMMYT, Mexico introduced the 'Norin' genes in Sonora 63, Sonora 64, Lerma Rojo 64A and Mayo 64. Similarly Dee-Geo-Woo-Gen genes for dwarf stature of rice from China were used to breed dwarfer rice varieties. Exotic vegetable genetic resources have been introduced which could be used to improve various attributes. For example, EC 380989-92 of watermelon with yellow flesh and early maturity, EC 310299 of tomato with slow ripening gene, pungent type of chilli EC 391073, high yielding and bold clove garlic EC 367655-58 have been introduced (Gautam *et al.*, 1998). Gynoecious lines of cucumber (EC 382737-39) have been received from U.S.A.

Collection through introduction from within the country and from other countries is being taken care by various institutions in the country, including the National Bureau of Plant Genetic Resources, New Delhi. For example, germplasm collection of 42,147 lines of wheat are in I.A.R.I., NBPGR, PAU, GB Pant University of Agriculture and Technology etc. (Tandon and Mathur, 1988). At the International Crops Research Institute for Semi-Arid Tropics (ICRISAT), Hyderabad, a little more than 18,000 accessions of pearl millet, representing 42 countries have been assembled and India has provided 11,140, genetic stocks (Harinarayana *et al.*, 1988). About 20,000 genetic resources, belonging to *G. arboreum, G. herbaceum, G. hirsutum, G. barbadense* and Wild species have been assembled at various cotton centres in India (Bhale and Narayanan, 1988). During 1985-86, 2320 exotic germplasm of different temperate, tropical and sub-tropical vegetable crops were introduced by N.B.P.G.R. from 34 countries (Seshadri, 1988). According to Gautam *et al.* (1998) 1,65,986 accessions are in *ex-situ* seed repository of National Gene Bank. It includes 70,414 lines of cereals and pseudo-cereals, 16,587 millets and minor millets, 24,857 oilseeds, 26,614 pulses, 9284 vegetable and spices, and many others.

7.2 PLANT INTRODUCTION AND QUARANTINE

An effective means of plant introduction of new and desirable plant material is via expeditions to the unexplored parts of the world. Nevertheless, with the establishment of International Board of Plant Genetic Resources (I.B.P.G.R.) at Rome and the National Bureau of Plant Genetic Resources, with headquarters at New Delhi, meeting the demands of various plant genetic resources has become much easier. The objective of I.B.P.G.R. is to encourage and assist the countries to conserve and use plant genetic resources for the benefit of present and future generations. The main objectives may be cited as (Bhagmal, 1998):

(i) Assist the countries, particularly in the developing world, to assess and meet their requirements for conservation of plant genetic resources.
(ii) To strengthen and contribute to the international collaboration in the conservation and use of plant genetic resources (PGR).
(iii) To develop and promote improved strategies and technologiés for the conservation of PGR, and
(iv) Provide information service on PGR.

The I.B.P.G.R. has offices in different regions of the world, viz., Regional Office for Europe at Rome; for Asia the Pacific and Oceania at Serdang (Malaysia); for Central and West Asia and North Africa at Aleppo (Syria); for sub-Saharan Africa at Nairobi (Kenya) and for America at Cali (Columbia). In addition it has offices for South Asia (Sub-regional) at New Delhi and for East Asia at Beijing (China) and for West Africa in Mlamey (Niger).

The N.B.P.G.R. has base centres at nine places in the country (Akola, Cuttack, Delhi, Hyderabad, Jodhpur, Ranchi, Shillong, Shimla, and Srinagar) while regional Stations are established at Akola/ Amravati, Bhowali, Jodhpur, Shillong, Shimla and Trichur. It has an experimental farm at Issapur (Paroda, 1988).

The scientists who are engaged in plant improvement in India and require certain plant material have to submit their requirements to NBPGR. These may be through the State Agricultural Universities, or through the Project Coordinators or Project Directors of All India Coordinated Projects or Directors of National Research Centres. Scientists working elsewhere, other than those mentioned above, may approach NBPGR directly for their requirements. Whenever the required plant material is not in the active collection of the NBPGR or NRCs or the Coordinated Projects it has to be procured from other country, which may be received as gift or exchange. But in any case it has to pass through quarantine procedure.

Germplasm exchange certainly helps to conserve valuable genetic resources and develop superior varieties of various crops. Nevertheless, such exchanges are sometimes disastrous due to inadvertent introduction of or spread of exotic disease or insect pest or weeds, leading to heavy losses. For example, apple scab (*Venturia inaequalis*) became very serious after its introduction in our country along with varieties like Red Delicious, Golden Delicious etc. from other countries. It is found in Kashmir Valley, Himachal Pradesh and other Himalayan ranges. Recently it has been reported from Nilgiris (Bhaskaran and Jeyarajan, 1985). The European Corn Borer (*Pyrausta nubilalis*) introduced into U.S.A. through broomcorn shipments from Italy/Hungary in 1908 was responsible for loss of 349 million dollars in 1949 (Rajak, 1988).

In 1906 the Government of India ordered compulsory fumigation of all the incoming cotton bales at the ports of entry to prevent introduction of Mexican cotton boll weevil (*Anthonomus grandis*). Subsequently the Government of India enacted the Destructive insects and Pests Act of 1914 which was the earliest measure taken to empower the then Federal and State Governments to frame the necessary rules regulating import and movement of certain plant parts and plant material into India and among the states within the country. The Act after its promulgation in 1914 has undergone a number of amendments and corrections, the latest being Plants, Fruits and Seeds (regulation of Import into India) Order, 1984. The current DIP Act provides:

(i) No consignment of plants or parts thereof or any plant product shall be imported into India without valid "import permit" issued by the Plant Protection Adviser to Government of India.

(ii) All consignments shall be imported to India only through prescribed ports of entry, where plant quarantine and fumigation facilities have been established (Rajak, 1988). These are: **Sea ports:** Kolkata, Mumbai, Chennai, Bhavnagar, Tuticorin, Cochin, Rameshwaram, Vishakhapatanam and Nagapatanam, **Airports:** Amritsar, Mumbai, Kolkata, Chennai, New Delhi, Trichurapalli, Trivendram, Patna, Varanasi and Hyderabad, **Land frontiers:** Attari-Wagah Border, Amritsar Rail Head, Attari Rail Head, Bongaon, Gede, Kalimpong, Sukhiapokhri.

(iii) All consignments on arrival at quarantine station shall be inspected and, if necessary, fumigated disinfected/disinfested by the authorized officials.

(iv) No consignment shall be imported into India unless accompanied by a valid phytosanitary certificate issued by the authorized official or quarantine personnel at source.

(v) The seed/planting material requiring isolation growing shall be grown in post-entry quarantine facility approved by Plant Protection Advisor.

It is well known that the quarantine inspections are not full proof and neither can they bring out all the hidden pathogens that may be present at the time of inspection. There is no single treatment that will eliminate the pathogen present in germplasm. It is, therefore, essential to grow the imported germplasm in isolation for a reasonable period of time to ensure that imported material is free from exotic pathogen. The Directorate of Plant Protection, Quarantine and Storage has established two screen/glass houses at Faridabad for post-entry quarantine examination.

The authority for enforcing plant quarantine in India basically rests with the Directorate of Plant Protection, Quarantine and Storage. However, the Government of India has also authorized three other national institutions to handle quarantine clearance of seed and plant material imported for research. These are National Bureau of Plant Genetic Resources (NBPGR), New Delhi, for germplasm material of Agri-horticultural crops; Forest Research Institute, Dehradun for forestry crops and Botanical Survey of India, Kolkata for plants of botanical interest (Nath *et al.*, 1988).

After the introductions are received they should be properly catalogued, i.e. proper serial number, information regarding the source from which it was received, name of variety, species, whether wild relative, weedy species etc., the plant characters, etc. should be recorded.

In order to assess the potential of the introduced material its evaluation over locations and if possible over years should be carried out. If the material has been requisitioned for insect/disease resistance, it should be evaluated in artificially created epiphytotic/artificial infestation with reared insects.

References

Allard, R.W. (1960). *Principles of plant breeding.* John Wiley and Co., New York.

Anonymous (1970). 150th Anniversary Souvenir. Royal Agri-Horticultural Society, Calcutta.

Anonymous (1984). Potato cultivars released by the Central Potato Research Institute. Eds. P.C. Mishra, P.C. Gaur and N.M. Nayar. Central Potato Research Institute, Shimla.

Bhagmal (1998). Plant Genetic Resources Activities in South Asia. Paper presented at short training course on "PGR Principles, Procedures and Approaches in collection, evaluation and conservation". NBPGR, New Delhi, 27th July to 5th August, 1998.

Bhale, N.L. and S.S. Narayanan (1988). Need for augmenting genetic resources in cotton — Indian perspective. In *Plant Genetic Resources — Indian Perspective*. Eds. Paroda *et al.* NBPGR, New Delhi, pp. 220-231.

Bhaskaran, R. and R. Jeyarajan (1985). Apple scab disease spreads to the Nilgiri district. *TNAU Newsletter*, 15(5): 2.

Briggs, F.N. and P.F. Knowles (1967). *Introduction to plant breeding*. Publ. Reinhold Publishing Corporation, New York.

Burton, G.W. (1958). Cytoplasmic male-sterility in pearl millet (*Pennisetum glaucum* (L.) P.Br.). *Agron.J.*, 40: 230-235.

Cuzin, V.F. (1976). Soviet Great Encyclopedia (3rd ed.). Moscow Publishing Company. *Soviet Encyclopedia*, 24: 281-282.

Datta, S.K. (1996). *Chrysanthemum*. NBRI Bulletin No.1. National Botanical Research Institute, Lucknow, p. 26.

Frankel, O.H. (1957). The biological system of plant introduction. *Indian Journal of Genetics and Plant Breeding*, Vol. 17(2): 336-342.

Gai, J. (1996). Soybean breeding. In J. Gai (ed.) *Plant breeding*. Çrop Science, China Agriculture Press, Beijing.

Gautam, P.L.; A. Kumar and B.M. Singh (1998). Germplasm collection, maintenance and utilization in improvement of vegetable crops. Presented at the summer School on "Advance Technologies in the improvement of vegetable crops including cole crops" at IARI, New Delhi, 4 to 24th May, 1998.

Getinet, A. and S.M. Sharma (1996). Niger (*Guizotia abyssinica*, (L.F.) Cass.) Promoting the conservation and use of underutilized and neglected crops. IBPGR, /Rome, p. 60.

Giles, W.F. (1941). Cauliflower and broccoli-what they are and where they come from. *J.Roy.Hort.Soc.*66: 265-277.

Gill, K.S. and D.S. Virk (1995). Progress and prospects of pearl millet hybrids in India. In "Hybrid Research and Development" Proc. of Seminar, 1993. *Indian Society of Seed Technology*, New Delhi. Eds. M.Rai and S. Mauria, pp. 45-60.

Gopal Sut (1991). Dahlia. In "Baliraja" Krishi Vidnyan Prakashan, 768 Sadashiv Peth, Pune: 411-430, pp. 117-119. (In Marathi)

Harinarayan, G.; S. Appa Rao and M.H. Mengesha (1988). Prospects of utilizing genetic diversity in pearl millet. In "Plant Genetic Resources, Indian Perspective" Eds. Paroda *et al.* NBPGR, New Delhi, pp. 170-182.

Hays, W.B. (1957). *Fruit growing in India.* Publ. Kitabistan, Allahabad

Horne, F.R. (1954). Winter cauliflower: History and breeding in the South West. *Sci.Hort.*, 11: 128-139.

Hymowitz, T. (1970). On the domestication of soybean. *Econ.Bot.*, 24: 408-421.

Jain, K.B.L. (1989). Cereals: wheat and barley. In *Plant breeding — theory and practice*, Ed. V.L. Chopra. Oxford and IBH Publishing Co. Pvt. Ltd., New Delhi, pp. 77-99.

Krapovickas, A. (1969). The origin, variability and spread of groundnut (*Arachis hypogea*). In *The domestication and exploitation of plants and animals*, Eds. P.J. Ucko and I.S. Falk Publ. Gerard Duckworth Co Ltd., London.

Krishnaswamy, N. (1951). Origin and distribution of cultivated plants of South Asia. Millets. *Indian Journal of Genetics and Plant Breeding* 11: 67-74.

Kumar, L.S.S. (1957). Scope and importance of plant introduction in South Asia. *Indian Journal of Genetics and Plant breeding*, 17(2): 362-369.

Mukherjee, S.K. and P.K. Majumdar (1966). Genetics and breeding of tropical and sub tropical fruits. *Indian Journal of Genetics and Plant Breeding*, 26A: 155-167

Mukhopadhyay, A. (2002). Gladiolus. Publications and information Division, I.C.A.R., New Delhi 110-012, p. 83.

Murdock, G.P. (1959). Staple subsistence crops of Africa. *Geographical Review*, 50: 521-540.

Nath, Ram, V.K., Mathur and B.R. Verma (1988). Plant quarantine activities in Indian context. In *Plant Genetic Resources — Indian Perspective*. Eds. Paroda *et al.* NBPGR, New Delhi, pp. 470-480.

Nieuwhof, M. (1970). Personal communications. Quoted by Vishnuswarup and S.S. Chhatterjee (1999).

Paroda, R.S. (1988). Genetic Resources Activity in India — Some considerations. In *Plant Genetic Resources — Indian Perspective*. Eds. Paroda *et al.*, NBPGR, New Delhi, p. 17-27.

Poehlman, J.M. (1987). *Breeding Field Crops*. Third Edition. AVI Publishing Co. Inc. Westport, Connecticut.

Pushkarnath (1964). Potato in India, varieties. Indian Council of Agricultural Research, New Delhi, p. 466.

Rahudkar, V.B. (1991). Bharatiya Udyane. Baliraja Publ. Krishi Vidnyan Prakashan, 768, Sadashiv Peth, Pune 411 030, pp. 25-29 (In Marathi).

Rajak, R.L. (1988). Plant quarantine regulations in relation to Exchange of Plant Genetic Resources. In *Plant Genetic Resources — Indian Perspective*, Eds. Paroda *et al.*, NBPGR, New Delhi, pp. 49-58.

Ramachandram, M. and M.V.R. Prasad (1995). Hybrid castor research and development in India. In "Hybrid Research and Development". Proceedings seminar, 1993. *Indian Society of Seed Technology*, New Delhi, Eds. M.Rai and S. Mauria, pp. 94-103.

Rao, U.P.; M.V.S. Sastry, J.R.K. Rao; G.S.V. Prasad and M.B. Kalode (1995). Breeding varieties resistant to major insect pests in rice. In "Genetic Research and Education: Current trends and the next fifty years" *Indian Society of Genetics and Plant Breeding*, New Delhi, Eds. B. Sharma *et al.* pp. 223-227.

Sandhu, G.S.; B.R. Sharma; B. Singh and J.S. Bhalla (1974). Sources of resistance to jassid and white fly in okra germplasm. *Crop Improvement*, 1: 77-81.

Seshadri, V.S. (1988). Genetic Resources and their utilization in vegetable crops. In *Plant Genetic Resources — Indian Perspective*, Eds. Paroda *et al.*, NBPGR, New Delhi, pp. 335-343.

Seshadri, V.S. (1989). Vegetable crops. In *Plant Breeding. Theory and Practice*, Ed. V.L. Chopra. Oxford and IBH Publishing Co. Pvt. Ltd., New Delhi, pp. 391-408.

Seshadri, V.S. and S.S. Chatterjee (2000). The history and adaptation of some introduced crops in India. *Asian Agri.History*, Vol.4 (3): 175-202.

Singh, B.D. (1983). *Plant breeding — principles and methods*. Publ. Kalyani Publishers, New Delhi.

Tandon, J.P. and H.C. Mathur (1988). Status of wheat germplasm in India and future needs. In *Plant Genetic Resources — Indian Perspective*. Eds. Paroda *et al.*, NBPGR, New Delhi, pp. 121-133.

Tedia, M.D.; P. Jain and S.I. Tanweer (2001). Success story of soybean in M.P. Information and management technology for soybean industries. In "Sharing in the Indian soybean sector" *Proceedings of India soyforum, 2001*" March, 17-18, 2001. Organized by Ministry of Agriculture, Govt. of India, Uni. Of Illinois, USA, ICAR, State Govt. M.P., Maharashtra, Rajasthan and SOPA. SOPA, Scheme 53, Malviyanagar, AB Road, Indore (India.).

Virupakshappa, K. (1995). Heterosis breeding in sunflower: Historical development, present status and future perspective. In *Hybrid Research and Development*, Indian Society of Seed Technology, New Delhi, Eds. M.Rai and S. Mauria, pp. 81-91.

Vishnuswarup and S.S. Chatterjee (1999). Origin and Genetic improvement of Indian cauliflower. *Asian Agri-History*, Vol.3(1): 5-19.

8

Selection: A Tool for Plant Improvement

Allowing certain plants from the population to produce offspring, while barring others from reproduction, comprises of selection. This permission for reproduction may be granted by the nature or by the plant breeder. In other words both, nature and breeder, exert selection pressure. Natural selection operates through differences of fertility among the adult individuals of parent generation as well as through differences of viability among the offspring generation. It is usually assumed that all the individuals in the population contribute equal number of offspring to the next generation. But individuals differ in their fertility, i.e. produce different number of gametes, and hence contribute different number of progeny. Proportionate contribution of offspring to the next generation is called as "fitness", "adaptive value" or "selective value". The force acting against is selection. Proportionate reduction in the gametic contribution of a particular genotype, as compared to the standard, which is usually the most favoured, is known as 'coefficient of selection', and denoted by s. Suppose genotype *AA* produces 100 offspring and *aa* only 95. The fitness of *aa* is reduced by 5. Thus selection operated against *aa*. When a gene is subject to selection its frequency is not same in offspring generation as compared with parental generation. Artificial selection, i.e., the selection resulting from action of breeder in the choice of parents, produces its change of gene frequency by separating the adult individuals of parent generation into two groups, viz., the selected and discarded, that differ in gene frequency.

In connection with selection it is important to recognize that:

(i) Selection can act effectively only on heritable differences, and

(ii) Selection can not create new variability but can act on variability that already exists in the population.

As pointed out at the beginning of this chapter selection tends to change the gene frequency. Effect of selection is to change the frequencies with which various genes or combinations of genes occur in the population. In case of cross-pollinated crops, selection can lead to assembly of new combinations of genes, not previously found in the population. According to Allard (1960) three phenotypically observable responses to selection are:

(i) Change in the proportion of previously existing genotypes, accompanied by shift in the population mean,
(ii) Appearance of new genotypes in the selected population, as compared to the base population, and,
(iii) Reduction in the variability for selected character in the selected population.

In the next two sections change of mean and change of gene frequency due to selection will be described.

8.1 CHANGE OF MEAN

Let us assume that a population of self-pollinated crop is made up of three genotypes, viz., *AA*, *Aa* and *aa*, with genotype frequencies 0.25, 0.50 and 0.25, respectively. Also assume that genotypes *AA*, *Aa* and *aa* are associated with phenotypic values 20, 15 and 10 units, respectively. We can assume that this is F_2 generation of a self-pollinated crop. The mean measurement of F_2 ($\bar{F}_2$) would be:

$$\frac{(0.25\times 20)+(0.50\times 15)+(0.25\times 10)}{1}=15 \text{ units}$$

Now suppose, selection operates against genotype *aa*, which are 25 per cent in the F_2 generation. The *aa* genotype is not allowed to reproduce and contribute progeny to next the generation. The mean measurement of F_3 generation ($\bar{F}_3$) could be got as follows:

Genotypes in F_3	*AA*	*AA* (From *Aa* in F_2)	*Aa* (From *Aa* in F_2)	*aa* (From *Aa* in F_2)	Total
Genotype frequencies (*f*)	0.25	0.50 (0.25	0.50	0.25)	–
	0.250	0.125	0.250	0.125	0.750
Values (*v*)	20	20	15	10	–

Now the $\bar{F}_3$ would be:

$$\frac{(0.25 \times 20) + (0.125 \times 20 + 0.25 \times 15 + 0.125 \times 10)}{0.750}$$

$= 16.66$ units

A reader may confirm that when genotype *aa* is not allowed to reproduce in F_3, the mean of F_4 generation would be 18.0 units. Thus selection against a genotype with smaller phenotypic value leads to increase in population mean.

Effect of selection on the mean could be described for a random mating population also. Assume that population in Hardy-Weinberg equilibrium, comprises of 3 genotypes *AA*, *Aa* and *aa* with genotype frequencies p^2, $2pq$ and q^2, respectively (see Chapter 5), such that $p^2 + 2pq + q^2 = 1$. Assume that the values associated with *AA* and *aa* are d and $-d$, respectively. These represent deviations from mid-point value i.e. ($\frac{AA + aa}{2}$), which in the present case is zero. Departure of heterozygote from mid-point is h. The population mean is then

$$\frac{(p^2 \times d) + (2pq \times h) + (q^2 \times -d)}{1}$$

$= p^2d + 2pqh - q^2d$

$= d(p^2 - q^2) + 2pqh$

$= d(p + q)(p - q) + 2pqh$, and because $p + q = 1$,

$= d(p - q) + 2pqh$ ...(8.1)

The expression (8.1) shows that populations mean depends on d, the magnitude of departure of either homozygote from mid-point, deviation of heterozygote from mid-point (h) and gene frequency. To elucidate the effect of selection on population mean assume that values of *AA*, *Aa* and *aa* are 20, 15 and 10, respectively. Hence, $d = 20 — 15$ (or $15 — 10$) $= 5$ and $h = 0$, because value of heterozygote is equal to mid-point $\left(\frac{20 + 10}{2} = 15\right)$.

Let us assume that the frequencies of *AA*, *Aa* and *aa* are 0.25, 0.50 and 0.25, respectively. Then the frequency of gene *A* (i.e. $\hat{p}$) would be

$$\frac{0.25 + ½(0.50)}{1} = 0.50$$

And likewise the frequency of gene *a* (i.e. $\hat{q}$) would be 0.50. Substituting values of *p*, *q*, *d* and *h* in (8.1) we get population mean as:

$$d(p - q) + 2pqh$$
$$= 5\,(0.5 - 0.5) + 2 \times 0.5 \times 0.5 \times 0$$
$$= 0$$

Because the values were expressed as deviations from mid-point, i.e. 15, the mean would be 15 + 0 = 15. Now suppose by selection the frequency of gene *A* (i.e. $\hat{p}$) has been increased to 0.60. What would be the mean of selected population containing 60 per cent *A* genes? Substitute new value of *p* in (8.1).

$$d(p - q) + 2pqh$$
$$= 5\,(0.60 - 0.40) + 2 \times 0.6 \times 0.4 \times 0$$
$$= 1.0$$

And the mean would be 15.0 + 1.0 = 16.0. Thus selection favouring gene *A*, leads to increase of population mean.

8.2 CHANGE OF GENE FREQUENCY

It was mentioned at the beginning of this chapter that the effects of selection are exerted by changing the frequencies with which various genes or combinations of genes occur in the population. Change of gene frequency resulting from selection can be shown for random mating population with convenience. While considering the effect of selection to change the frequency of gene, degree of dominance has also to be considered.

Let us assume that a population, which has attained Hardy-Weinberg equilibrium, has three genotypes *AA*, *Aa* and *aa* with genotype frequencies p^2, $2pq$ and q^2, such that $p^2 + 2pq + q^2 = 1$. Assume that coefficient of selection against *aa* is *s*. Contribution of 3 genotypes to the next generation, in terms of gametic contribution would be as follows:

Genotypes	*AA*	*Aa*	*aa*	*Total*
Genotype frequencies	p^2	$2pq$	q^2	1
Fitness	1	1	$1-s$	
Contribution of gametes	p^2	$2pq$	$q^2(1-s)$	$1-sq^2$

The total contribution of gametes is reduced after selection against genotype *aa*. The frequency of gametes *a*, and also the new frequency of gene *a* in progeny generation, would be (q_1):

$$\frac{\text{Total frequency of } aa + \text{Half the frequency of } Aa}{\text{New total}}$$

$$= \frac{q^2(1-s)+pq}{1-sq^2}$$

Which on simplification reduces to:

$$q_1 = \frac{(q-sq^2)}{1-sq^2}$$

The change of gene frequency due to selection can be obtained (Δq) as q_1 — q, where q is the gene frequency in original population. Thus Δq

$$= q_1 - q$$

$$= \frac{q-sq^2}{1-sq^2} - q$$

$$= \frac{q-sq^2-q+sq^3}{1-sq^2}$$

$$= \frac{-sq^2(1-q)}{1-sq^2} \qquad \text{...(8.2)}$$

The change is negative. Change of gene frequency depends, not only on the intensity of selection(s), but also on the initial gene frequency.

The change of gene frequency following selection could be elucidated with an hypothetical example. Remember, it is difficult to follow change of frequency for individual genes governing inheritance of quantitative characters because effect of each gene is small similar to one another and mimicked by environmental agencies. But the underlying cause for change of mean or variance is change of gene frequencies. Suppose the frequency of the gene *a* in a population is 0.40 and coefficient of selection against *aa* is 0.30. The new frequency would be,

$$q_1 = \frac{q - sq^2}{1 - sq^2}$$

$$= \frac{0.40 - (0.30 \times 0.40^2)}{1 - (0.30 \times 0.40^2)}$$

$$= 0.369.$$

And the change of gene frequency is 0.369 – 0.400 = - 0.031. Alternatively, substitute appropriate values, in 8.2 to get Δq. Thus:

$$\Delta q = \frac{-0.30 \times 0.40^2 (1 - 0.40)}{1 - (0.30 \times 0.40^2)} = -0.030.$$

The change of gene frequency resulting from selection when there is no dominance i.e. gene action is additive, can also be shown.

When there is no dominance, i.e. gene action is additive, the heterozygote is intermediate in phenotype as compared to both the homozygous parents. Suppose coefficient of selection against *aa* is *s*, then obviously coefficient of selection against *Aa* would be ½ s. The contribution of gametes by the *AA*, *Aa* and *aa* to the next generation would be got as shown below, when dominance is absent:

Genotypes	*AA*	*Aa*	*aa*	*Total*
Genotype frequencies	p^2	$2pq$	q^2	1
Fitness	1	1-½ *s*	1-*s*	
Contribution of gametes	p^2	$2pq - pqs$	$q^2 - sq^2$	1-*sq*

The total contribution of gametes to the next generation would be:

$$p^2 + 2pq - pqs + q^2 - sq^2$$

$$= 1 - pqs - sq^2, \text{ and because } p + q = 1, p = 1 - q$$

$$= 1 - (1 - q)qs - sq^2$$

$$= 1 - qs + sq^2 - sq^2$$

$$= 1 - sq.$$

The frequency of gametes *a* (and also the frequency of gene *a* in the next generation) would be (q_1):

$$\frac{q^2 - sq^2 + pq - \frac{1}{2}pqs}{1 - sq}$$

Which on simplification becomes:

$$\frac{-sq^2 + q - \frac{1}{2}sq + \frac{1}{2}sq^2}{1 - sq}$$

Therefore, change of gene frequency (Δq) would be $q_1 - q$

$$\frac{-\frac{1}{2}sq(1-q)}{1-sq} \qquad \ldots(8.3)$$

Expression (8.3) also shows that change of gene frequency, due to selection, is a function of intensity of selection (s), initial gene frequency and the degree of dominance. Although (8.2) and (8.3) were shown for a single locus with two alleles, reasoning holds true for any number of loci, in the absence of non-allelic interaction.

Because the components of genotypic variance, such as additive genetic, dominance and epistatic or non-allelic interaction variance, depend upon gene frequencies, components of variance also change as the gene frequencies change due to selection. Before we proceed to learn about change in genotypic components of variance due to selection, a word about directional selection.

Usually a plant breeder exercises directional selection, i.e. involves selection for only one level of a character, for example, in favour of tall plants or early maturity or intense colour of flowers, etc. This leads to fixation of genes and usually the extreme phenotypes are favoured while others are eliminated (of course in the absence of reverse mutation and migration). There is a form of selection called disruptive selection, in which case selection is made for more than one level of a particular character and extreme types are mated together. Thoday (1964) exercised selection in *Drosophila melanogaster* for both low and high bristle number in each generation and in each generation mating was permitted between some of the flies from low bristle selections and some of the flies from high bristle selections. Consequently there would be exchange of genes between the two groups. Thoday discovered that using disruptive selection, he could develop a polymorphic population in a relatively few generations. This polymorphic population continued to diverge until it became two separate populations, one with low abdominal bristle number and the other with high abdominal bristle number. Disruptive selection led not only to a polymorphic population but the polymorphism could be maintained, even with deliberate crossing together of high and low lines. An interesting feature was that genetic variability was much increased under disruptive selection. From the results Doggett (1970) concluded that disruptive selection leads to:

(i) Divergence without isolation,
(ii) Rapidity of crop development, and
(iii) Maintenance of very high variability.

When selection favours either homozygote, complete selection leads to fixation of that genotype and elimination of all other genotypes. However, when heterozygote is favoured, both the homozygotes remain in the population.

8.3 GENOTYPIC VARIABILITY AND SELECTION

Selection acts effectively on heritable variation and selection can not create variability. The genotypic variance has three components (see chapter 5) viz., additive genetic, dominance and non-allelic interaction. Among these components additive variance is known to be the chief cause of resemblance between relatives and the chief determinant of observable genetic properties of the population and hence response to selection. The additive component, as was shown in chapter 5 is described by symbol *D* for self-pollinated crops. According to Mather and Vines (1952) "*D* depends on difference between individuals distinguished by being homozygous for two allelomorphs of various genes". On the other hand, H, the dominance component depends upon departure of heterozygote from the mean of two comparable homozygotes (Mather and Vines, 1952). It is therefore obvious that variation attributable to differences between homozygotes is fixable. However, superiority of heterozygote dissipates as the magnitude of heterozygosity reduces with self-fertilization. Hence it is not fixable component of genotypic variance. Even in case of cross- pollinated crops, additive genetic variance is regarded as fixable component although in these crops additive variance does not arise only due to additive gene action (Expression for additive variance of random mating populations is $2pq[d+(q-p)h]^2$). Only when $p = q = ½$, additive genetic variance of random mating populations is free from dominance component. Additive variance will not contain dominance component when $h = 0$. Selection leads to change of gene frequency and therefore components of genotypic variance. This can be very simply verified by substituting the values of p and q in the expression of additive variance. Suppose departure of either homozygote from mid-point (i.e. d) is 5 units and departure of heterozygote from mid-point is 3 units. Suppose in the initial population, frequency of desirable gene, say A, is 0.20. The additive variance in this population would be:

$$\sigma^2 A = 2pq\,[d+(q-p)h]^2$$
$$= 2 \times 0.2 \times 0.8\,[5 + (0.8-0.2)3]^2$$
$$= 14.79$$

If the frequency of gene A is increased by selection, to say 0.60, the additive genetic variance will be utilized and hence lead to decline in magnitude of additive variance. Let us substitute value of gene frequency in the expression for additive variance and verify if additive variance declines. Thus,

$$\sigma^2\text{A} = 2pq\,[d+(q-p)h]^2$$
$$= 2 \times 0.6 \times 0.4\,[5+(0.4-0.6)3]^2$$
$$= 9.29$$

The additive variance is now lower in magnitude. This is also true for self-pollinated crops.

If all the genotypic variability were due to additive gene action, improvement theoretically should continue until all the desirable alleles have been fixed in the population. The rate of which selection leads to fixation and hence to exhaustion of genetic variability depends upon gene number and gene frequency. When the number of genes is large, change of gene frequency will be very small for each gene due to selection, despite the fact that gene frequency may be nearer to 0.5. Change of gene frequency will be even more smaller when gene frequency is lesser than 0.5. Hence it is very difficult to fix all the desirable genes and achieve complete fixation when gene number is larger and gene frequency is less than 0.50. Slow, steady progress for several generations can be expected when selection pressure is exerted for a trait whose inheritance is governed by several genes with small effects, acting in additive fashion.

8.4 GENETIC ADVANCE

The objective of most, if not all the breeding programmes, is to select individuals with superior phenotypic values from the genotypically variable base population and utilization of these individuals for creation of new population to be employed either as a potential new commercial variety or as a base for new cycle of selection. The phenotypic value of an individual can be linearly expressed as:

$$P = \mu + G + E + GE, \qquad ...(8.5)$$

where μ is the general population mean, G is the genotypic value which is conferred on individual by the genotype it carries, E is the environmental deviation which tends to cause a + or – shift from the general population mean, depending upon its nature of influence, GE is the interaction effect. Usually it is assumed that a specific change of environment has similar effect on different genotypes. When a specific change in the environment has different effects on different genotypes, interaction is said to exist between genotypes and environments.

It should be obvious from equation 8.5 that genetic advance under selection, as represented by improvement in G, the genotypic value, in the selected population as compared to the base population, depends upon:

(i) The amount of genetic variability, i.e. differences in magnitude of G among different individuals in the base population,
(ii) The intensity of selection, i.e., proportion of individuals chosen from the base population, and
(iii) The masking effect of environmental and GE interaction components of variability on genotypic variability.

Now suppose a breeder has grown a large number of homozygous lines of a self-pollinated crop at several locations for a number of years. Average yields of all the lines are available from the information so collected. When the mean yields ($\bar{X}_1$) of number of these lines are plotted on a graph paper and points joined by a line, the picture will be a bell shaped curve which is symmetric about the overall mean ($\bar{\bar{X}}_1$). Suppose the overall mean of this population is μ_0 (This is certainly a phenotypic mean value). From this population we decide to select certain individuals, say q in number, deviating from μ_0, to be the parents of next generation. The mean value of selected group of individuals is μ_1. The difference between the mean phenotypic value of individuals selected to be the parents and the entire base population, before selection, is known as selection differential (Figure 8.1). Allard (1960) has shown that genetic advance under selection can be predicted from equation shown below:

$$Gs = (k)\,(\sigma\, p)\,(h^2) \qquad \ldots(8.6)$$

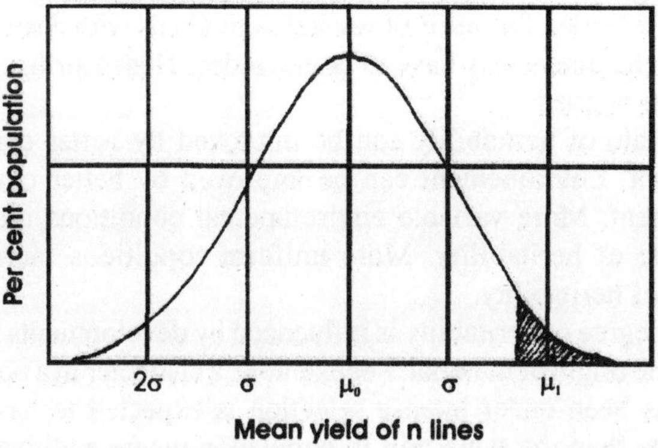

Fig. 8.1: Distribution of mean yields of n original lines

In this equation, *Gs* denotes the genetic advance under selection, i.e. difference between the mean genotypic value of *q* selected lines and the mean genotypic value of *n* original lines. σp is the phenotypic standard deviation of mean yields of the *n* original lines and h^2 is the heritability. Heritability, in this case is the ratio of additive genetic variance to the total phenotypic variance (comprising of $\sigma^2G + \sigma^2E$, where σ^2G is the genotypic variance and σ^2E the environmental variance. $\sigma^2G = \sigma^2A + \sigma^2D + \sigma^2I$). *k* is the selection differential, which takes into account the mean phenotypic value of *q* selected lines the mean phenotypic value of all the *n* original lines being evaluated, phenotypic standard deviation (σp) and the intensity of selection, i.e. proportion selected from the base population, *q/n*. Thus if highest yielding 1 per cent individuals are selected, $k = 2.64$. On the other hand, *k* shall be 2.06 and 1.76 if highest yielding 5 and 10 per cent individuals, respectively are selected. Heritability alone does not influence the genetic advance. Thus even if $h^2 = 100$ per cent, progress would not be spectacular unless the base population has larger of genotypic variability. Hundred per cent heritability suggests that phenotype provides a perfect correspondence with genotypic value.

A careful study of (8.6), reveals that genetic advance can be improved in certain ways. More stringent selection, i.e. lesser the proportion chosen from base population, is likely to improve genetic advance. However, this has to be executed very carefully. Smaller number of selected individuals is likely to lead to genetic drift. In case of cross-pollinated crops, reduced population size may result in loss of vigour and fecundity due to inbreeding. Moreover, a breeder has to improve several characters simultaneously. While exerting stringent selection for one trait several individuals with desired levels for other characters may have to be discarded. Hence stringency may have to be relaxed.

Estimate of heritability can be improved by better conduct of experiment. Environement can be improved by better conduct of management. More variable environmental conditions reduce the magnitude of heritability. More uniform conditions increase the estimate of heritability.

The degree of heritablity is influenced by developments that take place in the origin of material. For example, a character in a population which has been under intense selection is expected to have lower heritability than the same trait in population where additive genetic variance has not been reduced by selection pressure and both populations are evaluated in identical environmental conditions. For example, heritability of yield may be higher in early segregating generations than the later generations.

Phenotypic variability in the base population also determines the amount of genetic advance that can be expected. If the two populations that are subjected to selection show 1 and 2 units of standard deviation respectively, similar intensity of selection may lead to larger selection differential for population with larger standard deviation. Consequently, relatively larger genetic advance is expected for population with greater standard deviation.

8.5 RESPONSES TO SELECTION

Change produced by selection in which plant breeder is primarily interested is change of population mean. The difference between the mean phenotypic value of offspring of selected parents and whole of the parental generation before selection is known as response. The measure of selection applied is the average superiority of individuals chosen as parents. The mean phenotypic value of the selected individuals, expressed as deviation from mean of the whole population before selection is called as selection differential. On the basis of observations, Allard (1960) classified observed responses to selection into five categories, viz., initial rapid gain followed by a long period of slow progress, slow steady progress for a long period, slow progress leading to a plateau during which there is no response, little or no response to selection and lastly rapid gain followed by a period when there is no response followed by another period of rapid gain. According to Allard the differences in responses to selection pressure owe to differences in genotypic organization of populations subjected to selection. Evolutionary factors also influence the slow release and storage of genotypic variability.

When the inheritance of a character is governed by a number of genes, some of them may have large effects on the character and rest smaller effects. Change of gene frequency is rapid for genes with larger effects. The response to selection continues till all the genes with favourable effects have been fixed. Subsequently, genes with small effects come into play. But in this case rate of change of gene frequency is relatively slower. Rapid gain in the initial stages followed by slow progress for a considerable time is shown by traits whose inheritance is controlled by such a genetic organization.

Slow steady long continued response to selection is observed for characters whose inheritance is possibly governed by a large number of genes with small effects. Because the number of genes is large, several genes are spread over different chromosomes. At the Illinois Agricultural Experiment Station in U.S.A. selection was carried out for several generations for high oil contents in the germ of maize seed. In the first few generations of selection, selected lines were developed which contained more than 6.0 per cent oil. Selection in subsequent generations led to development of lines which exceeded the oil content than the base material. This possibly resulted from recombination of genes responsible for oil content present in the selected material. This increase in oil content continued for several generations without reduction in genetic variability. Selection was also being carried out in three other lines for low oil content, high protein and low protein in kernel. Because oil in maize resides in germ, after about 20-25 generations of selection for low oil content, progress was slower. It is expected that reduction in oil content means reduction in size of germ. There is probably a physiological threshold beyond which the germ size cannot be reduced. Hence progress towards low oil content was slower. Because a number of genes are spread on different chromosomes, intermingled with genes for other traits, selection for high oil content or low oil content led to simultaneous, inadvertent, changes in genes governing inheritance of other traits, such as height of ear, size of kernel, plant height, yield of kernel, etc. The four lines, after about 50 generations of selection, had characteristic features. The four lines differed in maturity, yield, ear height, tiller number etc.

A selection experiment carried out by Mather and Harrison (1949) on *Drosophila melanogaster* for abdominal chaetae is much educating about strorage and release of variability. These workers carried out selection for increasing number of abdominal chaetae in D*rosophila*. Chaetae number increased upto about 20 generations of selection but was accompanied by reduction in reproduction. Hence, selection was relaxed for a few generations, which resulted in restoration of fertility but decline in the chaetae number. Reduction in chaetae number stabilized at a higher number as compared to the original population. Upon resumption of selection the chaetae number continued to increase, without decline in fertility.

Mather (1943) believed that best adapted individuals in the population are those which manifest a harmonious combinations of all the characters leading to maximum fitness in the environment in which population exists. Wright (1951) also proposed that the best adapted form in the species is one which is close to average in all the quantitatively varying characters. This suggests that natural selection will act against extreme phenotypes. If reproductive wastage is to be avoided, it is to the advantage of a population to achieve uniformity of phenotype and avoid production of extreme forms. But environment is not constant, it keeps on changing. In order to meet the long term evolutionary changes, population should also be genotypically flexible, i.e. should have genetic variability to adapt to changing circumstances.

Successful species have resolved this conflict between immediate fitness and long range flexibility to meet the needs of future environmental changes. According to Mather (1943), by building up a system of linked genes variability may be 'free' or 'potential'. For example, in case of a single locus with two alleles, there will be 2 kinds of homozygotes, *viz*., *AA* and *aa*. Variability observed between the two homozygotes is free. But variability is not free in heterozygote, *Aa*. Heterozygote must be sib-mated or self-pollinated before the variability becomes free and is available for selection. Hence variability in *Aa* is potential. When the character is governed by two pairs of genes (*A*, *a*, *B*, *b*), homozygotes *AABB* and *aabb* will produce extreme phenotypes and nature is likely to act against these genotypes.

However, if the two genes *A* and *B* have reinforcing effects, the variability may be in 'bound' form in homozygotes *AAbb* and *aaBB*. Before the release of free variability, the two genotypes have to hybridize followed by recombination between *A*, *a* and *B*, *b* to produce extreme forms like *AABB* or *aabb*.

In case of abdominal bristle number, the release of variability upon resumption of selection after about 25 generations probably resulted from segregation and recombination among genes governing inheritance of chaetae number.

8.6 LIMIT TO RESPONSE

Response to selection can not be expected to continue indefinitely. Sooner or later it is to be expected that all the favourable alleles originally segregating will be brought to fixation. As the genes approach fixation, the genetic variance should decline and the rate of response to selection diminish. When the fixation is complete, response should cease. The population should fail to respond to selection in the opposite direction also. Further response to selection is only possible when new genetic variation arises due to mutation, or recombination following hybridization or out crossing.

Not only the minimal limit of selective advance is determined by the number of genes governing the inheritance of the character under consideration but also the speed of selective advance. For example, when only one gene pair, say *A*, *a* governs the inheritance of a trait under improvement, about 25 per cent of the individuals in the F_2 generation will be either *AA* or *aa*. Selection of any number of individuals upto 25 per cent from the F_2 population with desirable phenotype will lead to immediate fixation of the desired homozygous type. The progress will be achieved immediately. When two gene pairs govern the inheritance of a trait and environmental variation does not influence the phenotype of the character, about 6.25 per cent individuals in F_2 generations will be homozygous for each of the extreme genotypes, i.e. will be either *AABB* or *aabb*. The chances of selecting extreme homozygous types are thus reduced with increase in gene number. Some genetically less extreme types may have to be accepted alongwith desirable types. With 10 gene pairs, 5 generations of self-fertilization is likely to result in about 74 per cent homozygous population. With 20 gene pairs heterozygous in F_1, about 54 per cent population may attain homozygosity after 5 generations of selfing

(Table 8.1). After about 8-10 generations of self-fertilization the population attains homozygosity for most of the loci, in the absence of tight linkage. When the number of genes governing inheritance of a trait is large, selection for homozygous types at all the loci and their immediate fixation becomes difficult.

Table 8.1: Per cent homozygous plants in the population after self-fertilization

After generations of self-fertilization	*F_1 heterozygous at loci*			
	1	*10*	*20*	*100*
1 (F_2)	50.000	0.097	0.000	0.000
2 (F_3)	75.000	5.630	0.317	0.000
5 (F_6)	96.870	73.740	54.370	4.150
8 (F_9)	99.600	96.000	92.270	66.950
10 (F_{11})	99.900	99.900	98.000	90.430

Additive component of genotypic variance is the only component, which is fixable. Because gene frequencies influence the additive and dominance components of genotypic variance, heritability changes with change of gene frequencies. Because gene frequencies are different in different populations, estimates of heritability are also different for the same character.

The speed of selective progress, is determined by following factors, as suggested by Mather and Jinks (1971):

(i) Intensity of selection.
(ii) Number of genes or effective factors governing the inheritance of character to be improved.
(iii) Variation in the magnitude of '*d*' increments of various genes governing the inheritance of trait.
(iv) Variation in the magnitude of '*h*' increments of various genes involved.
(v) Linkage among various genes.
(vi) Magnitude of masking effect of non-heritable agencies.

8.7 PURE LINE SELECTION

It was a Danish biologist, W.L. Johannsen (1903, 1926) who explained the genetic basis of pure line and pure line method of selection. In the seed lot purchased from market, Johannsen observed seeds of various sizes in Princess variety of beans. Beans are self-pollinated and hence are expected to comprise of homozygous individuals in the commercial seed lot. On the basis of size Johannsen selected nineteen seeds. When the progeny rows of the 19 seeds were raised he observed that each seed lot manifested a characteristic mean seed weight. For example, line number 1 had average seed weight of 64 centigrammes while the mean seed weight of line 19 was 35 centigrammes. All intergrades were observed among lines from 1 to 19. Second feature of the variability observed was that there was little variability for seed weight within each line but the magnitude of variability was not comparable with that observed in the original commercial seed lot. Johannsen believed that little variability observed for seed size within each of the lines was due to various environmental factors, which affected individual plants within the lines to varying degrees. He confirmed the role of non-heritable agencies for little variability within the lines via two kinds of experiments. In the first experiment, Johannsen separated each line into the seed classes of 10 centigrammes each, and grew the seeds separately. For example, in line number 2, he selected seeds of weights 40, 50, 60 and 70 centigrammes. These were called mother seeds. Progenies of these mother seeds had average seed weights of 57.2, 54.9, 56.5 and 55.5 centigrammes, respectively. The variability within line was, therefore, not genetic in origin. In another experiment he selected in each of the line seeds in both plus and minus direction for six successive years. That is he developed two sub-lines within each line, one with larger seed and the another with smaller seed. Selection in opposite directions within the lines was not effective in separating the line into a line with relatively larger seed and line with smaller seed. For example after 6 years of selection for larger and smaller seed in line number 1, the progeny of larger seed had average seed weight of 67.66 centigrammes and offspring of smaller seed had mean seed weight of 69.07 centigrammes.

Johannsen proposed that the original commercial seed lot of Princess bean was composed of several pure lines and hence it was genotypically variable. According to him pure line is the progeny of a single, self-fertilized homozygous individual. But after the nineteen pure lines were established, variability within each of the lines `was non-genetic in origin and hence did not lead to permanent change following selection. This strongly points out that non-heritable variation does not lead to permanent improvement in desired direction.

Genotypic variability among the pure line arises due to mutation, natural hybridization among genotypically different forms residing adjacent to each other and mechanical mixtures.

The conventional vicw is that the inbreeders are composed of homozygous or little variable lines while the outbreeders are made up of a mass of different heterozygous individuals. Jain and Allard (1960), however, found evidence of heterozygote advantage at five out of eight loci while working with one of the barley composite crosses. According to them results strongly suggest permanent maintenance of some degree of hybridity in self-pollinated crops. Though barley is regarded as rigorous inbreeder, the barley used in this experiment was 1 to 2 per cent out-pollinated. Jain and Allard (1960) argued that a low degree of outpollination, coupled with some heterozygote advantage, could maintain hybridity indefinitely in inbreeders. Hence conventional view of population structure of self-fertilized plants may be inaccurate and there may be more variability in heterozygous phase than we think (Simmonds, 1962).

Pure line selection is an ancient method for improvement of self-pollinated crop species. Because the land races comprise of genetically diverse types of plants, selection from these variants leads to considerable improvement. There are two methods of simple selection, which are advocated to be employed for improvement of self-pollinated crop plants. Once again, it should be pointed out that by virtue of mating system in self-pollinated crops, the varieties are homozygous at most, if not all, the loci. Moreover, the varieties may not be genotypically homogeneous (see Chapter 5). The end product of pure line selection is a homogeneous population with individuals homozygous at nearly all the loci. The variety developed via pure line selection is made up of progeny of a single plant selected and evaluated for superiority. In mass selection, several plants which mature at about same time, have same plant height and similar for several other traits are selected, harvested and threshed and seed composited. Consequently, mass selection leads to development of a variety of a self-pollinated crop, which comprises of several pure lines.

Mass selection is also a selection method used for improvement of cross-pollinated crops. But the consequences of mass selection in cross-pollinated crops are different than in self-pollinated crops as will be shown later.

Pure line method of selection was developed as a result of fundamental studies of scientists like Vilmorin, Mendel, Johannsen, and others whose field experiments led to conclusion that progeny of an individual plant selection of self-pollinated crops may be expected to breed true, for most part.

Although selection for desirable plants was carried out by several workers, in a bid to improve the economic potential, contribution of Louis de Vilmorin of France has led to better understanding about the usefulness of selection in self and cross-pollinated crops. He selected best plants in each generation for 50 consecutive generations in 4 varieties of wheat, and raised progeny of each selected plant. But he was unable to detect improvement in yielding ability. On the other hand, selecting best plants and growing their progeny from 1850 to 1862, improved the sugar content in sugar beets so that sugar beet industry because economically important. This method of selecting the best plants from genotypically variable population and raising progeny rows to evaluate genetic worth of selection was known as 'Vilmorin method' or 'Vilmorin Isolation Principle'.

Pure line breeding essentially involves three steps. In the first step, a large number of superior plants are chosen from the population. The number of selections made may be determined by the finances available, experimental space that is available, the magnitude of genotypic variability for the characters to be improved. However, in case of cereals, pulses, oilseeds, a few thousand selections could be made. However, in case of vegetable plants, space is a limitation to raise large number of progeny rows and that too with wider row to row and plant to plant spacing. Fageria *et al.* (2001) suggested selecting 200 to 3000 phenotypically superior plants from genotypically variable population of vegetable crops. Seeds are harvested from individual plants and kept separate.

The number of plants that are selected from genotypically variable base population is very much important. Once the plants are rejected, there is hardly any chance of recovering the variability that is lost by rejecting the plants. This is the main difference between self-pollinated and cross-pollinated crops. In self-pollinated crops, a selection of heterozygous, superior appearing plant leads to homozygosity at a rapid pace and ultimately to fixation of homozygous types, with negligible heterozygosity.

The second step consists of growing progeny row of each selection, for observational purpose. The idea underlying progeny testing is to detect how much correspondence is there between phenotype and genotype. If the selected plant was superior by virtue of its genetic architecture the mean performance of progeny will be superior. Lines with obvious defects should be immediately discarded.

The progeny rows in subsequent years may be exposed to artificially created epiphytotics or insect populations, in order to evaluate their reaction to diseases and/or insects prevalent in the region where the variety is to be ultimately released for general cultivation. Drastic reduction should be made during progeny row evaluation. Probably three years of progeny row evaluation is adequate.

The last and final step involves, multi-location testing of superior pure line selections. In this testing, established commercial variety should also be included as a check or control. The length of time required for multi-location testing depends upon circumstances, but 3 years of evaluation can be regarded as adequate. A variety that is found statistically significantly superior than the check for character of economic importance may be released for general cultivation in the area of evaluation. Multi- locational evaluation over years takes care of genotype × location, genotype × year and genotype × location × year interaction.

The pure line method of selection was very useful breeding method in India, when the variability existing in land races was not exploited. But now the genotypic variability in land races has be utilized fully through single plant selection. Hence, it may not be more useful now.

The varieties evolved by pure line method may remain genotypically pure so long as the individual plants are not cross-fertilized, recurrent mutations do not cause change in genotypic composition and further care is taken during seed increase that mechanical mixture does not contaminate the genetic purity and identity of variety.

In the earlier period, pure line selection was utilized in India to evolve new varieties of several crop species. NP 4, a best quality wheat variety in the world was a pure line selection from local material called Mundya (means awnless). NP 6, NP 12 and NP 52 are *aestivum* wheats evolved via pure line selection. Motia, Gulab are *durum* wheats developed by this breeding method. Sujata, an excellent quality *aestivum* wheat arose as a pure line selection from C 306. NP 11 and 12 of linseed; BDN 1, C 11, Gwalior 3, Khargone 2, Number 148, UPAS 120 of pigeonpea; Jyoti, Pragati, TMV 1 and TMV 2 of groundnut; Vidisha 60-1, Maldandi 35-1, Ujjain 6, Aispuri of sorghum were evolved by this method.

In the early stage of rice breeding in India, pure line selection was widely practiced for improvement. Madhukar, Jalmagna, Jalgaon 5, Janki, etc. are some varieties developed recently by this method.

Several varieties of vegetable crops were bred via pure line selection in India. *NP* 46A, Sindhur, *G* 1, *G* 2, *G* 3 and *G* 4, Patna Red, *NP* 34 of chilli; Arka Basant, Arka Mohini and Arka Gaurav of capsicum; Pusa purple long, Pusa purple round, Pusa purple cluster, Pant Samrat, Arka sheel, Arka Shirish of eggplant; Pusa Himlata and Pusa Komal of French beans; Sharad Bahar of cluster beans; Arka Chandan of pumpkin; Arka Rajhans, Arka jeet and MH 11 of muskmelon; Pusa summer Prolific long of bottlegourd are products of pure line selection method. Pusa drum head of cabbage and Hisar 1, Dania of cauliflower were evolved by this method (Fageria *et al.*, 2001).

Pure line selection can not be regarded as desirable method for improvement of cross-pollinated crops, because:

(i) The populations of cross-pollinated crop species are made up of heterozygous individuals. Pure line selection, without self-fertilization or sib-mating can not lead to homozygosity and fixation of the genotypes. Moreover, because individuals are heterozygous, progeny does not breed true.
(ii) Selection of few genotypes and isolation for maintaining genotypic purity may lead to inbreeding and hence to inbreeding depression.
(iii) Selection of a few individuals may lead to random genetic drift.

Consequently, pure line method is not used for evolving high yielding or economically desirable varieties of cross-pollinated crop plants. In case of brassicas, varieties Type 151, Benoy, Type 42, K 88 of yellow sarson (*B. juncea*), presently grown varieties of mustard, *viz.*, Sekhar, BR 40, Seeta, RH 30, Durgamani were evolved through pure line selection (Rai, 1989).

8.8 MASS-SELECTION

As against pure line selection where several single plants are selected but the one whose progeny is identified as superior makes up a variety, the mass selection is a breeding method used to evolve a variety that is made up of several phenotypically similar pure lines. A number of plants are selected from genotypically variable population, especially the land races in self-pollinated crops, which are phenotypically similar for characters like plant height, maturity, kernel colour etc. These plants are harvested and seed composited after threshing. The selected plants are not further evaluated through progeny test before compositing of seed. Hence the effectiveness of this method of selection to improve quantitative characters depends upon the magnitude of heritability. Why? Because in case of qualitative characters, such as colour of flower, presence or absence of awns, etc. the selected plant breeds true because in self-pollinated crops the individuals are homozygous at all the loci. But for quantitative characters, which are governed by several genes with small, similar and cumulative effects and are also influenced by environmental factors, the selected plants may not breed true. In other words, there may not be a perfect correspondence between phenotype and genotype of selected plants. Number of genes governing the inheritance of traits of economic importance, the gene frequencies and heritability determine the success of mass selection, without progeny test.

In case of vegetable crops 200 to 2000 phenotypically superior looking plants are selected from the population. Seeds of selected plants are composited (Fageria *et al.*, 2001). Mass selection was a useful method of plant improvement in India till the genotypic variability existing in the indigenous varieties was not fully exploited. But now mass selection is useful for purification and multiplication of the varieties of self-pollinated crops. For purification of existing varieties a few hundred to few thousand plants are selected from field of varieties grown from authentic seed source. Plants are selected on the basis of distinguishing characters known for a given variety. In the next year, progeny rows are grown from part of the seed. Remnant seed is stored safely to meet the requirement in case of crop failure. The 'on' lines, i.e., lines which meet the distinguishing features in field, are tagged and harvested separately. After threshing the seed is compared with known distinguishing characters and only 'on' types of seed is bulked/composited. Rest of the seed is rejected.

Mass selection in cross-pollinated crops does not differ from the procedure employed in self-pollinated crops. Desirable individual plants are chosen from genotypically variable population.

Selected plants are harvested and seed is composited, without progeny testing. To understand the difference between mass selection in self and cross-pollinated crops, consider two crop plants, viz., wheat representative of self-pollinated crops and maize representing cross-pollinated crops. Plants of wheat variety are homozygous because for generations there is self-fertilization. The selected plant will breed true to type. In case of maize, the male flowers (tassel) are separated from female flowers (silk/ear) in space, i.e. there is monoecy. Moreover, there is protandry, i.e. male flowers mature earlier than the female flowers on the same plant. Consequently, the female flowers on a given plant are mostly cross-fertilized. The source of pollen parent is not known. Hence the seeds of maize within a given ear are half-sibs and progeny does not breed true. The pollens may carry superior genes if coming from plants with superior genotype or poor if produced by plants with poor genotypes. In cross-pollinated crops, mass selection is very effective for increasing the frequency of genes of those characters whose heritability is high, i.e. there is perfect correspondence between phenotype and genotype. But mass selection is not very effective for increasing the yield of adapted varieties of cross-pollinated crops. This is probably because:

(i) Superior genotypes can not be identified on the basis of phenotypic appearance of single plant.

(ii) Pollination is not controlled. Consequently selection is carried out only on the basis of phenotypic appearance of maternal parent, i.e. the plant that is selected. There is no control over paternal parent. And lastly

(iii) Strict selection leads to reduced population size and consequently leads to inbreeding depression.

Growing a small progeny of say 20 to 50 plants, per selection, is the most efficient way to distinguish among plants which are superior due to their genetic makeup from those superior due to superior environment around them. Conducting multilocational, replicated trials, can enhance the accuracy and authenticity of progeny row testing. The procedure involving compositing of seed of superior selections identified on the basis of progeny row testing is called line breeding. Adequate number of lines are selected and seed composited to avoid chances of inbreeding due to small population size.

The environmental variation influences the phenotypic expression of plants. Heterogeneity in the fertility status of the soil and the microenvironment around individual plants is one such environmental factor. In order to overcome the influence of this agency Gardner (1961) proposed a modification of mass selection for maize. This method is called 'grid method' of mass selection or 'stratified mass selection'. In the grid system the isolated field of maize, from which selection was to be made, was subdivided into approximately square plots of 40 plants each, leaving an un-harvested border around the field. Planting rate was very low. Gardner (1969) used a rate of planting of 7841 plants per acre to permit near maximum expression of genotype and minimum competitive effects between plants. Field was isolated from other maize fields to avoid contamination and taking advantage of selection differential. At harvesting 5 to 8 plants from each plot were selected. On the basis of superior appearance and harvested and kept separate. The selected plants were dried to a constant moisture level before shelling and weighing. Four highest yielding plants were then chosen from each plot to contribute equal numbers of seeds to the next generation. Using grids, to reduce the environmental variation between plants, Gardner (1969) reported 2.93 per cent gain per year in ten cycles of mass selection in maize variety Hays Golden.

Varieties Bikaneri Nerma, Narmada, H 777, SRT 1 and F 414 of cotton were developed by employing mass selection (Koutu and Mandloi, 2002). Mass selection has been used in spinach, beet root, cole crops, bulb crops, cucurbits and sweet potato to develop improved varieties (Kalloo, 1988). In spinach mass selection is very common method to develop new varieties. Female plants are more productive than male plants therefore selection of female plants leads to increase in proportion of female plants to male plants. Mass selection has increased the outer leaves in cauliflower. The outer leaves in cauliflower protect the curd and enhance the curd quality. Onion varieties Arka Kalyan, Arka Niketan and Arka Pragati were developed through mass selection from IIHR 145, IIHR 153 and IIHR 149, respectively (Kalloo, 1988). Agrifound Light Red, Agrifound Dark Red of onion were evolved by mass selection (Fageria *et al.*, 2001). Varieties of toria (*Brassica campestris*) like Type 9, Type 36, ITSA, TL 15, PT 30, BR 23, M 27, TS 29 were developed by mass selection (Rai, 1989). BS 20, BS 70 of brown sarson (*B. campestris*) and T 27 of taramira (*Eruca sativa*) also originated from local germplasm through mass selection.

Mass selection is used for the maintenance and seed multiplication in case of cross-pollinated vegetable crops (Seshadri, 1989). In case of onion and cabbage selected bulbs or heads of a variety are planted in isolation. Intermating of selected plants produces seed. Population, however, should be large enough to avoid random drift.

8.8.1 Ear-to-Row Method of Selection

C.G. Hopkins proposed this breeding method in 1897, to improve the oil and protein contents of maize kernel. This method involves selection of a number of phenotypically desirable plants from genotypically variable population grown in isolation. The seed of selected plants are not composited after harvesting and threshing, as is usually done in case of mass selection. Instead threshed seed of each plant is kept separate. Part of the seed is used to plant ear to row progeny the next year in a bid to evaluate the genetic potential of selected plants. Remnant seed is stored safely. Based upon the performance of progeny rows, remnant seed of superior selections is composited to initiate next cycle of selection. Although this method was found superior for improvement of oil content in the germ of maize kernel and protein in maize kernel, it was not effective for improving characters with low heritability, especially the yield.

According to Mukherjee (1989) lack of response to selection could be attributed to:

(i) Inefficient and inadequate field plot techniques, which failed to distinguish superior progenies from their single row performance. Single row plots had to be grown due to limited quantity of seed,

(ii) Continued ear-to-row selection results in inbreeding (since population size is small) and consequently loss of vigour,

(iii) Lack of consistency in performance from one generation to the next, and

(iv) Low heritability for characters for which improvement is sought.

Lonnquist (1964) proposed a modification of ear-to-row selection scheme.

References

Allard, R.W. (1960). *Principles of plant breeding*. John Wiley and Sons, New York.

Briggs, F.N. and P.F. Knowles (1967). *Introduction to plant breeding*. Reinhold Publishing Co., New York.

Doggett, H. (1970). *Sorghum*. Longmans, London.

Eberhart, S.A. (1972). Techniques and methods for more efficient population improvement in sorghum. In *Sorghum in seventies*. Eds. N.G.P. Rao and L.R. House. Oxford and IBH Publishing Co., New Delhi, pp. 197-213.

Fageria, M.S.; P.S. Arya and A.K. Choudhary (2001). *Vegetable crops, breeding and seed production*, Vol. I, Kalyani Publishers, Ludhiana.

Falconer, D.S. (1960). *Introduction to Quantitative Genetics*. Oliver and Boyd, Edinburgh.

Gardner, C.O. (1961). An evaluation of effects of mass selection and seed irradiation with thermal neutrons on yield of corn. *Crop Sci.*, 1: 241-245.

Gardner, C.O. (1969). The role of mass selection and mutagenic treatments in modern corn breeding. Proc. of the 24th Annual Corn and Sorghum Res. Conf., pp. 15-21.

Jain, S.K. and R.W. Allard (1960). Population studies in predominantly self-pollinated species. I. Evidence for heterozygous advantage in a closed population of barley. *Proc. Nat. Acad.Sci.Wash.*, 46: 1371-1377.

Johannsen, W.L. (1903). Ueber Erblichkeit in populationen und in reinen Leinen. Gustav Fischer, Jena.

Johannsen, W.L. (1926). *Elemente der exacten Erblichkeitslehre*. Gustav Fischer, Jena.

Kalloo, G. (1988). *Vegetable breeding*, Vol.I. CRC Press, Florida, pp. 92-93.

Koutu, G.K. and K.C. Mandloi (2002). Cotton. In *Genetic improvement of field crops*. Eds. Singh, C.B. and D. Khare, Scientific Publishers (India), Jodhpur, pp. 247-253.

Lonnquist, J.H. (1964). Modification of the ear-to-row procedure for improvement of maize populations. *Crop Sci.*, 4: 227-228.

Mather, K. (1943). Polygenic inheritance and natural selection. *Biol.Rev.*, 18: 32-64.

Mather, K. and B.I. Harrison (1949). The manifold effects of selection. *Heredity*, 3: 1-52, 131-162.

Mather, K. and A. Vines (1952). The inheritance of height and flowering time in a cross of *Nicotiana rustica*. In *Quantitative inheritance*, Ed. Reeve, E.C.R. and C.H. Waddington. H.M.S.O., London, pp. 49-79.

Mather, K. and J.L. Jinks (1971). *Biometrical genetics. The theory of continuous variation*. Chapman and Hall Ltd., London.

Mukherjee, B.K. (1989). Concepts and methods. Breeding procedures for cross-pollinated crops. In *Plant breeding-theory and practice*, Ed. V.L. Chopra, Oxford and IBH Publishing Co. Pvt. Ltd., pp. 199-212.

Rai, B. (1989). Brassicas. In *Plant breeding-theory and practice*. Ed. V.L. Chopra, Oxford and IBH Publishing Co. Pvt. Ltd., New Delhi; pp. 159-170.

Seshadri, V.S. (1989). Vegetable crops. In *Plant breeding. Theory and practice*. Ed. V.L. Chopra, Oxford and IBH Publishing Co. Pvt. Ltd., New Delhi, pp. 391-408.

Simmonds, N.W. (1962). Variability in crop plants, its use and conservation. *Biol.Rev.*, 37: 422-465.

Thoday, J.M. (1964). Effects of selection for genetic diversity. *Genetics Today*, 3: 533. Pergamon Press.

Wright, S. (1951). The genetical structure of populations. *Ann.Eug.*, 15: 323-354.

9

Breeding Methods for Self-Pollinated Crops

The primary objective of earlier plant breeders was to increase yield of economic traits of various crop species. In these endeavours, superior varieties being grown elsewhere in the country or other countries of the globe were introduced or selections were made from the land races using either pure line or mass selection method of improvement. However, after realizing that the genetic variability existing in the land races was utilized by simple selection methods, the breeders turned their attention to hybridization between different varieties. The improved land races were crossed, in this effort, to another improved types or to some introduced exotic types. The idea behind crossing of two or more genotypes was to combine in a single genotype the desirable genes present in two or more genotypes, and not to produce hybrid.

In addition to breeding for higher yield of economic produce, a breeder is required to incorporate genes for resistance to some disease and/or insect pest. For example, tomato varieties need to be resistant to leaf curl virus, okra varieties have to be resistant to yellow vein mosaic virus, sorghum hybrids have to be bred for resistance to shoot fly, grain mold and so on. Quality of the commercial produce also needs improvement in several crops before they are released. For example, superior fruit quality in case of bell pepper, comprising of pleasing flavour, high sugar/acid ratio, high pigment content and vitamin C has to be given attention while breeding for yield. Capsaicin, an alkaloid, determines the pungent flavour of red pepper (chillies), is an important quality attribute. Curd quality is an important character in case of cauliflowers. Non-ricey, compact, bract free protected curds with retentive white colour is desirable in case of cauliflowers. Heat tolerance of cabbage and cauliflower varieties is also being recognized as a desirable quality attribute, which will permit cultivation of these cole crops in peninsular India.

Prior to 1961, exchange or procurement of varieties grown in other parts of the country, what to say about exchange from other countries, was difficult. Breeders usually did not want to part with and/or share the germplasm and segregating material developed by them. This was a great hurdle in the improvement programmes. Moreover evaluation of varieties across various locations in the country was also practically impossible. Consequently, varieties bred at a given station in the state were specifically adapted to the location and some area around it. After the launching of All India Coordinated Commodity Improvement Projects by the Indian Council of Agricultural Research (I.C.A.R.), New Delhi, exchange of germplasm is free and multilocational testing has become easier. Moreover, each coordinating centre is staffed with an entomologist, plant pathologist, in addition to breeders, agronomist, physiologist etc. Consequently, the experimental varieties entered for multilocational evaluation by the breeder from each coordinating unit, are also assessed by the entomologist and pathologist for their reaction to prevalent insect pests and diseases, respectively. Access to world germplasm has also become easier due to the efficient functioning of National Bureau of Plant Genetic Resources. The Project Coordinators of the All India Coordinated Projects can arrange for supply of desired germplasm to the breeders. Because of the possibility of multilocation evaluation over several locations in the country identification of a widely adapted (a variety which gives consistently superior performance over a wide range of environments) and specifically adapted genotypes has become possible.

The segregating material produced by crossing two (or more) varieties of self-pollinated crops can be handled either by pedigree selection or bulk population method. Pedigree selection and back-cross methods were widely used for improvement of inbred lines of maize. Moreover, Hallauer *et al.* (1988) expressed that pedigree selection will always be an important component of modern maize breeding programme. According to Bauman (1981) about 17 per cent of the total effort for development of inbred lines of maize can be ascribed to backcross resources. Pedigree method has been recommended for cross-pollinated vegetables like cucurbits, cole crops (cabbage, cauliflower and knol khol) carrot, radish, turnip and onion and often cross-pollinated okra (Swarup, 1977).

In the earlier days, a variety developed by selection from the segregating material, generated by crossing of two (or more) varieties, was called as hybrid variety. For example, wheat varicties Hybrid 65 and Hy 633 were developed from segregating materials of crosses (Gabo × A 115) and (EB 76 × Ex 73), respectively. This is a misnomer. A hybrid variety is F_1 between two true breeding lines (even 3 or 4), which exhibits hybrid vigour and is highly heterozygous (see chapter 11). This F_1 is given for commercial cultivation.

Between the pedigree method and bulk population method of handling the segregating material, employed to develop a desirable variety, pedigree method appears powerful. Pedigree method utilizes progeny row testing for identification of a genetically superior plant. It may be mentioned here, once again, that the two plants may appear superior, but whether the superiority is due to genetic reasons or better environment around the plant can be distinguished by progeny test. Secondly, the breeder has adequate opportunity to exercise his skill to identify and select desirable family and plant within family. In the bulk population method, selections are not made till the material reaches F_5 generation. This method is based on the assumption that by bulking seed from all the plants in segregating generations most, if not all, genetic variability can be maintained and by the F_5 generation most of the individuals attain homozygosity at majority of the loci. As a result, without having the botheration of maintaining pedigree records till F_5 generation, segregating material can be advanced, and selections could be made when there is least chance of segregation. In this method predominant force of selection is natural selection. Nevertheless, artificial selection could also be exerted if thought necessary and useful.

Single seed descent is a modification of bulk population method. In this method a single seed (or pod, or a fruit) from each plant is taken in F_2 generation to raise F_3 generation. And the same procedure repeated to grow F_4 and F_5 generation. Selections are made in F_5 generation and carried forward according to steps advocated for pedigree method. This procedure may be suitable for vegetable crops, in my opinion. The reasons are: (*i*) seed from all the plants in F_2 generation could be advanced upto F_5 generation and hence maximum genetic variability is available to a breeder in F_5 for selection of superior plants, (*ii*) the plants can be closely spaced for advancing in F_3 and F_4 without adverse effect on their expression in F_5, when plants are space planted. Even glass house facility could be availed to reduce time for advancing generations upto F_5.

9.1 CHOICE OF PARENTS

Making of a cross is not very much difficult. Any two varieties of a crop which belong to the same species, do not have genetic abnormalities could be hybridized. Even two genotypes belonging to two different species of same genus, and two genotypes belonging to different genera from same family could be crossed. Nevertheless, success varies. For example, in potato a wild species *Solanum demissum* has been crossed with *Solanum tuberosum* to incorporate resistance against late blight. Similarly *Lycopersicon pimpinellifolium, L. peruvianum* and *L. hirsutum* have been crossed with *L. esculentum* to incorporate resistance against *Fusarium* wilt and leaf mould, tobacco mosaic virus and root knot nematode and T.M.V. and *Septoria* leaf spot disease, respectively (Swarup, 1977).

Success of crossing depends upon the choice of parents. And the choice of parents depends upon the objective. But if the parents are rightly chosen, the segregating material may comprise of a wealth of desirable segregants. When the objective is to improve yielding ability of variety which is being cultivated by the farmers in a given region and is also popular, the parent for crossing should be chosen which is superior for this attribute. In the days prior to 1961, the year before launching of coordinated projects, a breeder had to collect germplasm from various sources, evaluate the lines and then identify desired parent. But since most superior entries are evaluated in the advanced varietal trials, across locations and years, organized by the coordinated projects identification of second parent should not be difficult. A variety from advanced varietal trial, which performs consistently well over locations and years, could be selected as second parent. Even the two varieties from advanced trial, which are superior, could be crossed to generate a genetically rich segregating population. Alternatively the breeders of coordinated projects have access to a rich germplasm. Record on yield, quality attributes, maturity, plant height, reaction to diseases and insects averaged over several locations in the country is available to him. On the basis of this information he can choose an ideal parent.

Objectives of programmes could be different such as breeding for disease resistance or insect resistance, frost tolerance, heat tolerance, quality attributes, etc. A suitable parent has to be chosen on the basis of the data available to the breeder. One thing needs to be kept in mind by the breeder that for whatever character he chooses a second parent; the parent should manifest this trait with high intensity because some intensity is usually lost during transmission and in new genetic background.

For qualitative character, which is usually governed by one or two genes, selection of parent may not pose a serious problem. But for quantitative traits, such as yield of grain, or fruits or lint, etc. heredity of which is governed by numerous genes with small, similar and usually cumulative effects and are also influenced by environmental agencies, the choice becomes tough. It may be desirable to evaluate several lines showing desirable attribute, for their general combining ability. Combining ability is the relative ability of a genotype to transmit desirable genes to its progeny, produced by crossing with other lines. And general combining ability is the average performance of a strain in a series of crosses. A good general combiner usually transmits genes with superior performance for a character under consideration, to its progeny. Diallel mating design (Griffing's 1956 approach) and line × tester mating design enable evaluation of true breeding lines for their general combining ability (See Dabholkar, 1999 for more information).

The second parent should be genetically diverse. If the second parent possesses several genes other than those existing in the variety to be improved, transgressive segregation could be expected. Although Mahalanobis D^2 statistic is proposed for detecting genetic divergence, it may not be truly so. It is also assumed that varieties from different geographic origins are genetically diverse. However, this may not be all the times necessarily true. The genetic diversity is not quantitatively assessed and is taken as synonymous with phenotypic or even geographic diversity (Murty and Arunachalam, 1967). Moll *et al.* (1962) could not find any direct relationship between geographical distribution and genetic diversity as evaluated in terms of heterosis between crosses in maize. Timothy (1963) observed considerable phenotypic divergence between Mexican, Brazillian and Andean maize collection but little genetic diversity between them as measured by performance of crosses between them. Geographical distribution and genetic diversity could not be directly related in Brassica, linseed, wheat, sorghum and tobacco by Murty and Arunachalam (1967).

While selecting a parent for crossing it is important to find out from the catalogue that undesirable characters are not linked with desirable attributes to be transferred. In pedigree and bulk method the variety to be improved and the second parent are crossed only once. If undesirable genes are linked with genes for desirable traits chances of recovering desirable recombinations is difficult with the population sizes that are usually grown. However, with the back-cross programme chances are bright for eliminating undesirable linked genes because recurrent parent is crossed again and again to desirable segregants.

Theoretically the variability for general combining ability is attributed to additive genetic variance and additive × additive kind of non-allelic interaction. These are fixable components of genetic variance. Lines superior in *per se* performance are usually good general combiners.

Occasionally improvement is sought in several attributes. In such a case the second parent may not be superior for all the characters to be improved, but for one. Hence a third and even a fourth parent may be desired. In this case two single-crosses may be crossed to produce segregating population. For example A is variety to be improved, and *B*, *C*, *D* are desired donor parents, then A may be crossed with *B* and *C* with *D* to produce single-crosses and $(A \times B) \times (C \times D)$ may be crossed. But such a cross may result in complex segregation. A very large segregating population has to be grown to isolate an individual desirable for all intended attributes. Instead, improvements could be made in steps.

There is adequate evidence, which suggests that several self-pollinated crops also exhibit heterosis (Ashton, 1946). Predominant component of genetic variance observed in self-pollinated crops is additive genetic (Matzinger, 1963). Smith (1952) has shown that in case of *Nicotiana rustica*, a variety could be isolated from segregating material, of two identified parents, which surpass the better parent and even the F_1 in performance. Homozygous lines of tomatoes can also be selected from segregating materials of certain crosses, which yield at par with hybrid combination from which they are derived. Gardner (1972) is of the view that it should be possible to develop varieties of sorghum that can compete favourably with hybrids. Many organizations like to promote hybrids because farmers should come back to them for seed each year. There is no doubt that F_1 hybrids have several advantages such as stability in performance over years and locations, uniformity of character expression, etc. but hybrid seed is costly which marginal farmers may not afford.

Results of experiments by Jayasekara and Jinks (1976) and Pooni and Jinks (1981) revealed evidence that heterosis for plant height in

Nicotiana rustica resulted from dispersion of genes, rather than over dominance. Dispersed genes could be manoeuvered to produce variety with performance at par with F_1.

9.2 MAKING OF A CROSS

For hybridization, irrespective of whether the crop is self-or cross-pollinated, artificial cross-pollination is to be accomplished. Crossing is usually performed by removing the stamens from hermaphrodite flowers from plants of variety to be employed as seed parent. Pollen grains from the anthers of variety to be used as pollen (male) parent are collected at correct stage and transferred on the stigmatic surface of emasculated flowers. The process of removal of anthers from flowers of variety being used as seed parent is called emasculation. As will be described later a breeder can accomplish emasculation mechanically, by using hot water, by using polythene bags or by suction. Emasculation is not required in case of monoecious and dioecious species, but the precaution that is essential is protection of female flowers from contamination. The actual steps involved in emasculation differ from one crop to another. In following a few examples of emasculation procedures in various crops will be described.

The first requirement that is important for crossing/hybridization between two genotypes is that both should synchronously flower. When the two parents differ in days required to initiate flowering, staggered sowing may have to be done. For example, an improved indigenous variety of sorghum, viz., Vidisha 60-1, requires about 85 to 90 days to commence 50 per cent flowering while SPV 475 requires about 70 days for 50 per cent flowering. For accomplishing hybridization of Vidisha 60-1 with SPV 475, the latter should be sown 5, 10 and 15 days later than Vidisha 60-1. Late sown sorghum succumbs to attack by shoot fly (*Atherigona soccata*) and hence the emerging seedlings have to be protected from this pest.

In wheat the spikelets for emasculation are selected when there is 50 per cent heading. The secondary (central) floret is removed and then the glumes are clipped back at the tip with scissors. Subsequently anthers are removed with the help of forceps. A label is tied at the base of spike containing information about name of female variety, date of emasculation, name of pollen parent and subsequently the dates of pollination. After about one or two days of emasculation ripe anthers are taken from the male parent and placed on the stigma of emasculated floret. The emasculated spike should be enclosed inside a glassine or paper bag in order to protect stigmas from contamination by unwanted pollen (As a matter of fact wheat is cleistogoamous and chances of contamination are remote).

Select those plants in rice for emasculation where panicles have emerged 50 to 60 per cent out of boot leaf. Emasculation is performed around 3 p.m. Remove young florets from bottom of panicle. Give a slant cut to the floret so that about one third to one half of the floret is removed, without injury to pistil. Remove the anthers from each floret with the help of forceps, which are six in number. Cover the emasculated panicle with butter paper bag and put a label at the bottom of panicle. Stigma of emasculated floret remains receptive for 3 to 7 days after emasculation. But pollination done next day after emasculation gives good results. Panicles of male variety are cut and placed in earthen cups containing water. These cups are placed inside room at high temperature. When the anthers extrude from glumes they are ready for pollination. Shake the panicle of male plant over the emasculated floret.

Emasculation is done in sorghum, by removing anthers from few selected florets (because there are several thousand spikelets in a panicle of sorghum, a few are retained and rest are clipped off). Pollen can be collected in pollen bags, in case of sorghum, and dusted on florets of a panicle, which have been emasculated a day or two earlier.

The base of the flower of cotton is surrounded by three leaf like triangular bracts forming what is commonly called as 'squares'. The cotton flower contains several stamens, forming a tube like staminal column around the style, which is united with the inside base of corolla. The pistil is formed of 3 to 5 carpels. For emasculation those flower buds are selected which are expected to open the next day. Anthers are removed by cutting the petals with the help of nails of index finger and thumb. Emasculation is accomplished between 3 to 6 p.m. The emasculated flower bud is enclosed inside a red coloured tissue paper bag. A label is hung at the base of flower bud after emasculation, inscribed with appropriate information. On the day following emasculation of flower buds of the female parent, pollination of emasculated buds is performed between 8 to 12 a.m. Flower buds from male parent are collected (of appropriate stage), the anthers are exposed, after removing the cover of petals, to sunlight. This causes bursting of anthers. One flower bud is taken and the anthers are brushed to stigma of 4 to 5 emasculated flower buds. The pollinated buds are then enclosed inside white coloured tissue paper bags. A thread is tied at the pedicel of pollinated bud as an identification mark (Koutu and Mandloi, 2002). These days only a tag is put at the base of crossed flower bud.

The flower buds of pulses are in general very small in size. Appropriate flower buds (likely to open the next day) are selected and emasculated in the evening. Other buds, which are not selected for emasculation, are removed. For emasculation the young flower bud is held between the index finger and thumb and stamens are removed with the help of forceps. The anthers (9 + 1, diadelphous) are covered by the keel (see Figure 9.1). The anthers are removed by pushing very gently the keels apart. Emasculated buds are then enclosed inside paper bag and label put on pedicel. Ripe anthers are collected from flowers of male variety and gently pressed against stigma of emasculated bud. Pollinated buds are again enclosed inside paper bag.

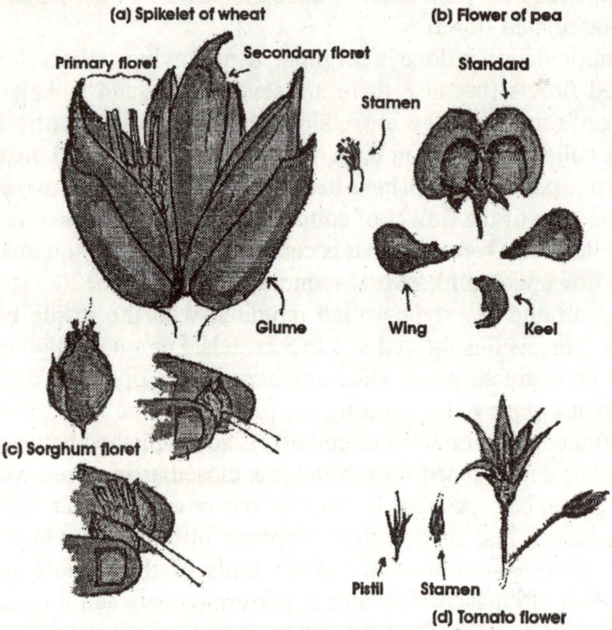

Fig. 9.1: Flowers and emasculation

Most of the cucurbits are monoecious (male and female flowers separated from each other on same plant in space) and bear large sized flowers. As mentioned in chapter 2, most varieties of muskmelon are andromonoecious, and others are monoecious. The staminate flowers of *Luffa, Cucumber* and *Cucumis melo* occur in clusters, but in other cucurbits such as bottle gourd (*Lagenaria siceraria*), pumpkin (*Cucurbita moschata*), etc. they are solitary. The pistillate flowers in cucurbits are borne singly in the leaf axis. The staminate and pistillate flowers on the male and female varieties are enclosed in butter paper bags about 24 hours before they open. The corolla of the male flower buds is often tied with a string or very thin wire to prevent opening. On the following day, male flowers are collected and their pollen dusted on female flowers of variety chosen as female parent. After pollination, the flower of female variety is enclosed in the butter paper bag and tag placed at the pedicel. The tag contains information about female and male parent and dates of pollinations. Paper bags may be removed from female flowers four to five days after pollination for better fruit set. The flowers of onion are borne in a simple umbel at the terminal end of flowering stalk. The inflorescence of carrot is a compound umbel. The entire umbel of the plant of variety used as seed parent is enclosed with muslin cloth bag. The opened flowers are removed every day for few days until peak flowering. After this flowers are not removed. The flowers on the umbel are then emasculated. After emasculation of sufficient number of flowers, remaining flowers are removed. The umbel of the pollen parent, which is also enclosed inside muslin cloth bag, is cut off from the plants of male variety, placed in a glass bottle containing water, which is fastened to a bamboo stake close to the plants of female parent. Umbels of the male and female parents are commonly enclosed within same muslin bag for about a week. In order to facilitate pollination, male umbel is gently rubbed over the emasculated umbel of female parent. Similar crossing procedure is followed for onion and carrot.

The cabbage, cauliflower, turnip and radish are usually self-incompatible. Bud pollination is successful way of selfing, and crossing (if the second variety has also the same genes for self-incompatibility). Cabbage is biennial and requires two seasons to produce flowers. In the first season heads are produced and in the following season flowering and seed formation takes place. The main crop may be allowed to over winter and produce flowers in its original position, that is, where they are planted in the seedling stage. Alternatively the crop may be allowed to fully mature in the first season. The heads are kept intact and a cross-cut is given to facilitate emergence of stalk bearing flower buds. Those flower buds are emasculated which are expected to open in one to two days. The emasculated flower buds are immediately pollinated with pollen from the male parent. According to Swarup (1977) seed set is better when buds are pollinated two days before opening of the flower. After pollination buds are enclosed inside butter-paper bags and properly labeled. For pollination, the pollen is collected from flowers of male parent, which are enclosed in the paper bags before opening. Pollen from freshly collected anthers is taken and placed on stigmatic surface, with the help of camel hair brush.

In tomatoes those flower buds are selected for emasculation, which are expected to open in a day or two. The most suitable stage for emasculation is when the corolla has just opened and forms an angle of 45° with respect to flower axis. At this stage self-pollination is not expected because the anthers do not burst open. All other flowers, which are not to be emasculated, or those which are expected to be pollinated already, are removed. The flower bud is held in left hand between the thumb and index finger. With the help of thumb and index finger of right hand corolla are held near the base and with one upward pull the corolla are taken out along with stamens (Ram, 1998). Alternatively emasculation may be accomplished with the help of forceps. One arm of forceps may be inserted between the anther cone and sepals and the other between anther cone and pistil. Forceps is rotated gently around flower axis and the cap is removed (Kalloo, 1988). Emasculation is executed similarly in chillies.

Three kinds of flowers are observed in brinjal, viz., long styled, medium styled and short styled. Fruit setting is better in long styled flowers (Ram, 1998). Emasculation of the flower buds is carried out similar to tomato with the help of forceps.

In case of tomato, brinjal and chilli pollination of emasculated flower bud is carried the next day morning after emasculation. Pollen for this purpose is collected from flowers of male parent, which are in full bloom. The pollinated flower buds enclosed in paper bags are properly labeled.

Hot water treatment has been advocated for rice (Chandraratna, 1964; Jodon, 1938) and sorghum (Stephens and Quinby, 1933). This method takes benefit of differential sensitivity of anthers and the pistil. It has been observed that the rice anthers are killed if immersed in hot water (42 to 44 degree centigrade) whereas it does not have adverse effect on the pistil. The rice panicle is therefore immersed in thermos flask containing water with temperature at 42 to 44°C. A technique of mass emasculation was devised for sorghum by Stephens and Quinby (1933). The panicle is immersed (before the anthers ripe and start shedding pollen) in water stored in thermos flask at 48°C for about 10 minutes. This temperature kills pollen but not the pistil. Sorghum anthers are sensitive to temperature around 47°C. Plastic bags are also used to emasculate sorghum florets (Freeman *et al.*, 1962). Enclosing the sorghum panicle at appropriate stage inside plastic bag and fixing at the base causes rise in temperature inside plastic bag and kills the anthers.

9.3 PEDIGREE METHOD

In this method of handling the segregating material, resulting from cross of 2 genotypes (or even more), superior plants are selected in successive segregating generations, and a record is maintained of parent-progeny relationships. The pedigree record comprises of notes taken regarding the relationships among families, which have been grown. It is important to take note of the fact that pedigree record does not establish a retraceable route enabling development of variety by repeating the same cross. The record provides basis for retaining selections or rejecting them on the basis of their origin from same family.

General steps that are followed in this method to develop a variety are presented in the following paragraphs. But these are not strict guide lines. Changes can be made depending upon the crop, the number of genes governing inheritance of character being transferred, the space and finances available, and the like.

Usually about 10-15 F_1 plants are grown. How many plants should be grown depends upon the size of F_2 population that can be grown and the quantity of seed produced by each plant. For example, a single plant of tobacco, sorghum, pearl millet, cucurbits produces good amount of seed. Hence relatively few plants may be grown in F_1 generation. But urd, mung, gram, soybean etc. produce relatively fewer seeds per plant. Hence more plants of F_1 have to be grown of these crops.

In case of vegetables, commercial crop of cucurbits is planted at relatively wider spacing (2 m × 2 m), hence about 2500 plants are grown in an hectare. On the other hand, okra, cabbage, cauliflower and knol khol are grown in rows 60 cm apart and within rows plant to plant distance of 50 cm. As a result there are about 33,000 plants per hectare. Onion is grown with close spacing (0.3 × 0.1 m) and hence there will be about 3,33,000 plants per hectare. On the basis of growth habits and ability to produce seed by each plant about 1 to 2 rows of carrot, radish and onion are sown in F_1 (with about 35 70 plants). Swarup (1977) recommends raising 2 to 3 rows four feet long in case of cucurbits, 2 to 4 rows of okra and 3 to 5 rows of cabbage and cauliflower. About 12-18 plants of cucurbits will be raised in 2-3 rows with 2 plants per hill. Approximately 18 to 36 plants of okra, 25 to 35 plants of cabbage and cauliflower need to be grown in F_1 generation.

In case of cereals about 1/20th of an acre or about 0.02 hectare is assigned to the F_2 generation. Seed rate for commercial planting of wheat is one hundred kilogrammes per hectare. Hence 2.0 kilogrammes F_2 seed shall be needed to plant 0.02 ha. Since half the seed is stored in reserve to meet the contingency of crop failure, 4.0 kilogrammes F_2 seed shall be required. Plan number of rows of wheat F_1s to be sown so that about 5.0 kilogrammes seed is available.

Tigchelaar (1986) suggested growing 10-15 plants in F_1 generation to obtain sufficient seed of tomato to sow an F_2.

It is desirable to grow both the parents along side the F_1. This will facilitate confirmation of whether F_1 plants are truly hybrid in origin or are selfs. The F_1 plants are selfed or sib-mated to produce F_2 seed. Any plant growing in F_1 row, which does not conform to the character requirement of a hybrid should be immediately rouged out. Each plant in F_1 is threshed separately or in case of fruit vegetables the seed is extracted from each F_1 plant separately. After examination of seeds, with respect to seed health, genuinity, seeds from all the F_1 plants from same cross are composited and stored safely.

From the Table 5.2 it is evident that maximum genetic variability for a given cross can be expected only in F_2 generation (at least in case of self-pollinated crop). The magnitude of genetic variability depends upon the genetic width, i.e. the number of gene differences between the two parents, number of characters being considered and linkage relationship among genes. Relatively more genetic variability will be observed in F_2 generation of a cross involving genetically diverse parents than the F_2 of genetically less diverse parents. In the self-pollinated crops genetic variability diminishes in magnitude in generations subsequent to F_2. At the end of 7 to 8 generations of selfing, in the absence of selection, population will comprise of 2^n different kinds of homozygous individuals. For example, if F_1 was

AaBb, the end product would be *AABB*, *AAbb*, *aaBB* and *aabb* in equal proportions. Maximum opportunity is offered to the breeder by F_2 generation for exercising his skill of selection. Practically every plant in F_2 is genetically different. If a genotype is rejected in F_2, there is practically no chance of regaining it in subsequent generations. It is lost. Hence efforts should be made to grow a large F_2, as far as possible. The importance of size of population in F_2 can be recognized from an hypothetical example. Assume that two gene pairs govern the inheritance of a trait, say *A*, *a* and *B*, *b*. F_1 received these genes in repulsion phase, i.e. *Ab/aB*. One plant out of 16 in F_2 generation is expected to be *AABB*, (in the absence of tight linkage). When F_1 is heterozygous at 10 loci, about 1 plant shall be homozygous for all 10 dominant genes out of 10 lakh plants in F_2 generation. Raising such a large F_2 is practically not possible. Raising 2000 to 5000 plants in F_2 in case of cereals, pulse, oilseeds etc. appears reasonable. In wheat about 2000 plants are grown in F_2 (Mishra *et al.*, 2002).

According to Sneep (1977) if 21 independent loci are considered and F_1 is heterozygous at 21 loci, then a random sample of 1,684 F_2 plants shall not suffer an irreparable loss of desirable alleles. Shebeski and Evans (1973) are also of the view that if the parents differ for 25 important genes, one out of 1,330 plants in F_2 are likely to contain the best 25 important genes in either homozygous or heterozygous state.

The size of F_2 population for cereals depends upon number of F_3 that could be handled by the breeder (Allard, 1960). The ratio of F_2 individuals to F_3 progeny families ranges from 10:1 to 100:1. The ratio could be wider when it is possible to discard F_2 plants with ease. According to Harrington (1952) the total number of plants selected from F_2 should not be less than 5 per cent, in case of cereals. A check variety should be planted after every 10 rows of F_2.

Swarup (1977) recommended growing 120 to 210 plants of cucurbits, 225 to 450 plants of okra, 350 to 490 plants of cabbage and cauliflower, 1,850 to 3,700 of carrot, radish and turnip and 1,110 to 2,220 plants of onion in F_2 generation. In the light of information given in Table 5.2, it appears that growing only 120 to 490 plants in F_2 generation may not be satisfactory population size. Even when it is assumed that most of the characters of importance are simply inherited, the two parents are likely to differ at more than 5 loci. And hence at least 1000 plants should be grown in F_2 generation. Because row to row distances in most of the vegetable crops are around 60 centimetres, this should not be difficult. But raising a large F_2 provides more opportunities to isolate superior segregants. For tomato, Tigchelaar (1986) recommends growing of 400 to 500 plants in F_2 generation. Closer spacing could be adopted for vegetable crops so that more plants could be accommodated in a given area (Kalloo, 1988). Close planting according to him did not have adverse influence on selection of desirable plants. Plants observed promising when grown at closer spacing were also superior at wider spacing. Singh (2002) recommends growing 1000 plants of tomato, okra, chilli and eggplant in F_2 generation.

F_2 generation provides first opportunity to a breeder to select desirable plants. As mentioned earlier, if superior plants are not selected in F_2, they are lost forever, because in self-pollinated crops after F_1 heterozygosity decays at the rate of $(½)^n$, where n is the number of generations of self-pollination. This rate of approach towards homozygosity is very fast indeed.

A number of plants selected on the basis of their superior phenotype are heterozygous at many of the loci. Their superior performance could be due to this heterozygosity. But as the generations advance homozygosity increases and the superior performance may diminish. This point should be borne in mind by the breeder.

Attitude of a breeder should be to discard plants in F_2 generation, which appear undesirable. Several times he is tempted to select a plant with the expectation that it will throw desirable segregants in F_3 or F_4 generation. It is imperative to overcome this temptation but at the same time genuinely superior plants should not be discarded. Because breeder has to take close look at each and every plant in F_2, wider row to row distance will be desirable. Selection of plants in F_2 for characters of low heritability, such as yield, may not be much fruitful. But selection of plants for plant height, colour of grain or fruit, presence or absence of pubescence, etc. could be useful.

Selection based on single plant yield in early segregating generations has been highly ineffective in barley (Bell, 1963), wheat (McGinnis and Shebeski, 1968) and chickpea (Byth *et al.*, 1987). According to Green *et al.* (1981) factors causing low heritability of yield in early generations include environmental variation, non-additive gene effects in heterozygous plants and high genotype-environment interaction. Inter-plant competition may be the chief factor of environment responsible for high variance of pure lines, as suggested by Hamblin (1977). Inter-plant competition could be eliminated by wider plant spacing.

Screening of selections for their reaction to diseases and insect pests as well as to abiotic stresses (such as frost, heat, salinity, drought etc.) should be deferred till F_3 generation. This is because an F_2 plant may be desirable for several characters but may be eliminated from further propagation if destroyed by some disease (exposed to epiphytotic) insect (exposed to artificially reared insect) or some abiotic stress.

Only those plants should be selected in F_2, which exhibit combinations of desirable characters from both the parents. When adequate population is grown in F_2, there may be several plants, which show superior phenotypes for quantitative characters. It is certain that these plants do not carry all the desirable genes for characters under consideration. In F_3 and subsequent generations there is no chance that various desirable genes for a quantitative (or even qualitative) traits will recombine. Whatever genes are borne by different superior looking F_2 plants will become homozygous and get fixed in subsequent generations, in case of self-pollinated crops. What a breeder can do to overcome this undesirable feature is select two superior plants and cross them together. In this way several bi-parental crosses could be made in F_2 for converging desirable genes. Of course, population improvement method is the appropriate approach to achieve this aim.

The plants selected in F_2 generation are assigned serial numbers in pedigree method of selection. For example, first selection in F_2 of cross (*SPV* 475 × *SPV* 462) would be written as (*SPV* 475 × *SPV* 462)–1. Other selections will receive numbers in sequence (Figure 9.2).

The plants selected in F_2 are harvested separately. After threshing, in case of crops other than vegetables, the seed of each selection is kept separately in paper packet or cloth bag. Seed may be treated with insecticide or fumigated to protect from damage in the store. Before storing, the seed should be examined in laboratory. Damaged, improperly developed seed may be discarded. In case of vegetable crops like tomato, brinjal, etc. seed is properly extracted from each plant selected in F_2, dried well, examined in laboratory and undamaged, well developed, healthy seed should be stored in paper bags or cloth bags. Seeds of cucurbits are also extracted separately from each F_2 plant. Same thing holds true for cabbage, cauliflower, radish, carrot, onion etc. Identity of each selected F_2 plant should be maintained.

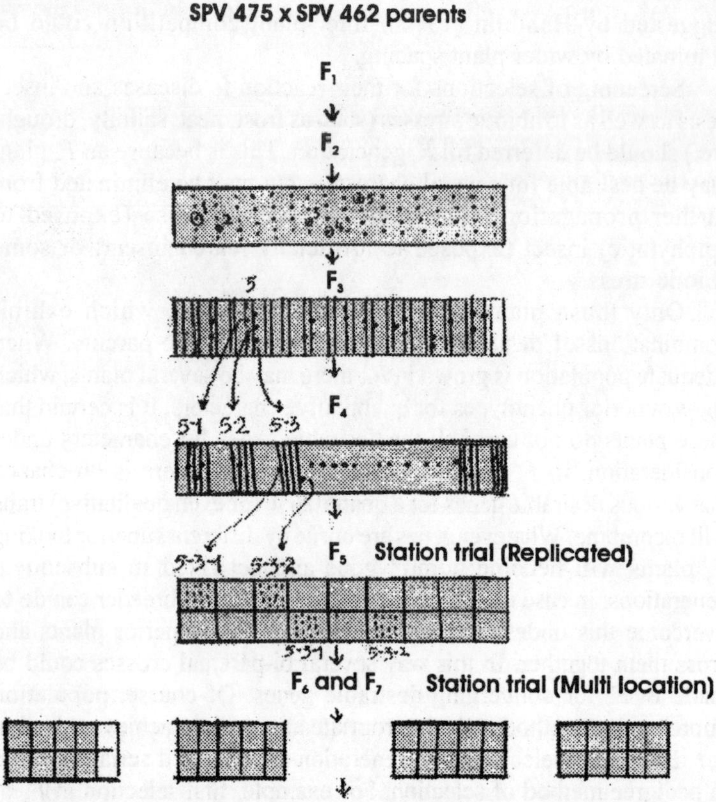

Fig. 9.2: Pedigree selection

It was mentioned earlier that for 10 to 100 plants of F_2, 1 plant is selected which will provide F_3 seed to plant F_3 progeny rows.

If the two parents have nicked well to produce desirable progeny, several selections may be made in F_2. However, if the parents did not complement each other the F_2 may be largely made up of undesirable plants. In this case entire F_2 population may be discarded. In wheat about 2000 plants are grown in F_2, so are grown in other crops such as sorghum. About 20 to 200 selections could be made in F_2. After selection in field, some selections are discarded during laboratory screening. About 3 to 4 rows, 5 m long, of each F_3 progeny could be grown. Number of rows and the length is the discretion of breeder which depends upon the quantity of F_3 seed, number of F_3 selections, space and funds available. After each of the 10 F_3 selections one standard check should be grown for comparison.

Singh (2000) suggested that about 100 progeny rows of F_3 in case of vegetable crops such as tomato, brinjal, chilli, okra may be planted. However, Tigchelaar (1986) recommended growing of about 100 F_3 progenies in 5 row plots of 10 m length, in case of tomato.

The ratio of F_2 plants to F_3 progeny families to be grown was suggested by Swarup (1977) for cucurbits, okra, cabbage and cauliflower, radish, carrot and onion. The ratios suggested were 10: 1 for cucurbits, 20:1 for cabbage, cauliflower, knolkhol and okra, while 50:1 was recommended for carrot, radish, turnip and onion. He proposed growing 10 to 15 plants per family of cucurbits, 20 plants per family of okra, cabbage, cauliflower and knolkhol, 30 to 40 of onion and 40 to 50 plants of carrot, radish and turnip. The number of families raised in F_3 could be 20 to 40 of onion, 35 to 70 for carrot, radish and turnip, 18-25 of cabbage, cauliflower and knolkhol, 12-20 for cucurbits and 10-15 of okra.

Family differences start appearing in F_3. Hence, it is desirable that at least 10 to 30 plants per progeny should represent each F_3 selection. The potential of a cross will be reflected from the appearance of F_3 progeny rows. Breeder should take a close look at F_3 progeny rows and identify superior progenies. Subsequently superior plants from superior progenies are selected. Individual plants appearing exceptionally good in otherwise undesirable progeny rows should not be discarded.

When the object of a breeding programme is resistance against a disease and/or insect pest, the usual procedure is to make two sets of F_3 progeny families. One set of F_3 progeny families is exposed to a disease epiphytotic created artificially (or in case of insects a counted number of larvae, reared artificially, are placed on each plant) and other set evaluated for agronomic attributes. At CIMMYT, Mexico, the F_1s are evaluated for agronomic attributes as well as reaction to disease but major selection pressure is exerted from F_3 generation onwards. F_1s with poor agronomic features and highly susceptible to diseases are discarded (Rajaram *et al.*, 1995). Advanced generation lines are evaluated at one or more locations not only for disease reaction but also for agronomic features. Agronomically acceptable entries with high degree of resistance to diseases under Mexican conditions are then distributed world wide for international multilocational testing. F_2 populations of crosses of chickpea involving parents resistant to *Helicoverpa armigera* were screened in pesticide free fields and those F_2s involving parents resistant to both *Helicoverpa* and wilt were screened in wilt sick plot at ICRISAT, Hyderabad. Single plant selections were made, based on visual observations and actual borer damage, in F_2 and subsequent generations upto F_5/F_6 by Singh and Colleagues (1995). Based on technique developed to screen for grain mold in the field, F_1 and F_2 generations of sorghum were evaluated at the International Crops Research Institute for Semi-Arid Tropics, Hyderabad (Mukuru, 1992).

The total number of selections made in F_3 generation should not exceed the total number of families grown in F_3 generation. In early generations breeder is concerned with single plant performance but in later generations family performance is also taken into consideration. Selection of more number of plants means more progeny rows in the next generation. This involves more space, more labour and money. If many of them turnout to be undesirable, time, money, space and labour are wasted. About 2 to 5 plants per F_3 family could be selected in vegetables like cucurbits, okra, cabbage, cauliflower, radish, carrot and onion (Swarup, 1977).

The F_4 selections are also grown in 3 to 5 row plots, about 5 m long in most of the cereals. They are still grown unreplicated as in F_3. A standard check is planted after every 10 progeny families. Many families attain homozygosity by F_4, nevertheless, much genetic diversity remains within families. Single plant selection remains the basis of selection in this generation also. By F_4, genetic differences among families are much greater than differences among individual plants within families. Consequently, this generation provides opportunity to eliminate inferior families. This is done both on visual basis as well as on the basis of the pedigree record. One of the two or three families descending from same F_2 plant, showing superiority amongst themselves, should be retained and rest discarded.

Swarup (1977) suggested that since carrot and onion manifest much inbreeding depression after two to three generations of selfing, the seed should be bulked F_4 generation onwards. On the other hand, seed of selected families of cabbage, cauliflower and radish is bulked in F_5 generation because these crops show less inbreeding depression. According to him evaluation of temperate vegetables is essential both in research station in plains and the research station located in hills. F_4 and onwards, replicated trials should be organized in plains as well as hills. This will enable identification of a variety suitable for hills as well as plains. Seeds of some of the temperate vegetables are produced in hills. It is, therefore, imperative to ensure the genetic purity and uniformity of the material at different locations in plains, because genotype-environment interaction can influence selection of superior types in hills and plains.

According to Swarup (1977) comparison should always be made among varieties with similar features. For example, early varieties should not be compared with late maturing varieties. Similarly large fruited varieties of tomato and capsicum should not be evaluated along with varieties having small or medium sized fruits. Varieties of onion bred for storage and table purpose should be tested separately.

Evaluation of selections should begin from F_4 onwards for quality characters.

Plants attain considerable homozygosity by F_5 generation. F_5 progeny families are grown in replicated trials for their systematic evaluation. In case of cereals the plot size may comprise of 4 to 6 rows 5 m long. The replicated trial is usually grown at the breeding station itself, but if the seed is available and facilities permit two to three locations within the state would be desirable. This will provide information about genotype-environment interaction. About 25-30 families belonging to a cross reach F_5 stage. Selections may be made from superior families. On the basis of yield evaluation trial of F_5 progeny rows; selections of superior progeny families may be bulked. In F_6 generation, the superior progenies are evaluated along with a check (usually the dominating variety of the region) in replicated multilocation trials. The locations are once again within the state in which the breeder is operating. The plot size will depend upon the seed available, the crop with which the breeder is dealing with and the number of replications of the trial. For instance, if seed is limited and at least 3 replications per trial have to be maintained then 4 row plot 5 m long may be satisfactory for sorghum. But if adequate seed is in stock then number of replications may be four and plot size would be 6 rows 5 m long. The primary objective of this evaluation is to identify one or 2 superior selections, over locations, as compared to the check. In F_7 generation the evaluation trial may be repeated.

The F_6 and F_7 progenies should be, if possible, simultaneously evaluated for their reactions to prevalent diseases and insects in the region. If the breeding programme exclusively aims at some quality character, simultaneous evaluation for this trait should be done. In F_8, generation the seed of superior entries (may be 2-3) is bulked and entered in coordinated testing in the All India Commodity Projects.

The two or three superior progenies, identified on the basis of station trials over locations and years (F_6 and F_7), now assume the form of experimental varieties. In F_8 generation the bulks of each of the 2 or 3 progeny families are uniform in height, maturity and most of the agronomic characters. These entries are submitted for coordinated testing. As the procedure is, in most of the coordinated projects, the freshly received entries are (accepted on the basis of record of station trial) evaluated in preliminary varietal trial over several locations in the country. The check is usually the best variety, which is widely adapted, high yielding, moderately resistant to the prevailing diseases and insects.

Breeders from various states contribute their entries to coordinated testing. The Project Coordinator gives one set of the entries, which makes up a preliminary varietal trial, to the entomologists and plant pathologists working at different coördinating centres. Those experimental varieties, which show excellent performance over locations, and are desirable with respect to a disease and/or insect pest, are promoted to advance varietal trial. Size of advanced varietal trial is larger than the preliminary trial. The number of locations is more for advanced testing. At least 2-3 locations in each of the coordinating states are chosen for advanced varietal trial. Usually, about 20 to 30 locations are available for advanced testing of cereals, pulses, oil seeds etc. The number of locations may be less for vegetables.

The breeder also grows F_8 and subsequent generations at his station. If superior plants appear within the experimental varieties, selections are made and after bulking the selections from same experimental variety evaluated in F_9, F_{10} and so on. Process is repeated till superior plants appear.

Three years of evaluation of experimental varieties in advanced varietal trial are supposed to be adequate to take decision about release of a variety at the national level (widely adapted) or to a specific state (specifically adapted). A variety, which is consistently superior for 3 years than other varieties, including the check, and is not susceptible to prevalent diseases and insects, is considered for release.

If a variety is *at par* with check for yield and other agronomic traits but highly resistant to prevalent disease and/or insect pest or to salinity or heat or frost or some such adversities, or superior for quality attributes, it may be considered for release. For example, a sorghum variety resistant to grain mold or shoot fly and/or stem borer may be considered suitable for release even if it yields *at par* with standard checks.

In rice, varieties are classified into five categories on the basis of length and shape of grain, viz., long bold, long slender, medium slender, short bold and short slender. In India, Pakistan and West Asia long slender grains fetch a premium price in market (Sharma, 1989). Hence if a variety has long slender grain it may be considered for release, provided it yields comparable to the checks.

High vitamin C content in cabbage, tomato, high sugar content in muskmelon and watermelon, better keeping quality in onion, ability to withstand long transportation and better shelf life in case of muskmelon, watermelon and tomato are the characters which may be considered before discarding a variety on the basis of yield alone.

9.3.1 Varieties Developed Through Pedigree Method

Varieties of various crops have been developed in India by employing pedigree method to handle segregating material. A few examples will given.

Pusa Ruby variety of tomato was developed from cross Sioux × Improved Meeruti. This is also an example where an exotic variety Sioux introduced from United States of America was used to develop superior variety. Tomato varieties Pusa Early Dwarf, Punjab Chhuhara, Selection 2, HS 101, HS 102 were also developed via pedigree method. A variety of okra, viz.; Pusa Sawani was developed from cross Pusa Makhmali × Bengal local. Pusa Kranti, H 4, HS, H 8 of brinjal, Pusa Dofasli of cowpea, Pusa Sharbati and Punjab Sunehari of muskmelon were evolved by using this method of breeding (Kalloo, 1988). Selection 8 of cabbage was developed from segregating material of cross EC 10109 × EC 24855, at Katrain. Pusa Himani, Pusa Rashmi of radish are the outcome of selection from crosses Black radish × Japanese white and Green top × deshi, respectively.

Pedigree method has been responsible for evolution of chilli varieties Andhra Jyoti (G 2 × Bihar variety), Pusa Jwala (NP 46A × Puri Red), Pusa sadabahar, (Pusa Jwala × IC 31339), Pusa

C1 (NP 46A × Kandhari), Punjab Lal (Perennial × Long Red) and Jawahar 218 (Kalipeth × Pusa Jwala).

Pedigree method is also a popular and often used breeding method for evolving varieties of pulse crops. Prabhat, ICPL 151, ICPL 87 of pigeonpea; Kabuli varieties L 144, L 550 and ICCC 32 of chickpea were developed through pedigree method (Asthana and Chandra, 1995). Deshi chickpea variety H 208 was developed from a three-way cross, viz., (S 26 × G 24) × C 235. Variety C 235 is tolerant to blight, wilt and pod borer and still being cultivated in India.

Cotton varieties Digvijay, Khandwa 3, MCU 6 and MCU 8, LH 900, LH 1134 and G.cot 14 were evolved through pedigree method (Koutu and Mandloi, 2002). This method was used to develop a widely adaped, high yielding variety of sorghum CSV 13 (= SPV 475). Sorghum varieties Jawahar Jowar 741, 938, 1041 and 1022, CSV 10 (SPV 346), CSV 11 (SPV 351), were also developed via pedigree method.

Durum wheat varieties NP 404 (Gaza × EKD 6), NP 406 (Gaza × EKD 6), HI 8498 (Raj 6070 × Raj 911), and *aestivum* wheat varieties NP 832 (E 145 × Pissi local), NP 839 (GB-Aus × Niphad 4), HI 1418 (HI 999 × HI 601), UP 301 (LR 64 × Sonora 64), Arjun are the result of pedigree method. Jaya and Padma varieties of rice were developed from cross Taichung Native 1 × T 141, by this method. Bala, Cauveri, Karuna, Ratna etc. are other rice varieties resulting from pedigree method. The list of crops and varieties developed by employing pedigree method is very large.

9.4 BULK POPULATION METHOD

When the two parents, which are crossed differ at *n* loci, the F_1 will be heterozygous at *n* loci. With each generation of self-pollination after F_1, heterozygosity reduces by one half at each locus. At the end of selfing, the population comprises of 2^n kinds of homozygotes. But each of the homozygote is different in genetic constitution. The bulk population method of handling the segregating population makes use of the fact that change from homogeneous heterozygosity of F_1 to heterogeneous homozygosity of F_{10} occurs naturally during the period of 10 years in a normally self-pollinated crop. Of course, this assumes absence of tight linkage, mutation and natural selection. It is assumed that after ten generations of self-fertilization nearly all the individual plants in the population shall be homozygous at all the loci, although the individuals may differ from each other in their genetic constitution.

In the bulk population method the F_2 population is grown as large as possible. While Singh (1993) suggested growing as large as 30,000 to 50,000 plants from F_2 to F_6 generation, Harrington (1952) advocated growing F_2 on 1/200th to 1/100th of an hectare. If the Harrington's suggestion is taken into consideration, then for wheat about 10 and 20 thousand plants will occupy 1/200th to 1/100th of an hectare, respectively. Planting rates and cultural practices are usually those recommended for commercial cultivation of the crop. At maturity the plot is harvested in bulk and seed obtained after threshing is used to plant similar plot the following season. The method, according to Allard (1960), is best suited for crops such as cereals, pulses and oilseeds. However, it is most unsuitable for fruit-crops and most of the vegetable crops.

The bulk-population method of breeding self-pollinated crops was developed by Nilsson-Ehle in 1908 to combine the winter hardiness of Square head variety of winter wheat in Standup variety of winter wheat. In contrast to the pedigree method of selection, in this breeding method the early generations are grown as bulk populations. During this bulking period, natural selection plays significant role in shifting the gene frequencies towards better forms. This selective force is important if bulking is continued for large number of generations. Nevertheless, there is no need to depend entirely on natural selection during bulking. Artificial selection can also be exercised at any time during the period of bulk-propagation. How many generations should the segregating material be bulk propagated? The decision regarding this differs with the breeder and the nature of cross. Those breeders, who consider absolute homozygosity of the population as unnecessary, initiate single-plant selections in F_5 or F_6 generation. Others prefer to bulk propagate till the F_6 generation and start selecting in F_7 or F_8 generation. Poehlman (1987) and Briggs and Knowles (1967) advocate bulk propagation upto F_5 and initiate selections in F_5 generation, while Harrington (1952) suggested starting selection in F_{10} or F_{11} generation.

Various steps involved during handling of segregating material via bulk population method are presented in the following. However, modifications could be made depending upon the crop, circumstances and the objectives of improvement programme.

After effecting a cross of two varieties grow sufficient plants of F_1 to produce enough seed required to raise F_2 generation. A large plot of F_2 generation is grown in isolation. Commercial planting rates are adopted. Harvest all the plants in F_2 and after threshing store the seed safely to plant F_3 in the next season (Poehlman, 1987; Briggs and Knowles, 1967). Harrington (1952) suggested harvesting entire F_2 but after threshing at least 20 per cent seed should be taken for sowing an F_3 generation, in an area of 1/20th to 1/15th of an hectare. Bulk harvesting is carried out in F_4 generation. Although Briggs and Knowles (1967) suggest maintaining the same number of plants in F_4 as in F_2, i.e. 5000, Harrington (1952) recommends taking a random sample of at least 15 per cent seed from the harvest of F_3. He further advocates growing of 1 hectare plot from F_5 to F_{10} generations. Space planting is adopted in F_5 or F_6 generation. Single plant selections are then made (Figure 9.3). About 300 to 500 plants with desirable agronomic features are selected. Subsequently the selections are handled as in pedigree method. F_6 or F_7 selections (if selected in F_5/F_6) are grown in four row plots 5 m long with a standard check after every tenth F_6/F_7 selection. Usually within-family variation will be least. Hence select superior families in F_6/F_7 and then superior selection in superior families. Subsequently the selections could be evaluated in station trials over locations within the state. After about two years evaluation best selections, now experimental varieties, are entered in the coordinated evaluation programme.

At present the bulk population method is used to obtain homozygous lines with least efforts and expenses. The average percentage of homozygosity, as mentioned earlier, is very high by F_6 generation of selfing, and approaches approximately 100 per cent by F_{10} generation, in spite of the number of genes segregating. Only when the heterozygotes are superior than homozygotes, return to homozygosity at the rate $(\frac{1}{2})^n$ will be delayed. Hence, in those crops, where there is very negligible cross-pollination, selections made in F_6 and subsequent generations will be almost homozygous at most of the loci.

There is inadequate evidence for critical evaluation of long continued bulk propagation as a method of handling segregating material of self-pollinated species. Nevertheless, a few plant breeders prefer to continue bulking for a number of generations, in some cases for 25-30 generations or more. The logic behind bulking for extended generations is that the homozygous types in F_6 to F_8 generation are composed of nearly poorer types with respect to agricultural value. This is true irrespective of whether artificial selection is practiced or not. The proportion of homozygotes with agricultural defects are reduced by both natural and artificial selection in subsequent generations.

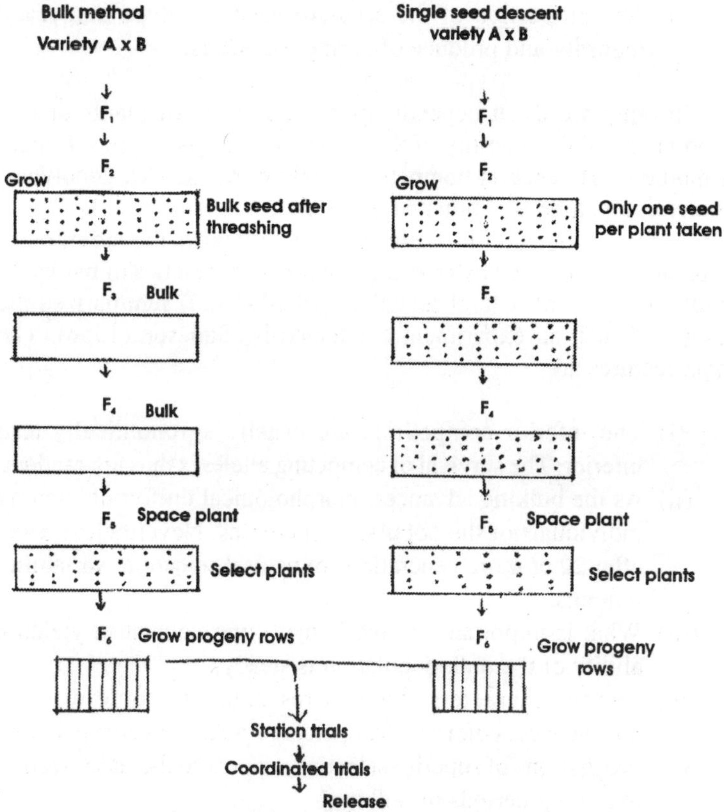

Fig. 9.3: Bulk method and SSD

9.4.1 Role of Natural Selection in Bulk Populations

It is a very well recognized fact that natural selection played significant role in the evolution of cultivated plant species both before domestication and even after domestication. But in nature the selection pressure is associated with survival. When the plants of two genotypes compete against one another in a bulk plot, large enough to avoid drift (Change of gene frequency due to sampling in small population is genetic drift. Genetic drift leads to change of gene frequency) survival depends upon two factors:

(i) The number of seeds produced by plant of each genotype, and

(ii) The proportion of the seeds of each genotype that reach maturity and produce offspring, i.e. seeds.

In other words, it depends upon the fertility of plants of two genotypes and the viability of the offspring of the two types. If there is marked difference in competitive abilities of the two genotypes, poorer competitor is eliminated promptly.

A few experiments were carried out with the composite populations, produced by crossing a number of varieties of barley, to study the effect of natural selection. Allard (1960) summarized the results of such an experiment conducted by Suneson (1956). The main features are:

(i) The inferior competitors are usually agronomically also inferior. The survival of competing alleles is thus not random.
(ii) As the bulking advances, morphological uniformity among individuals of the population increases. Nevertheless, even after 20 or more generations of natural selection, variability remains.
(iii) What is important to note is that improvement in yielding ability of the bulk population is steady.
(iv) Certain characters, such as resistance to some disease, pubescence, colour of fruit, etc. are neutral in survival, and
(v) Proportions of superior selections appear to increase steadily over long periods of bulking.

It is the Vth aspect which led some breeders to continue bulking for extended number of generations. Moreover, because some characters appear neutral in competition, such as disease resistance, which is an important character, there is scope for artificial selection also during bulk propagation. Selection by breeder is also required during bulk propagation for characters like colour of fruit, pubescence, quality attributes which are important in commercial varieties but neutral in survival.

9.4.2 Single Seed Descent (SSD) Method

Goulden (1939) proposed a breeding system in which selection is eliminated for certain number of generations and in each generation one or two seeds are taken from each plant to produce next generation.

This method is known as Single Seed Descent (SSD). Brim (1966) outlined a similar modified pedigree method for use in soybeans. This method was found to be effective for breeding high yielding oats (Kaufmann, 1971). Knot and Kumar (1975) compared the pedigree method and single seed descent method for improvement of spring wheat. The single seed descent method was found to have considerable merit. A brief out line of the steps adopted for SSD are as follows:

F_1 generation	:	Grow adequate number of F_1 plants.
F_2 generation	:	Space plant about 2000-3000 F_2 seeds. From each plant, one seed (or a pod or a fruit) is taken at harvest. Identity of individual F_2 plants is not maintained. Tigchelaar (1986) suggested growing 400 to 500 plants of tomato in F_2. These are space planted. Selection of plants is made on the basis of growth habit, reaction to diseases, maturity duration and characteristics of fruits.
F_3 generation	:	Seeds collected from each of the F_2 plant are sown to raise F_3 generation. At harvest single seed is taken from each plant. This will be F_4 seed. In tomato, 100 individual progeny families are grown from plants selected in F_2. Each progeny family is grown in 5 row plots 10 m long. Selections are made initially for superior F_3 families and subsequently for individuals within these families. Three to 4 seeds are collected from each selected plant and bulked (Tigchelaar, 1986).
F_4 generation	:	F_4 seed collected in previous season is sown. Once again one seed is taken from each F_4 plant. Identity of individual is still not maintained. In case of tomato, as suggested by Tigchelaar (1986), seeds collected in F_3 generation are sown without maintaining identity. No selection is made.
F_5 generation	:	The F_5 seed obtained from F_4 is space planted. Selections of desirable plants are made. Identity of each plant is maintained. Seed of each selected plant is kept in separate packets.

In case of tomato the bulk is grown. Selection of desirable plants is made. Consideration is also given for yielding ability.

F_6 generation : About 3 to 5 progeny rows, 5 m long, are grown from selected plants. A standard check is also sown after each of the 10th selection. Selection is made initially of better looking progeny and subsequently superior individual in better progeny. Identity of each selection is maintained. Seed of selections is kept separately.

Progeny rows of selections made in F_5 tomato progenies are grown. Superior plants are selected from superior families and seed of selections kept separately.

F_7 onwards : The selections should be handled in the way suggested for pedigree method. For instance, station trial is organized and then superior experimental variety is entered in coordinated programme and subsequently superior variety is released if found suitable.

Single Seed Descent method was originally proposed as an effective means of maintaining all possible descendents from F_2 generation. In other words, the objective of this method is to conserve all possible genetic variability expected in F_2. Because of some limitations it is not possible to do so in pedigree method. Since optimum phenotypic development of plants is not necessary in the early generations, close planting is done to save the space and money. The generations could be advanced in glass house with closely spaced plants. If off-season nursery facility is available this can be taken advantage of to advance the generation and thus gain time required for movement from one generation to the next. Record is not to be maintained of plants upto F_5 or F_6 generation. These are some advantages of the single seed descent method.

Boerma and Cooper (1975) compared early generation testing, pedigree method and single seed descent methods for improvement of soybeans. Single seed descent emerged as the most efficient breeding method because it required less selection efforts than early generation testing and pedigree method. The single seed descent allowed rapid advancement of early generation segregating populations and did not use expensive yield testing until later generations, when yield testing is more efficient. According these research workers single seed descent method should be used in plant improvement where isolation of high yielding lines is the primary objective.

The author is of the opinion, which should be verified through experiments, that single seed descent would be an efficient improvement procedure in case of vegetables like tomato, brinjal, chilli, peas, okra, radish, carrot and cowpea. Relatively large F_2 and subsequent generations, upto F_5 or F_6, could be raised than are being presently grown, so that maximum genetic variability is conserved till F_5 or F_6 when breeder will initiate selections.

Single seed descent can not be employed for improvement of cross-pollinated crops. The reason is that it is assumed in SSD that while advancing generations, from F_2 to F_5 or F_6 by taking single seed from each plant, homozygosity increases. In F_6 most of the plants are homozygous upon which selection if exercised the progeny will breed true. In case of cross-pollinated crops advancing from F_2 to F_6 by taking a single seed from each plant will not lead to homozygosity. Moreover, because of sampling, resulting from taking a single seed from each plant, gene frequency will change. Selfing of plants in case of cross-pollinated crops will lead to inbreeding depression. Maize, carrots, alfalfa, show more drastic effects of inbreeding.

9.5 BACK-CROSS METHOD

When a superior variety, agronomically desirable for most of the characters and popular among the farmers and consumers, is found defective for certain trait, such as susceptibility to a disease or an insect or to some abiotic stress, it can be improved without much change in its genetic constitution. A variety or an entry from germplasm, or an introduction, which exhibits the character lacking in superior variety in most intense form is chosen as a donor or non-recurrent parent. If the gene for the trait to be transferred from the donor is dominant in nature, the F_1 between the superior variety and donor is crossed back to the superior parent to produce back-cross 1 seed. The back-cross-1 (BC 1) progeny is again back-crossed to the superior parent to produce BC 2. In this way 5 to 6 successive back-crosses will result in reconstitution of entire genotype of superior variety but the gene transferred from donor will be additional and in heterozygous condition. One generation of selfing makes this gene homozygous. In this process, selection is made for the trait being transferred from donor. Because the superior variety is back- crossed again and again to the desirable segregates with an objective to recover its genetic constitution, it is called the recurrent parent.

This method is very much useful. Arkel is a very popular variety of garden peas but is susceptible to powdery mildew (*Erysiphe polygoni*). Through back-cross breeding method gene for resistance to powdery mildew could be incorporated in this variety. Several resistant sources are known, such as EC 326, T 10, P 185, PC 6578, IC 4604 etc. (Ram, 1998).

Back-cross method was mostly used in the past to incorporate genes for resistance to diseases or insect pests. But these days back-cross method is being used to transfer genes for male-sterility, non-photosensitivity, heat tolerance, etc. to the popular and desirable varieties/suitable genotypes. The male-sterile lines 2219A, 2077A, 296A, AKMS 14A, and IMS9A of sorghum were developed through back-cross method or some modification of back-cross. The milo cytoplasm and kafir nuclear genes from CK 60A were mostly utilized.

The tropical sorghum varieties are tall and late maturing. In U.S.A. after the tropical varieties were collected from Africa they were converted into early, dwarf and day length neutral varieties adapted to temperate climate (Rana and Rao, 1989). In sorghum, wide range of maturity lengths are observed. Genes controlling duration of growth are located at four loci, viz., Ma_1, Ma_2, Ma_3 and Ma_4. Photoperiod has considerable influence on duration of growth of many sorghums. Lateness is dominant to earliness. The stem internode length is determined by genes at four loci, viz., Dw_1, Dw_2, Dw_3 and Dw_4. Tallness is partially dominant and the effects of four *Dw* genes are additive (Doggett, 1970). For a variety to be early, gene ma_1 should be present in homozygous state. The conversion of tropical sorghum varieties was done by substituting gene ma_1 for Ma_1 and recessive height alleles (dw_1, dw_2, dw_3 and dw_4) in the tropical varieties. The tropical variety was used as a recurrent parent in the conversion programme. Four back-crosses, coupled with selection, enabled recovery of over 98 per cent of tropical germplasm.

Multiline varieties are produced through back-crossing of the variety as recurrent parent, in separate back-cross programmes to different donors. The end product of such back-crosses is recurrent parent possessing diverse kinds of desirable genes from the different donors. Borlaug (1958) proposed development of component lines of a multiline by the back-cross method. Such lines are phenotypically similar but different with respect to genes for resistance to a disease. Multiline MLKS-11 was developed with Kalyansona as recurrent parent (Jain, 1989).

According to Briggs and Allard (1953) for a back-cross programme to be successful three aspects should be given due consideration, viz.,

(i) A satisfactory recurrent parent is required,
(ii) A worthwhile intensity of the character being transferred from donor to the recurrent parent be maintained during several back-crosses, and
(iii) It should be possible to recover entire genotype of the recurrent parent through sufficient number of back-crosses.

As mentioned earlier, F_1 is backcrossed to the recurrent parent to produce first back-cross. The F_1 carries 50 per cent genes of recurrent and 50 per cent genes of donor parent. After the first backcross, the BC_1 progeny carries 75 per cent genes of recurrent parent and 25 per cent of the donor. In this way genes of donor parent diminish by one-half with each backcross. The proportion of genes from donor will be $(\frac{1}{2})^n$ in the backcross progeny after n back-crosses. For example, after 6 back-crosses the backcross progeny shall contain $(\frac{1}{2})^6 = 0.156$ or 1.56 per cent genes of the donor not intended to be transferred. The backcross progeny therefore, progressively becomes like the recurrent parent in genotypic constitution, but possesses gene from donor.

Percentage of plants which are homozygous for genes of recurrent parent, entering with each backcross, can be calculated by using following formula:

$$\left(\frac{2^m - 1}{2^m}\right)^n \qquad \text{... (9.1)}$$

Here m is the number of backcrosses and n is the number of genes for which F_1 was heterozygous. Thus if 8 genes are involved ($n = 8$) the progeny after 6 backcrosses shall be homozygous for 88.1 per cent genes of the recurrent parent, provided selection was not practiced and genes were independent in transmission.

The strength of linkage does influence the rate at which genes of donor are eliminated during backcrossing. According to Allard (1960), if no selection is practiced for the genes, except the gene desired from donor, the probability with which a gene linked with the gene desired from donor shall be eliminated is given by formula 9.2.

$$1 - (1 - p)^{m+1} \qquad \ldots(9.2)$$

Where m is the number of backcrosses and p is the recombination frequency. New combinations of linked genes are called recombinations. And the proportion in which new combinations are formed between the two gene pairs as compared to sum of all the combinations is recombination frequency. Let us take an hypothetical example to understand this aspect. Certain common varieties of wheat, which have been derived from *Triticum timopheevi*, exhibit tight linkage between genes governing stem rust resistance and genes for late maturity. Obviously whenever efforts are made to incorporate gene for resistance to rust from such variety the gene for late maturity will automatically accompany it. Suppose rust resistant variety was used as donor. If we assume that recombination frequency between gene for rust resistance and late maturity is 20 per cent, then after five backcrosses, the gene for late maturity will be eliminated in about 74 per cent cases. Thus:

$$(1 - 0.20)^{5+1}$$

$$= 1 - (0.80)^6$$

$$= 1 - 0.2621 = 0.7379$$

or 73.79 per cent

However, when two desirable genes are linked, both the genes are transferred to the recurrent parent during backcross programme. For example in tomato, resistance to *Verticillium* and *Fusarium* is governed by genes V and F, respectively, which are very tightly linked (Fageria *et al.*, 2000). About 91 per cent plants in F_2 generation are resistant to *Verticillium* as well as *Fusarium* when these genes enter a cross in coupling phase. This is an advantage of tight linkage.

It has been said again and again that during the backcrossing, selection is made for the character being transferred from donor. Usually selection for characters of the recurrent parent is not necessary. The genes from recurrent parent enter with each back-cross. For accurate selection of the character of donor it is desirable that the character is highly heritable. It is easier to select plants for characters which are highly heritable, from segregating material. Simply inherited and highly heritable traits can be very comfortably incorporated by back-cross in the recurrent parent. Because highly heritable traits are least influenced by environmental agencies, there is correspondence between phenotype and genotype. The speed and precision with which a character is transferred is certainly determined by the number of genes governing inheritance of trait. A highly heritable character, even though governed by a relatively large number of genes, can be transferred to the recurrent parent with relative ease than a character with low heritability governed by fewer genes. The breeder should remember that while dealing with quantitative character, some intensity is lost in backcross programme. When the two characters have to be transferred to the recurrent parent, from two different donors, efficient plan is to transfer each character in separate programmes.

Backcross method is comparatively easier in cucurbits, brinjal, tomato, capsicum, etc. in which large number of seed is produced in fruit as compared to pea and beans which set few seeds per pod. Similarly back-cross programme can be easily executed in tobacco. In maize also back-cross programme can be comfortably carried out. The only caution that is required in execution of back-cross programme with cross fertilized crops is that sufficiently large number of plants of recurrent parent should be sampled so that maximum genetic variability of the parent is represented in gametes. Otherwise complete genotype of recurrent parent will not be recovered even after 5 or 6 backcrosses.

The success of backcross programme depends upon the number of backcrosses that are made to the recurrent parent. Three to ten backcrosses may accomplish the goal. When 10 backcrosses are made almost complete recovery of the genotype of recurrent parent is attained. According to Allard (1960) six back-crosses, together with rigid selection for character from the donor in the early generations, has been successful in larger number of back-cross programmes. He further suggested that selection for the type of recurrent parent, if based on moderate sized population, is equivalent to one or two additional backcrosses without selection. The population resembles the recurrent parent after third backcross.

Some breeders are of the opinion that three backcrosses are enough to reconstitute the genotype of recurrent parent in large part.

But much variability remains in the population upon which selection pressure could be exercised to improve adaptability and yield.

The varieties of self-pollinated crops are not always 100 per cent homogeneous. Several closely related pure lines may make up a variety. Hence even in case of self-pollinated crops gametes of several plants of recurrent parent should be sampled in backcross breeding.

Transfer of a dominant gene from the donor is relatively much easier. Five to six successive backcrosses of recurrent parent to the F_1 and backcross generations will reconstitute the genotype of recurrent parent plus a gene from donor. Of course linkage might delay the rate of reconstitution.

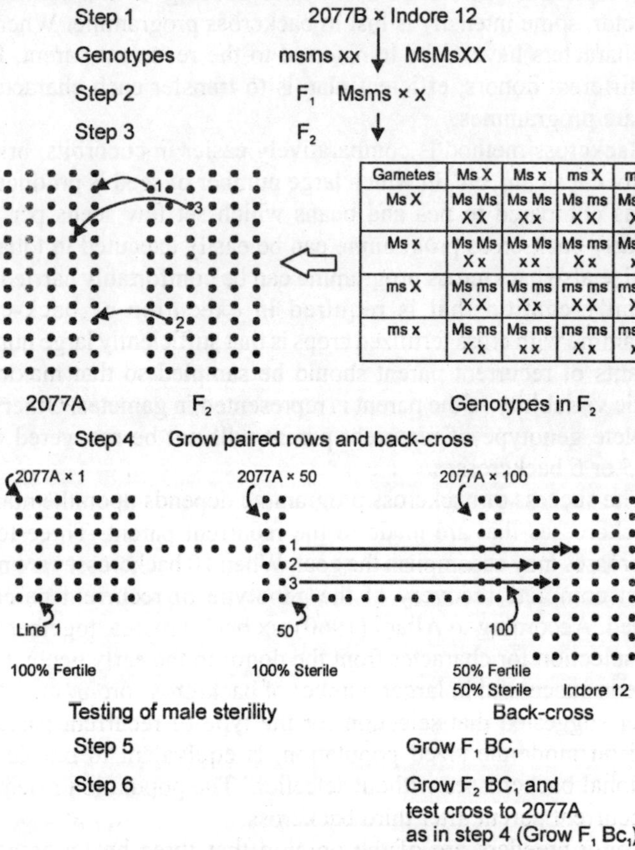

Fig. 9.4: Back-cross breeding for recessive trait

Transfer of a recessive gene from donor to recurrent parent is some-what arduous. F_2 generation of the cross has to be grown in order to identify plants, which have received the recessive gene. Phenotype of individuals carrying recessive gene can be observed only in homozygotes. For example, male-sterility in sorghum is governed by the interaction of recessive nuclear gene (ms ms) with hereditary factors in cytoplasm. A cross of parent with genes Ms Ms (Fertile) and parent carrying ms ms will result in F_1 with genotype Ms ms. The F_2 will comprise of 25 per cent Ms Ms (homozygous fertile), 50 per cent Ms ms (heterozygous fertile, because Ms is completely dominant) and 25 per cent ms ms (completely sterile). How can one detect the sterile plants? It is simple. Adjacent to F_2 population grow a few rows of cytoplasmic-genetic male sterile line, for example 2077A (see Figure 9.4). Choose desirable plants from F_2 (panicle of each plant should be enclosed by a sorghum pollen bag to protect from contamination by foreign pollen) and pollen from each plant is dusted on one panicle of 2077A (which are also covered by sorghum bags as soon as they emerge out of boot leaf). Suppose 50 plants are chosen from F_2, serially number them and put a label. Similarly put labels on 2077A, which received the pollen. For example, plant selected in F_2 is serialed 1, then the label on 2077A shall be 2077A × 1, and so on. Next season, grow the hybrids, ear to row. As soon as the panicles emerge from boot leaf cover them inside sorghum bag. Examination of each panicle within the rows, about a week after putting sorghum bags, will show that some lines have all the panicles sterile, these therefore received pollen of F_2 plant which carried ms ms genes and hence maintained sterility. Lines in which 100 per cent panicles are fertile are progeny of those F_2 plants which carried Ms Ms genes. There will be some lines in which about 50 per cent plants will be sterile and rest fertile. The pollen parents of these lines were Ms ms in genetic constitution. If resistance to a disease is governed by recessive gene, the F_2 should be exposed to epiphytotic to identify plants homozygous for resistance genes.

9.5.1 Backcross for Recessive Gene

Transfer of cytoplasmic-genetic male-sterility will be taken as example to describe the general steps undertaken to transfer recessive gene from donor through backcross. As mentioned in previous paragraph, male-sterility in sorghum is the result of interaction of nuclear genes, ms ms with genetic factors in mitochondrial DNA. Suppose Indore 12, a restorer of hybrid CSH 18, is to be converted to male-sterility. The maintainer line of 2077A (a male-sterile) is chosen as the seed parent. Any other *B* line could be taken. In the present illustration 2077*B* is taken. Why? Because 2077*B* carries on one of its chromosome pair genes ms ms. These genes have to be transferred to Indore 12, rather substituted for genes Ms Ms carried by it. The F_1 shall be *Ms ms Xx*. Where *X*, *x* represent genes other than those for male sterility transmitted to F_1 by the recurrent and donor parents (here 2077B is donor), respectively. As mentioned earlier, in order to identify those segregants, which carry genes ms ms in F_2, plants looking similar to Indore 12 are chosen. Each plant is labeled serially. Adjacent to the F_2, 2077A is also sown (on different dates so that the flowering of segregants coincides). Pollen from each of the selected F_2 segregant is used to pollinate a single plant of 2077A. The latter is appropriately labeled. The hybrids (i.e. 2077A × 1, and so on) are sown in the next season ear-to-row. Adjacent to the hybrids their corresponding pollen parents (the selected F_2 segregant) are also sown. As described earlier examination of panicles of hybrids will indicate which of the segregants carries ms ms genes (and hence could maintain the sterility of 2077A). Backcross the segregants carrying gene ms ms and phenotypically similar to the recurrent parent, i.e. Indore 12. In the subsequent season grow F_1 of BC_1 and in the next season F_2 of BC_1. Select plants from F_2 looking like Indore 12, label them serially, and pollinate individual 2077A plant. The latter is also appropriately tagged. Grow hybrid along with male parent (segregant from F_2BC_1). Find out which segregants maintained sterility of 2077A (Figure 9.4). Backcross the pollinator segregant to the Indore 12. This will be second backcross. About 5 backcrosses, in this fashion, will convert Indore 12 into a male-sterile line.

9.5.2 Convergent Improvement

This method was proposed by Richey in 1927 to improve the existing inbred lines so that hybrid could improve in its performance. This method is based on double backcross. The two inbred lines, says *A* and *B* are crossed to produce F_1. The F_1 is crossed separately to both the parents i.e. (*A* × *B*) × *A* and (*A* × *B*) × *B*. The idea behind backcrossing F_1 to both the parents is to improve both the inbred lines by careful selection for vigour and other desirable characters. If (*A* × *B*) is backcrossed to *A*, *A* shall be improved by the favourable genes contained in *B*. Likewise ($A \times B$) crossed to *B* is likely to improve *B* to the extent that desirable genes from A have been incorporated in *B*. After about three backcrosses selfing is practiced to fix the selected genes. The method is useful for improving vigour, resistance to diseases and/or insects and lodging.

Double backcross or convergent improvement is also recommended for improvement of vegetable crops. Swarup (1977) suggested use for double backcross to incorporate genes for large fruit size from one parent and earliness from other in tomato. The early and small fruited variety was crossed with a variety bearing larger fruits but late in maturity. The F_1 was backcrossed to early, small fruited parent and also to large fruited late parent, separately. In the former backcross early types were selected while in latter large fruit size was selected. After that both the backcross lines were crossed together, which resulted in plant with large fruit size and early maturity (Walkof, 1961).

9.5.3 Backcross Breeding and Cross-Pollinated Crops

As applied to the cross-pollinated crops, backcross method is practically similar to that employed for improvement of self-pollinated crops. But because the varieties of cross-pollinated crops are heterozygous and heterogeneous the number of plants of recurrent parent for back-crossing should large in number. However, this may not be so while backcross is used to improve some attribute of inbred lines of maize.

9.6 VARIETIES BRED THROUGH BACKCROSS METHOD

This breeding method has been used for improvement in almost all the crops. Its maximum use was made to transfer gene for resistance to a disease or an insect. Variety of safflower resistant to *Phytophthora*, viz., US 10 (Thomas, 1964) was evolved via backcross breeding. Variety UC 1, of safflower containing high oleic acid in the oil also emerged through backcross improvement (Knowles, 1968). Sabarmati and Jamuna varieties of rice were developed by backcrossing Taichung (Native) 1 with Basmati 370 (Sharma, 1989). Basmati rice is excellent in grain quality (such as intermediate amylose contents, intermediate gelatinization temperatures, soft gel consistency, high degree of grain elongation on cooking and excellent aroma) and they fetch high price in the international markets. Basmati varieties are tall "*indica*" types but are poor yielder. Backcross method was used to improve grain quality with high yield of semi-dwarf varieties.

Backcross method was used very widely in cotton breeding. It is being used to develop male-sterile lines. This method was employed successfully to transfer desirable characters of *Gossypium arboreum* and *G. herbaceum* to *G. hirsutum* (Singh, 1989). Varieties Deviraj and Devitej were developed through this effort (Bhat and Patel, 1956). Hairiness of wild species *G. tomentosum* was transferred at Indore to *G. hirsutum* to induce jassid resistance.

Wheat varieties C 286 and *NP* 890 were evolved through backcrossing (Jain, 1989). Varieties Kharchia, *NP* 852, *NI* 5439, *K* 8027 were also bred through this method. These are but few examples of varieties evolved by backcross method.

References

Adair, C.R. and J.W. Jones (1946). Effect of environment on the characteristics of plants surviving in bulk hybrid population of rice. *Jour.Amer.Soc.Agron.*, 38: 708-716.

Allard, R.W. (1960). *Principles of plant breeding*. John Wiley and Sons, Inc., New York.

Asthana, A. N. and S. Chandra (1995). Varietal improvement in pulse crops — current trends and future prospects. In "Genetic Research and Education. Current trends and the next 50 years". Ed. B. Sharma *et al. Indian Society of Genetics and Plant Breeding*, New Delhi, pp. 925-959.

Ashton, T. (1946). The use of heterosis in the production of agricultural and horticultural crops. *Imper.Bur.Plant Breed. And Genetics*, School of Agric. Cambridge, England.

Bauman, L.F. (1981). Review of methods used by breeders to develop superior corn inbreds. Proc. *Annu. Corn Sorghum Ind.Res. Conf.* 36: 199-208.

Bell, G.D.H. (1963). Breeding techniques. In "Barley Genetics" Wageningen, The Netherlands, Purdoc, pp. 285-306.

Bhat, N.R. and C.T. Patel (1956). Further progress in cotton improvement through interspecific hybridization at Surat. 6th Conference on Cotton Growing Problems in India. ICCC, Bombay, pp. 63-66.

Boerma, H.R. and R.L. Cooper (1975). Comparison of three selection procedures for yield in soybeans. *Crop Sci.*, 15: 227-229.

Borlaug, N.E. (1958). The use of multilineal or composite varieties to control air borne epidemic diseases of self-pollinated crop plants. Proc. *First*

Intern.Wheat Genet.Symp. The Public Press Ltd., Winipeg. Canada, pp. 15-27.

Briggs, F.N. and P.F. Knowles (1967). *Introduction to plant breeding*. Reinhold Publishing Corporation, New York.

Briggs, F.N. and R.W. Allard (1953). The current status of the backcross method of plant breeding. *Agron.J.*, 45: 131-133.

Brim, C.A. (1966). A modified pedigree method of selection in soybeans. *Crop Sci.*, 6: 220.

Byth, D.E.; J.M. Green and G.C. Hwtin (1980). ICRISAT/ICARDA chickpea breeding strategies. Proceedings International Workshop on Chickpea Improvement. ICRISAT, Hyderabad, 28 Feb to 2 March, 1979, pp. 11-27.

Chandraratna, M.F. (1964). *Genetics and breeding of rice*. Longmans. Green and Co. Ltd., London.

Dabholkar, A.R. (1999). *Elements of biometrical genetics*. Concept Publishing Company, New Delhi. 2nd edition.

Doggett, H. (1970). *Sorghum*. Longmans, London.

Fageria, M.S.; P.S. Arya and A.K. Choudhary (2000). *Vegetable crops. Breeding and seed production*. Vol.I. Kalyani Publishers, Ludhiana, pp. 83-103.

Freeman, J.E.; P.W. Watkins; H.H. Hadley and R.H. Hageman (1962). A simple method of hot water emasculation of sorghum. *Crop Sci.*, 2: 175-176.

Gardner, C.O. (1972). Development of superior populations of sorghum and their role in breeding programmes. In *Sorghum in seventies*, Ed. N.G.P. Rao and L.R. House. Oxford and IBH Publishing Co., New Delhi, pp. 180-196.

Goulden, C.H. (1939). Problems in plant selection. In "Int. Genet.Congr. Proc. 7th (Edinburgh)" Cambridge University Press. Ed. R.C. Burnett, pp. 132-133.

Green, J.M.; D. Sharma, L.J. Reddy; K.B. Saxena; S.C. Gupta; K.C. Jain; B.V.S. Reddy and M.R. Rao (1981). Methodology in progress in ICRISAT pigeonpea breeding program. In "Proceedings of the International Workshop on Pigeonpea Vol.I" 15-19 Dec., 1980. ICRISAT, Patancheru, India, pp. 437-449.

Griffing, B. (1956). Concept of general and specific combining ability in relation to diallel crossing systems. *Aust.J.Biol.Sci*. 9: 463-493.

Hallauer, A.R.; W.R. Russell and K.R. Lamkey (1988). Corn breeding. In "Corn and corn improvement" Agronomy Monograph, No.18, 3rd edition. Iowa Agric.Home Econ.Expt.Stn.Ames, Iowa.

Hamblin, J. (1977). Plant breeding interpretations of the effects of bulk breeding on four populations of beans (*Phaseolus vulgaris*, L.). *Euphytica*, 26: 157-168.

Harrington, J.B. (1952). *Cereal breeding procedures*, F.A.O. Development Paper No.28, F.A.O. of United Nations, Rome, Italy.

Jain, K.B.L. (1989). Cereals: Wheat and Barley. In *Plant breeding — theory and practice*, Ed. V.L. Chopra. Oxford and IBH Publishing Co. Pvt.Ltd., New Delhi, pp. 77-99.

Jayasekara, N.E.M. and J.L. Jinks (1976). Effect of gene dispersion on estimates of components of generation means and variances. *Heredity*, 36: 31-40.

Jodon, N.E. (1938). Experiments on artificial hybridization of rice. *Journal of the American Society of Agronomy*, 30: 294-305.

Kalloo, G. (1988). *Vegetable breeding* Vol.I. C.R.C. Press, Florida, pp. 27-28 and 79-81.

Kaufman, M.L. (1971). The random method of oat breeding for productivity. *Can.J.Plant Sci.* 51: 13-16.

Knott, D.R. and J. Kumar (1975). Comparison of early generation yield testing and single seed descent procedure in wheat breeding. *Crop Sci.*, 15: 295-299.

Knowles, P.F. (1968). Registration of UC 1 safflower. *Crop Sci.* 8: 641.

Koutu, G.K. and K.C. Mandloi (2002). Cotton. In *Genetic Improvement of Field Crops*. Eds. Singh C.B. and D. Khare, Scientific Publishers (India), Jodhpur, pp. 247-253.

Matzinger, D.F. (1963). Experimental estimates of genetic parameters and their applications in self-fertilizing plants. In *Statistical genetics and plant breeding*. Eds. Hanson, W.D. and H.F. Robinson, NAS — NRC Publication, pp. 253-279.

McGinnis, R.C. and L.M. Shebeski (1968). The reliability of single plant selection yield in F_2. In *Wheat genetic symposium proceedings*, 3rd pp. 109-114.

Mishra, P.C.; R.P. Sharma and S.S. Singh (2002). Wheat. In *Genetic improvement of field crops*, Ed. Singh, C.B. and D. Khare. Scientific Publishers (India), Jodhpur, pp. 1-10.

Mukuru, S.Z. (1992). Breeding for grain mold resistance. In *Sorghum and Millet diseases*. A second world review. Eds. de Milliano, W.A.J., R.A. Frederiksen and G.D. Bengston. ICRISAT, Patancheru, India, pp. 273-285.

Murty, B.R. and V. Arunachalam (1967). The nature of divergence in relation to breeding system in some crop plants. *Indian J.Genet.*, 26A: 188-198. Proc.3rd Int. Symposium.

Peter, K.V. (1998). *Genetics and breeding of vegetables*. I.C.A.R., New Delhi.

Poehlman, J.M. (1987). *Breeding Field Crops*, 3rd Ed. AVI Publishing Co. Inc. Westport, Connecticut, U.S.A., pp. 187-213.

Pooni, H.S. and J.L. Jinks (1981). The true nature of non-allelic interactions in *Nicotiana rustica* revealed by association of crosses. *Heredity*, 47: 253-258.

Rajaram, S.; R.P. Singh and M. van Ginkel (1995). CIMMYT's approach to breeding for durable disease resistance in bread wheat: A Global perspective. In "Genetic research and education — current trends and the next fifty years". Ed. B. Sharma *et al*. *Indian Society of Genetics and Plant Breeding*, New Delhi, pp. 325-338.

Ram, H.H. (1988). *Vegetable breeding — principles and practices*. Kalyani Publishers, Ludhiana.

Rana, B.S. and N.G.P. Rao (1989). Often cross-pollinated crops: Sorghum. In *Plant breeding — Theory and practice*. Ed. V.L. Chopra. Oxford and IBH Publishing Co.Pvt. Ltd., New Delhi, pp. 225-235.

Richey, F.D. (1927). The convergent improvement of selfed lines of corn. *Am.Nat.* 61: 430-449.

Sharma, S.D. (1989). Cereals: Rice. In *Plant breeding — Theory and practice*, Ed. V.L. Chopra, Oxford and IBH Publishing Co.Pvt. Ltd., New Delhi, pp. 101-120.

Shebeski, L.H. and L.E. Evans (1973). Early generation selection for wide range adaptability in breeding program. Proc.4th International Wheat Genetics Symposium, Columbia, M.O., pp. 587-593.

Singh, B.D. (1993). *Plant breeding: Principles and methods*. Kalyani Publishers, Ludhiana, pp. 215-255.

Singh, K.P. (2002). Personal communication.

Singh, M. (1989). Often cross-pollinated crops: Cotton. In *Plant breeding — theory and practice*, Ed. V.L. Chopra, Oxford and IBH Publishing Co.Pvt. Ltd., New Delhi, pp. 237-249.

Singh, Onkar; C.L.L. Gowda; S.C. Sethi and S.S. Lateef (1995). Inheritance and breeding for resistance to *Helicoverpa armigera* and pod borer in chickpea. In "Genetic Research and Education — current trends and the next fifty years" Ed. B. Sharma *et al. Indian Society of Genetics and Plant Breeding*, New Delhi, pp. 243-248.

Smith, H.H. (1952). Fixing transgressive vigour in *Nicotiana rustica*. In *Heterosis*, Ed. G.W. Gowen, Iowa State College Press, Ames, Iowa, pp. 161-174.

Sneep, J. (1977). Selection for yield in early generations of self-fertilizing crops. *Euphytica*, 26: 27-30.

Stephens, J.C. and J.R. Quinby (1933). Bulk emasculation of sorghum flowers. *Journal of the American Society of Agronomy*, 25: 233-234.

Suneson, C.A. and H. Stevens (1953). Studies with bulked hybrid populations of barley. U.S. Department of Agriculture, *Tech. Bulletin* No.1067, p. 14.

Swarup, Vishnu (1977). *Breeding procedures for cross-pollinated vegetable crops*. ICAR, New Delhi, p. 73.

Thomas, C.A. (1964). Registration of US 10 safflower. *Crop Sci.*, 4: 446-447.

Tigchelaar, E.C. (1986). Tomato breeding. In *Breeding vegetable crops*, Ed. M.J.Bassett, AVI Publ.Co., pp.138-171. Quoted by Ram, H.H. (1988).

Timothy, D.H. (1963). Genetic diversity, heterosis and the use of exotic stocks in maize in Columbia. In *Statistical genetics and plant breeding*. Symposium. *Raleigh*, 1961, pp. 581-591.

Walkof, C. (1961). Improvement of tomato fruit size and maturity by backcross breeding. *Canadian J.Plant Sci.*,41: 24-30. Quoted by Swarup, Vishnu (1977).

Williams, W. (1959). The isolation of pure lines from F_1 hybrids of tomato and the problems of heterosis in inbreeding crop species. *J.Agric.Sci.* 53: 347-353.

10

Breeding Methods for Cross-Pollinated Crops

The genetic organization of cross-pollinated crops is different from that of the self-pollinated crops because of differences in reproductive structure. Homozygosity is normal state at each locus for self-pollinated crops. Inbreeding does not lead to loss of vigour and fecundity and appearance of morbid and lethal forms. On the other hand, although some homozygous loci may exist in cross-pollinated crops, most of the loci are heterozygous. Natural populations of allogamous crops are heterogeneous. Practically every individual carries deleterious recessive genes shielded by their favourable dominant alleles. Upon inbreeding, these deleterious recessive genes become homozygous and manifest adverse effects on their carriers. The extent of adverse effects, upon inbreeding, is a function of number of harmful recessive genes carried by the plant before inbreeding. Because, in most of the cross-pollinated crops homozygous individuals are weaker, heterozygosity must be restored in the end product of any breeding programme. The end product could be homogeneous (as in case of single-cross hybrids) or heterogeneous (as in double-cross, three-way cross hybrids, synthetics and composites).

Mass selection is the oldest improvement method employed in cross-pollinated crops. Although this breeding method was successful for improvement of characters with high heritability, it was not effective for yield improvement. Selection of plants is made on the basis of phenotype, in mass selection. It is the phenotype of female parent alone that is taken into consideration because genetic constitution of male-parent is unknown. Moreover, there is very less correspondence between the phenotypic value and the genotypic value with respect to quantitative traits. Quantitative characters are governed by numerous genes each with small, similar and cumulative effects and these characters are influenced by environmental agencies to a greater extent than the qualitative traits.

Hopkins (1899) introduced ear-to-row breeding method in maize, which was based on Vilmorin principle. In this method, superior plants are selected from an open-pollinated population on the basis of superior ears. Next year part of the seed from each ear is sown in progeny row and remnant seed is stored safely. Superior progeny rows are identified on the basis of evaluation test. Remnant seed of selected entries is composited. Thus two generations are required per cycle. Gardner (1961) proposed, "griding" the field in order to reduce, if not entirely eliminate, environmental influence on phenotype. Lonnquist (1964) suggested modification of ear-to-row method.

On evolutionary grounds, complete dominance or overdominance may be responsible at some loci in cross-pollinated crops (Sprague, 1966). On the other hand, marked degree of dominance may be unexpected in self-pollinated crops.

In recent years exploitation of heterosis through hybrids has been the major objective of plant breeders dealing with improvement of cross-pollinated crops. As shown in Chapter 12, three basic steps are involved in the development of hybrids of cross-pollinated crops, viz.:

(i) Selection of desirable plants from heterozygous base population,
(ii) Progenies of selected plants are inbred for several generations to develop homozygous inbred lines, and
(iii) After assessing the general combining ability of inbred lines, isolate good general combiners and subsequently cross two, three or four inbred lines (after assessing specific combining ability of crosses) to produce single-cross, three-way cross or double-cross hybrid, respectively.

Initial success in the production of superior maize hybrids was achieved by development of inbred lines directly from the then existing open-pollinated varieties. As the hybrid maize programme advanced further tendency to isolate inbred lines from open-pollinated varieties was bypassed because of the low frequency with which superior inbred lines could be isolated subsequently. According to Eberhart (1972) a hybrid variety is nothing more than a genotype that can occur in a random mating population. But whereas the best genotype occurs in certain individuals of the random mating population, it is replicated in every individual of certain cross. Once a superior hybrid combination has been isolated from a population, selection of hybrids with greater superiority is difficult unless the population is improved as shown in Figure 10.1.

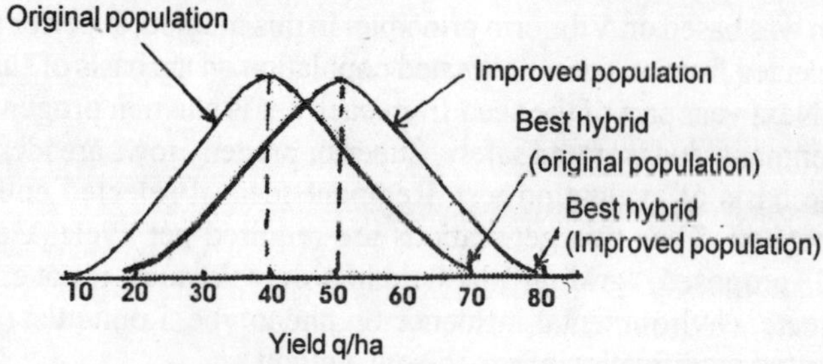

Fig. 10.1: Effect of population improvement on hybrid performance

10.1 RECURRENT SELECTION: METHOD OF OBTAINING SUPERIOR INBRED LINES

With the objective to improve performance of existing inbred lines, inbred lines with superior genotypes were crossed. Either pedigree method of selection or backcross method was employed to develop inbred lines with still superior genotypes. Inbred lines developed in this way were called as "second cycle inbreds".

Although the *per se* performance of second cycle inbreds was improved considerably, performance of their hybrids was not improved conspicuously as compared to hybrids of first cycle inbreds. While examining the standard system of hybrid development, two facts emerged, viz.,

(i) The proportion of inbred lines with superior genes or gene combinations that could be isolated from a population depends upon the extent of superior genes or gene combinations existing in the population, and

(ii) Selection during inbreeding to increase the frequency of superior genes or gene combinations was not effective.

Evidence available suggested that selection was not particularly effective in increasing the frequency of desirable genes, during inbreeding, except the genes with most easily observable characters. Proportion of superior genotypes in heterozygous base population depends upon frequency of superior genes in the population. Breeders realized long ago that a breeding scheme was needed which will maintain genetic variability in the population and at the same time increase the frequency of superior genes or gene combinations by providing for recombination among selected lines. These ideas were developed into a breeding system called as "recurrent selection". As a result of his experiments with early testing, Jenkins in 1940, described such a procedure. The steps were:

(i) Selection of superior plants from heterozygous source population. Evaluation of these plants for desirable character or characters and at the same time selfing the selected plants ($S_0 \rightarrow S_1$),
(ii) Growing ear-to-row progeny of selected plants, from selfed (S_1) seeds, and
(iii) Making all possible crosses among the progenies of selected and selfed (S_1) plants.

According Hull (1952) recurrent selection was meant to include reselection generation after generation with interbreeding of the selects to provide for genetic recombination. Selection among the isolates, inbred lines, clones was not recurrent selection unless the selects were interbred and a new cycle of selection begun. Depending upon procedure, on the basis of which desirable plants are selected and intermated, four types of recurrent selection methods have been suggested to develop superior inbred lines.

10.1.1 Simple Recurrent Selection

In this system, plants with desirable phenotypes are selected from heterozygous source population and also self-pollinated (S_1). Progeny rows are raised in the next season from selfed seeds of selected plants and all possible crosses are made among progenies of these selected and selfed (S_1) plants. In other words, selection is based on S_0 plants but mating is executed among S_1 plants. That completes C_0 cycle, because it was initiated in original random mating population. C_1 cycle will begin with selection of superior plants from seed composited at the end of C_0 cycle.

10.1.2 Recurrent selection for general combining ability

This recurrent selection method was developed as a result of Jenkin's (1940) experiments for early testing for general combining ability.

Early testing of inbred lines for general combining ability, proposed by Jenkins (1935), was founded on the assumption that the S_0 or S_1 plants differ in combining ability and these differences among the plants could be detected via top-cross although the S_0 or S_1 plants are heterozygous. According to him inbred lines acquire their individuality as parents of top-crosses very early in the inbreeding process and tend to remain fairly stable for combining ability thereafter. In this method of recurrent selection phenotypically desirable plants are chosen from the heterozygous source material and selfed (S_1). At maturity, plants are selected for some character or characters. Rest are rejected. In the next season, progeny rows of S_1 selected plants are grown alternating with some desirable heterozygous tester. The plants from S_1 progeny rows are used as seed parent, and in case of maize detasseled, and top cross seed harvested from S_1 progeny rows, individually. On the basis of evaluation test, those progeny rows, which yield better than the overall average, are regarded as superior. The remnant seed of superior S_1s selected on the basis of top-cross test is planted in the fourth season, ear-to-row, and progenies intermated in all possible ways. The seed from all crosses is composited at harvest. This forms base population for first cycle of recurrent selection (Figure 10.2).

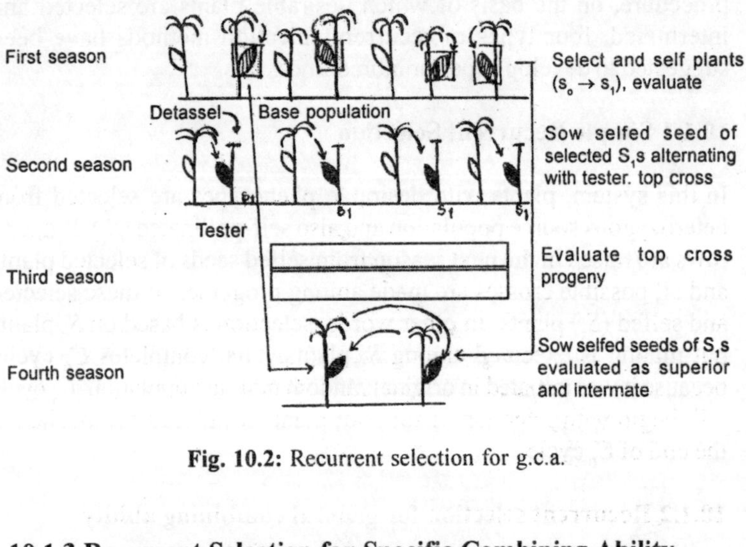

Fig. 10.2: Recurrent selection for g.c.a.

10.1.3 Recurrent Selection for Specific Combining Ability

This approach to recurrent selection ows to Hull (1945) who believed that in most cases heterosis results from overdominant gene action. In order to identify superior hybrid combinations, specifically combining inbred lines need to be identified. Unlike recurrent selection for general combining ability, the tester in this method is an inbred line. The objective of this method is to identify or produce lines, which combine well with the tester. Rest of the procedure is same as that for recurrent selection for general combining ability.

10.1.4 Reciprocal Recurrent Selection

Comstock, Robinson and Harvey (1949) proposed a breeding/ selection procedure to make maximum use of both general and specific combining ability. For this recurrent selection procedure two genetically divergent, heterogeneous populations, that are known to combine well together, should be chosen. They could be two varieties, two synthetics or two single-crosses involved in a successful double-cross hybrid. Let us denote the two populations as *X* and *Y*. A number of plants from population *X* are selected to function as male parents. Each male from population *X* is mated to several females from population *Y* (about 3 to 4). That is several plants from population *Y* are pollinated by a single male from population *X*. In this way a few hundred half-sib progenies are developed (Because one male pollinates several female plants, the progeny of common male is half-sib). Similarly several plants are selected from population *Y* and each male is mated with several females from population *X*. Several hundred half-sib progenies are developed in this way.

The male plants from both the populations are selfed. Thus S_1 seed of males is produced.

In the second season, the half-sib progeny families of males from *X* population and also from males of *Y* population are grown for evaluation (Figure 10.3). Superior half-sib progeny families are identified on the basis of performance for the character under consideration. In the third season selfed seed of males identified as superior (on the basis of half-sib progeny evaluation) from population *X* are planted ear-to-row in isolation and all possible crosses are made. Equal quantities of seed of each cross is taken and composited which forms the base population for first cycle of reciprocal recurrent selection. Selfed seeds of selected male plants from population *Y* are similarly planted and intermated in isolation. Bulking the seed of intermated plants is base for another cycle of recurrent selection. If we denote the initial populations as X and Y, then after one cycle the populations are X_0 and Y_0, respectively.

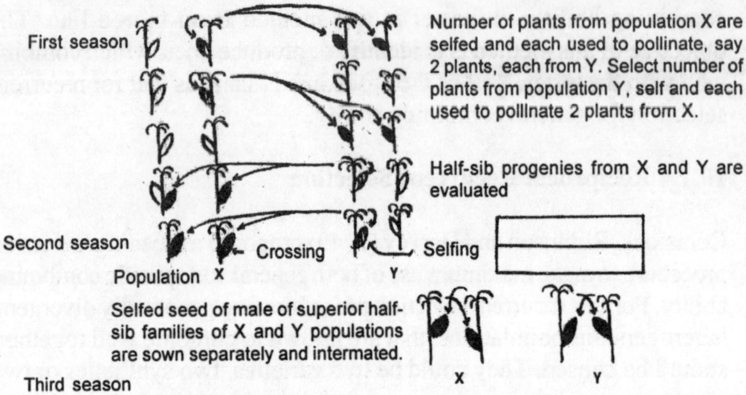

Fig. 10.3: Reciprocal recurrent selection

It is important to note that in all the four methods of recurrent selection, sufficient number of plants should be selected in each cycle to keep the level of inbreeding at its minimum.

Comstock *et al.* (1949) proposed reciprocal recurrent selection as an efficient means of improving the population cross of two broad genetic base populations. As mentioned earlier, recurrent selection methods aimed to increase the frequency of desired alleles for quantitative traits in the breeding populations and also maintain genetic variability for continued improvement.

Reciprocal recurrent selection was initiated in Iowa, U.S.A. in 1949 with two populations, viz., Iowa Stiff Stalk Synthetic (BSSS) and Iowa Corn Borer Synthetic No.1 (BSCB1). Objectives of the studies carried out with populations BSSS and BSCB1, by Martin and Hallauer (1980) were to determine changes in yield and correlated changes in other agronomic traits after seven cycles of reciprocal recurrent selection. To determine the effectiveness of reciprocal recurrent selection for yield improvement, they evaluated C_0, C_1, C_3, C_5 and C_7 of both BSSS(R) and BSCB1(R) and the $C_0 \times C_0$, $C_5 \times C_5$ and $C_7 \times C_7$ population crosses in six Iowa environments. The grain yield of the *BSSS*(*R*) × *BSCB*1(*R*) crosses increased 1.75 ± 0.37 quintals per hectare per cycle, but the yield of parental populations did not change significantly. Observed mid-parent heterosis for yield increased from 14.9 per cent for C_0 population cross to 41.7 per cent for C_7 population cross (Table 10.1). Estimates of genetic components of variance were determined for the yield from selection trials for each cycle. There was no evidence for decline in genetic variability in *BSSS*(*R*) and *BSCB*1(*R*) because of selection.

Populations produced after the reciprocal recurrent selection could be used for development of inbred lines from populations *X* and *Y* and subsequently crossed to produce hybrid. The populations could be used to develop varietal hybrids also.

Table 10.1: Means of populations and population crosses for yield of different cycles of reciprocal recurrent selection (Martin and Hallauer, 1980)

Entry	*Yield (q/ha)*
BSSS (*R*) C_0	50.0
BSSS (*R*) C_1	49.1
BSSS (*R*) C_3	51.5
BSSS (*R*) C_5	55.0
BSSS (*R*) C_7	52.4
*BSCB*1 (*R*) C_0	51.8
*BSCB*1 (*R*) C_1	51.3
*BSCB*1 (*R*) C_3	53.2
*BSCB*1 (*R*) C_5	50.4
*BSCB*1 (*R*) C_7	47.4
$C_0 \times C_0$	58.5
$C_5 \times C_5$	67.4
$C_7 \times C_7$	70.7
L.S.D. (0.05)	5.32

10.2 POPULATION IMPROVEMENT

According to Mather (1943) a big difference between a cross-pollinating crop and a self-pollinating crop is that dominance variance may be of considerable importance in the former while it is unlikely to be of much importance in the latter. But in both the types of crops additive genetic variance is important. Various recurrent selection methods have been described in section 10.1, which are used to develop improved inbred lines. But because additive genetic variance is also an important component of variability observed for yield, population improvement approaches have been proposed. Population improvement is a general term, which includes all those operations within a system in which the end product sought is sexually propagated improved type, either a random mating population or pure line (Sprague, 1966). Various population improvement schemes have been divided into two categories, viz., intra-population selection schemes and inter-population selection schemes. The object of former is to improve the population mean while the latter aims at improvement of variety cross (Empig *et al.*, 1972).

Initially population improvement schemes were proposed to develop populations of maize, superior in the frequencies of desirable genes and adequate genetic variability. However, this approach has been employed, with some modifications in sorghum (Gardner *et al.*, 1971; Ross *et al.*, 1971 and Doggett and Eberhart, 1968). Rai (1989) successfully employed it in rape and mustard. Singh (1989) reported use of population improvement in cotton. It was used in pearl millet also (Andrews *et al.*, 1985). Rachie and Gardner (1975) described an experimental approach for population improvement, utilizing genetic male-sterility and insect pollination in some grain legumes. Brim and Stuber (1973) applied recurrent selection in soybean by using male-sterility. The first composite variety of pearl-millet viz., *WC-C75*, was developed at *ICRISAT*, Hyderabad through recurrent selection (Andrews *et al.*, 1985).

Population improvement results from substitution of superior alleles for another in the population. What a breeder attempts to do is replace inferior genes with better ones at each locus and accumulate genes at different loci that interact favourably with one another. The force underlying desired change is selective force exerted by plant breeder.

Author is of the view that various population improvement schemes can be employed in multiflowered vegetable crops also where emasculation and pollination is relatively easy and a single pollination results in several seeds, for developing superior populations. The intra-population improvement methods are described in the following pages.

10.2.1 Mass Selection

This method was described at the beginning of this chapter. Classic mass selection study was conducted in 'Hays Golden' variety of maize. Gardner (1977, 1978) reported a realized gain of 3.00 per cent per cycle for grain yield as a linear response over first 15 cycles. The realized gain was in excellent agreement with expected response of 3.08 per cent per cycle. However, at cycle 13 of mass selection, responses tended to plateau, and started declining at cycle 17. It was probably because of changes in genetic variance of population. Non-additive genetic variance was probably more important at this stage. According to Mareck and Gardner (1979) although there was some decline in additive genetic variance with selection, more possible reason for decline in response was a series of stress environments experienced during later cycles of selection. According to Hallauer and Miranda (1981) mass selection is most effective improvement method for broad range of plant and ear characteristics of maize.

Selection for percentage of oil and protein in seed of maize was initiated in 1896 in United States of America in open-pollinated cultivar Burr's White. This is the longest selection programme executed in maize and the details of investigations after 76 generations of selection were reported by Dudley (1977). Divergent selection for oil and protein content was initiated by analyzing 163 open-pollinated ears. The seeds in ears represented half-sibs. The 24 ears with highest protein and oil and 12 with lowest protein and oil were employed to begin selection for strains with higher protein and oil and strains with lowest protein and oil. Selections were effective in all the instances. Gains after the completion of 76 generations were 292 per cent for high oil, 92 per cent for low oil, 133 per cent for high protein and 78 per cent for low protein.

Except for the characters with higher heritabilities, such as ear height, simple mass selection generally was not effective (Sprague, 1955). Because of the confounding effects of environment and pollen control, Gardner (1961) advocated modifications to increase efficiency of mass selection. Ear-to-row selection was designed to base selection on progeny rows rather than phenotype of individual plants (Hopkins, 1899). Nevertheless, experimental technique and pollen control limited its effectiveness. It was after the improvements suggested by Lonnquist (1964) and Compton and Comstock (1976) that the ear-to-row method of selection became an acceptable method for germplasm improvement (Hallauer *et al.*, 1988).

10.2.2 S_1 Progeny selection

Random sample of plants in the base population are selfed (S_1). The selfed seeds are grown ear-to-row in the next season, and progeny performance is evaluated. The remnant selfed seeds of those progenies are composited in equal proportions and grown in isolation, whose performance was superior. The random mated seed is the base for first cycle of S_1 progeny selection (See Figure 10.4). Alternatively the remnant selfed seed may be planted ear-to-row and manually intermated in all possible ways. The crossed seed is composited. Three generations per cycle are required for this method. It will be apparent from Table 10.2, that S_1 progeny selection makes maximum utilization of the additive genetic variation, which is responsible for the inheritance of quantitative character under consideration. Eberhart (1972) has shown that S_2 progeny selection utilizes more additive genetic variation than the S_1 progeny selection and rest of the population improvement schemes. But it requires 4 generations per cycle.

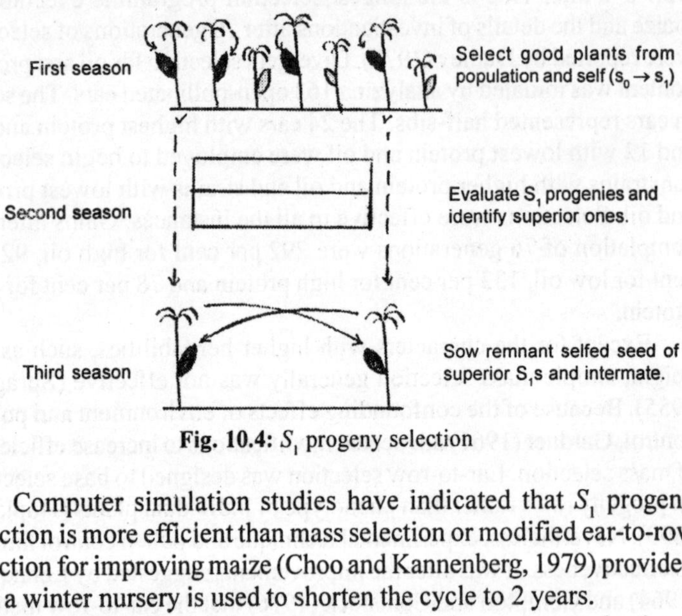

Fig. 10.4: S_1 progeny selection

Computer simulation studies have indicated that S_1 progeny selection is more efficient than mass selection or modified ear-to-row selection for improving maize (Choo and Kannenberg, 1979) provided that a winter nursery is used to shorten the cycle to 2 years.

S_1 progeny selection is similar to the simple recurrent selection.

Devey and Russell (1983) summarized the results of seven cycles of selection for improvement of stalk quality in open-pollinated maize cultivar Lancaster. Selection of plants during C_1 to C_4 cycles was on the basis of S_0 plant and S_1 progeny resistant to stalk rot following artificial inoculation with *Diplodia maydis*. The incidence of stalk rot significantly decreased over 7 cycles of selection. On a scale of 0.5 (no spread of disease) to 6 (premature death of plant), the level of stalk rot infection decreased from 3.27 for C_0 to 1.33 for the C_7, which compared favourably with a resistant check (1.44 for *B*14*A* × *C*103).

Table 10.2: Various intra-population improvement methods

Selection method	*Generations per cycle*	*Progenies* Evaluated	*Progenies* Used for recombination	*Expected progress**
		Intra-population selection schemes		
Mass selection	1	Individual plants	Individual plants	$\frac{k(\frac{1}{2})\sigma_A^2}{\sigma_{Pm}}$
Half-sib progeny test	3	Half-sibs	S_1's	$\frac{k(\frac{1}{2})\sigma_A^2}{\sigma_{Phs}}$
Half-sib test	2	Half-sibs	Half-sibs	$\frac{k(\frac{1}{4})\sigma_A^2}{\sigma_{Phs}}$
Full-sib test	2	Full-sibs	Full-sibs	$\frac{k(\frac{1}{2})\sigma_A^2}{\sigma_{Pfs}}$
S_1 progeny test	3	S_1's	S_1's	$\frac{k(\sigma_A^2 + C)}{\sigma_{Ps}}$

*Based on Empig *et al.* (1972) and Sprague (1966)

k = Selection differential in standard units

σ_A^2 = Additive genetic variance

$\sigma_{Pm}, \sigma_{Phs}, \sigma_{Pfs}, \sigma_{Ps}$ are the phenotypic standard deviations for mass selection, half-sib, full-sib and self progenies respectively.

$$C = \sum_{i=1}^{n} 2p_i q_i (p_i - \frac{1}{2}) h_i [d_i + (q_i - p_i) h_i]$$

Penny *et al.* (1967) reported results of S_1 recurrent selection for leaf feeding resistance to European corn borer, carried out in five synthetic varieties of maize. Manual infestation with artificially reared corn borer eggs of the plants was employed in each of the cycles of selection. After the completion of three cycles of S_1 recurrent selection, 100S_1 progenies were extracted from each of the cycles C_0, C_1, C_2 and C_3 and evaluated again with manual infestation with eggs of corn borer. Three cycles of S_1 recurrent selection were adequate to produce S_1

progenies with acceptable level of resistance to the first generation of European corn borer. On an average the S_1 ratings declined from 5.4 to 2.5 (on a scale 1 = resistant to 9 = susceptible). About 77 per cent of the progenies were within the resistant classes.

10.2.3 Half-Sib Progeny Test and Half-Sib Test

The term 'half-sib progeny test' has been used in place of the more conventional plant breeding term 'top-cross progeny test' in order to emphasize the genetic similarity in plant and animal breeding (Sprague, 1966). In the North Carolina Mating Design I also half-sib progenies are produces by crossing one male plant to several plants used as seed parents. The progeny families of a common male are half-sibs (because their female parents are different). Of course, the North Carolina Designs use F_2 population as base for producing half-sibs. In maize ear-to-row method also represents selection based on half-sibs, because a single ear of the plant selected is usually pollinated by several plants in the population.

In half-sib progeny test a number of plants from heterozygous population are selected before flowering. Each selected plant is selfed and used as male to pollinate to a number of other plants used as females or seed parents (about 3 to 4). In maize this is very easy. The pollen from tassel of one plant is used to pollinate ear on the same plant (selfing) and also to pollinate 3 to 4 other plants to produce half-sibs. In multiflowered vegetables, like tomato, this is possible. Some of the flowers are selfed and pollen from others used to pollinate emasculated flowers of other plants from same population. In the second season progenies are raised from crossed seed. These are evaluated for the character to be improved. In the third season selfed seed of male plants, selected on the basis of performance of their half-sib progenies, is planted and progenies intermated in isolation. The seed harvested and composited forms base population for first cycle of selection (Figure 10.5).

The half-sib test differs from the half-sib progeny test in that instead of sowing selfed seed of male parents for intermating, open-pollinated seeds of superior half-sib progenies in evaluation trial are composited in equal quantity and progeny inermated in isolation. It will be evident from Table 10.2 that half-sib progeny test utilizes more additive variation determining variability of the quantitative trait being improved, than the half-sib test. But the former also requires one additional generation per cycle. Half-sib progeny test could be utilized for developing inbred lines for synthetics.

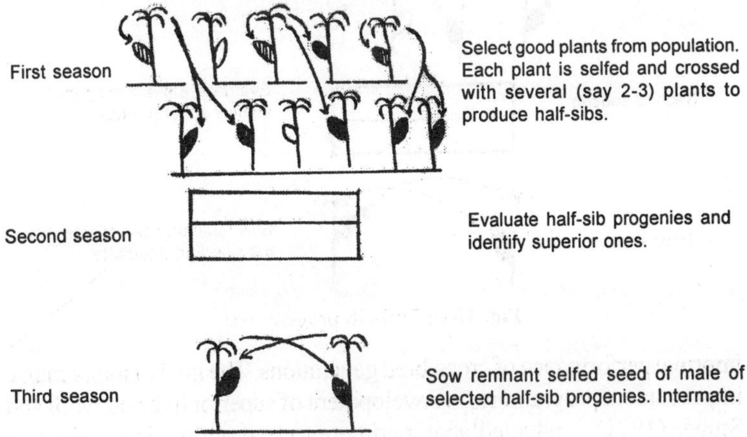

Fig. 10.5: Half-sib progeny test

10.2.4 Full-Sib Progeny Test

Full-sibs are a family of individuals having both parents in common. The full-sibs can be produced by selfing a plant or crossing two plants from same population. Because both the male and female plants from source population (of cross-pollinated crops) are heterozygous, progeny of full-sibs show considerable genetic variability. Individual plants within a full-sib family are closely related to each other as compared to individuals from another full-sib family. Production of full-sib families in cross-pollinated crops needs hand pollination (after emasculation in hermaphrodite flowered crops). In the first season a number of full-sibs are produced. One male is crossed to only one female plant. There is no need to maintain identity of male plant. In the next season half of the crossed seed is sown, ear-to-row, and progenies are evaluated for the character under improvement. In the third season, remnant seed of full-sib crosses, whose performance is superior, is sown in isolation and intermated. The intermated seed is composited which is the base for first cycle of full-sib progeny selection (Figure 10.6).

Various breeding procedures proposed for maize fall in two categories, viz., selection to improve population mean and selection to

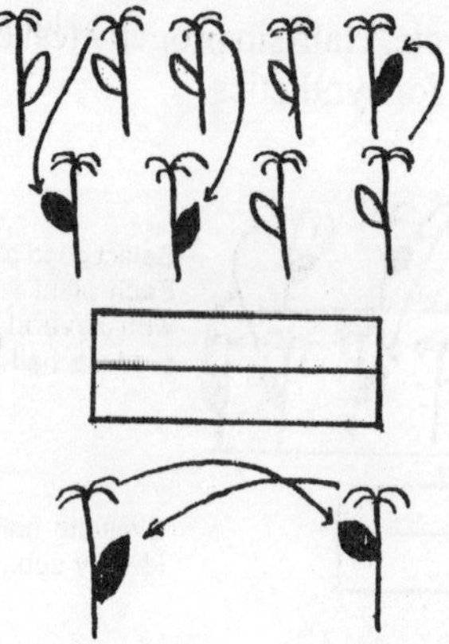

Fig. 10.6: Full-sib progeny test

improve performance of cross-bred generations. The goal of many maize improvement programmes is development of superior hybrids. Moll and Stuber (1971) conducted an experiment to find out, which of the two basic alternatives will lead to superior hybrids most efficiently. Comparisons were made between full-sib family selection and reciprocal recurrent selection for higher yield in maize following six selection cycles. Significant increases in yield were noticed after six selection cycles in six of the seven populations tested. Responses of varieties (Jarvis and Indian chief) to full-sib family selection were 2.4 times greater than their responses to reciprocal recurrent selection. Responses of variety hybrids to reciprocal recurrent selection were 1.3 times greater than response to full-sib family selection. Heterosis in variety hybrid increased markedly after reciprocal recurrent selection, but showed little change after full-sib family selection (Table 10.3).

Table 10.3: Heterosis in variety hybrid of maize before and after selection*

Selection cycles	*Kind of selection*	
	Full-sib family	*Reciprocal recurrent*
0	19.2	19.2
3	14.8	–
4	18.2	24.2
6	15.4	30.2

* Moll and Stuber (1971). Heterosis = $\left(\frac{\overline{F_1} - \overline{MP}}{\overline{MP}}\right) \times 100$

Full-sib family selection has not been used as extensively as half-sib family or inbred-progeny selection. In certain situations, full-sib family selection is effective and appropriate. Full-sib selection has been used effectively for developing improved *CIMMYT* corn populations and providing flexibility in the types of cultivars that can be used in national programmes, i.e. improved populations, improved full-sib families, hybrids and extraction of inbred lines (Pandey *et al.*, 1986).

The inter-population improvement schemes comprise of half-sib reciprocal recurrent selection and full-sib reciprocal recurrent selection.

10.2.5 Full-Sib Reciprocal Recurrent Selection

Hybrids are very popular these days as compared to the varieties for most of the crops. The genetic basis of heterosis is explained either as due to linked dominant factors with favourable effects on the character of interest or overdominance of genes concerned. Even genes with partial dominance with additivity over loci can be responsible for heterosis. Nevertheless, definitive evidence supporting either of the hypotheses has not yet been offered. In order to make use of additive gene effects as well as non-additive gene effects, presumably important for heterosis, Comstock, Robinson and Harvey in 1949 proposed reciprocal recurrent selection. This breeding procedure aimed to utilize both general and specific combining ability effects. Thus population mean is improved accompanied with improvement of variety cross. Because a random plant from one of the two heterozygous populations is used as male parent to pollinate 3 or 4 plants from another population, the resulting progeny is half-sib. In this way several half-sib progenies, in two populations, are produced for evaluation of male parents. After identification of superior male parents on the basis of performance of their half-sib progenies, selfed seeds of males of both the populations are separately intermated in isolation. This method is one of the inter-population improvement schemes and is also known as 'reciprocal half-sib recurrent selection'. The details have been described in section 10.1.4.

The principles of reciprocal full-sib recurrent selection are similar to those for reciprocal recurrent selection. The basic difference between full-sib reciprocal recurrent selection (*FSRRS*) and reciprocal recurrent selection (*RRS*) is that full-sib progeny families are produced in the former instead of half-sib families as in the latter, for evaluation of male parents. The reciprocal full-sib selection was described by Hallauer (1967), Lonnquist and Williams (1967) and Hallauer and Eberhart (1970). The reciprocal full-sib recurrent selection requires two eared maize plants, in at least one of the populations (Moreno-Gonzalez and Hallauer, 1982).

Original proposal of the full-sib reciprocal recurrent selection scheme requires maize populations that produce seed on two ears of individual plant. However, modifications can be made to conduct *FSRRS* with populations that include plants that produce seed on only one ear per plant (Hallauer, 1973).

In *FSRRS*, two populations, say *A* and *B*, are selected on the basis of their cross-bred performance (Jones *et al.*, 1971). As mentioned, this method requires two eared maize. Populations *A* and *B* are planted in alternate rows. In each row of *A* a few plants are selected and each is crossed to one plant from *B* to produce full-sibs and both parents are selfed. For example, self four plants from each *A* row and cross each to one plant from its paired *B* row. The female plants of *B* are also selfed. Similarly, self-pollinate four plants in each *B* row and cross each to one plant in its paired *A* row. The female plants in *A* are also selfed. Full-sib progenies are evaluated in replicated trials. Several full-sib progenies are developed. Plant S_1 seed of both parents of best families from populations *A* and *B* separately in isolation. The progenies in *A* and *B* are intermated separately.

The chief advantage of full-sib reciprocal recurrent selection is that superior families can be reproduced if they are found, because selfed seed from both parents is available. These will differ only in so far as recombination has occurred in selfing the parents. Thus, the cross between two populations can be improved at the same time superior hybrids can be released when they are obtained. In half-sib reciprocal recurrent selection seed from only male parents is available and hence superior families can not be reconstituted.

Secondly, because both parents of superior families are selected, more individuals can be evaluated with *FSRRS*. Thus for a similar amount of labour, higher intensities of selection and/or larger population sizes could be maintained.

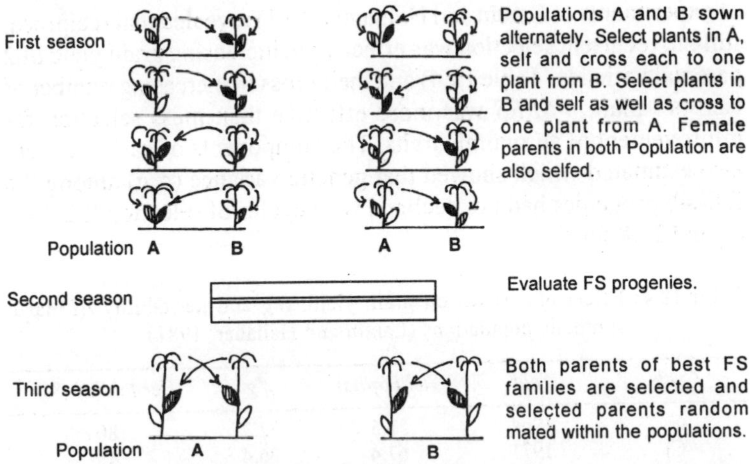

Fig. 10.7: Full-sib reciprocal recurrent selection

Jones *et al.* (1971) compared algebraically and with computer simulation the full-and half-sib reciprocal recurrent selection methods. Relative performance of two schemes depended upon the selection intensity and the environmental variance. The long term response of two methods according to these scientists, will depend upon effective population size which will be maintained. For any long term selection programme large populations are needed to avoid loss of genetic variation due to random drift and to a greater extent response to selection (Robertson, 1960). Full-sib reciprocal recurrent selection will enable a breeder to maintain a larger effective population size than the half-sib reciprocal recurrent selection. However, this will reduce higher selection differential that is possible. Lantin and Hallauer (1981) initiated reciprocal full-sib recurrent selection in two synthetic varieties of maize, viz., *BS* 10 and *BS* 11. *BS* 10 and *BS* 11 populations produce two ears per plant. Plants of each population were planted in alternate rows. Plant to plant crosses were made to produce full-sibs. For example, one plant was selected in *BS* 10 and one in *BS* 11. The selected plant in *BS* 10 was selfed at the same time its pollen used to pollinate plant from *BS* 11. This plant of *BS* 11 was also selfed. Similarly, plant of *BS* 11 was selfed and at the same time used to pollinate a plant from *BS* 10 (which was also selfed).

Full-sib progenies were evaluated in replicated trials. Selfed seed of both parents of superior full-sib progenies from populations *BS* 10 and *BS* 11 were sown separately and intermated in isolation. The results of experiments of Lantin and Hallauer (1981) revealed that reciprocal full-sib recurrent selection was effective in increasing grain yield of 2 parental varieties (Table 10.4) and their cross in increasing number of ears per plant. *FSRRS* was more effective than mass selection for improvement of agronomic traits. The components of variance were also estimated which showed that genetic variance ($\sigma^2 g$) among the full-sib progenies had not declined with cycles of selection (32.2 for C_0 and 32.8 for C_5).

Table 10.4: Effect of FSRRS on grain yield, $\sigma^2 g$ and heritability of maize synthetic populations (Lantin and Hallauer, 1981).

Cycle	*Year*	*Yield (q/ha)*	$\sigma^2 g$	*Heritability (%)*
C_0	1964	52.5	32.2	80
C_1	1971	67.4	26.4	45
C_2	1973	65.1	48.4	60
C_3	1975	70.4	33.9	60
C_4	1977	49.1	31.1	67
C_5	1979	73.3	32.8	57
Average		**63.0**	**34.1**	**62**

10.3 CHOICE OF POPULATION IMPROVEMENT SCHEME

Different selection schemes have been proposed for improvement of germplasm. Whenever objective is to improve the population mean choice is to be made from various intra-population improvement schemes while either of the two methods is chosen for improvement of variety cross. Mukherjee (1989a) suggested a few criteria such as operational simplicity, number of generations per cycle, gains per year, depletion of genetic variability, etc. for choosing a selection scheme from several intra-population improvement schemes.

According to Eberhart (1972) formula for gain from selection per year (G) predicts the rate of gain. The key formula for gain suggested is as follows:

$$G = c\ HD/y \qquad \ldots 10.1$$

In this equation *H* denotes heritability, *D* refers to selection differential. Heritability is the proportion of phenotypic variation (σ^2_{Ph}) among individuals or families evaluated that is due to additive genetic variance ($\sigma^2 A$), i.e. $\sigma^2 A / \sigma^2_{Ph}$. Selection differential is difference between mean phenotypic value of selected individuals or families ($\bar{X}_s$) and grand mean of population ($\bar{\bar{X}}$). If the population is normally distributed then selection differential can be expressed as $D = (\bar{X}_s - \bar{\bar{X}}) = k\sigma^2_{Ph}$. k is the intensity of selection, i.e. proportion of individuals selected from the population. c is measure of control over parents for material being selected and recombined. For example, in maize when a plant is selected on the basis of its phenotype in mass selection, value of c is equal to ½, because there is no control over the male parent. Any plant in the population pollinates the selected plant. On the other hand, in S_1 progeny selection the male and female gametes are from same plant, i.e. there is absolute control over the gametes. Hence, the value of c is 1. For full-sib progeny test also $c = 1$. Whereas the value of c is one for half-sib progeny test (because remnant selfed seed of selected plants is used for intermating) but the value of c is ½ for half-sib test. y denotes the number of years required to complete the cycle.

The gains from any one of the selection schemes, proposed for population improvement, could be enhanced by increasing the magnitude of additive genetic variation among the families being evaluated. One of the ways to achieve this objective is to develop genetically variable composite varieties by using superior material from breeding programmes in the centres of diversity. After the population has been built-up, the magnitude of additive genetic variance that is utilized depends upon the choice of selection scheme.

The choice among intra-population schemes depends upon several factors, a crude guide is ratio of the amount of additive genetic variance in the numerator of formula and number of generations per cycle (Empig *et al.*, 1972). It will be apparent from Table 10.2 that this ratio is highest for mass selection. This selection scheme requires only one generation per cycle and can utilize about ½ $\sigma^2 A$, which is responsible for character under improvement in the population. But mass selection will not be an efficient selection scheme if overdominance is the predominant type of gene action, as compared to other schemes of intra-population improvement.

As compared to mass selection, family selection methods are preferred because in the former the genetic worth is difficult to determine and the characters to be improved manifest low heritability. However, all the family selection schemes, which capitalize on the between family variance, utilize less of the $\sigma^2 A$ available in the base population than mass selection. The only exception being S_1 progeny selection. In the intra-population selection schemes that use another population as tester (recurrent selection for general and specific combining ability and top-cross) genetic gain can not be expressed in terms of σ^2_A of the base population, unless the gene frequencies in the two populations are equal.

While selecting for fecundity in *Drosophila pseudoobscura*, Kojima and Kelleher (1963) found that reciprocal recurrent selection was more effective than full-sib reciprocal recurrent selection.

The progress could be enhanced if the number of years required to complete one cycle is reduced. Reader will remember that y is one of the components of prediction equation 10.1, in the denominator. If off-season nursery is available, number of years could be reduced. For example, Eberhart (1969), was able to raise 3 generations in 2 years at Kitale, Kenya. Under such circumstances S_1 progeny test, which requires 3 generations per cycle, may be more efficient than mass selection, especially if mass selection is not able to efficiently utilize σ^2_A due to the problems of field management. Because environmental conditions of regular season and off season are usually not similar, mass selection cannot be carried out in off season.

Phenotypic variation (σ^2_{ph}) is also one of the components of equation for predicting genetic progress. Reduction of σ^2_{ph} can lead to increased gains. Phenotypic variation can be reduced via better field plot techniques. Crop rotation, use of proper machinery to prepare good seedbed, uniform plant stand per plot, use of higher rates of fertilizer, appropriate experimental design, are some means to reduce the estimate of σ^2_{ph}.

Selection intensity (the proportion of the individuals chosen from the entire population) is related to *k* value. For example, when 20 per cent of the individuals are selected from the population the value of *k* is 1.40, but when the intensity of selection is 1 per cent *k* is 2.64. Thus with stringent selection genetic progress could be enhanced. But too much reduction in the number of individuals chosen will lead to reduced population size and hence to genetic drift. Reader will recall that change of gene frequency due to sampling in small population is genetic drift. Minimum number of lines that should be retained is not very well known. Information on this aspect is scanty.

10.4 POPULATION IMPROVEMENT IN SORGHUM

Sorghum is often-cross pollinated crop, and usually the breeding methods advocated for self-pollinated crops are recommended and used for genetic improvement of this crop. However, after the reports came in that sorghum shows considerable heterosis and availability of cytoplasmic genetic male-sterile lines, hybrid varieties were developed initially in United States of America and subsequently in India and other parts of the world. Again like maize, for development of superior hybrids, the base population should possess superior genes and gene combinations in high frequency.

Traditional breeding of sorghum and even other naturally self-pollinated crops, generally involved crossing of two elite varieties followed by selection in F_2 and subsequent generations and fixation of desirable genes in homozygous lines via selfing. After a variety is developed and stabilized for desirable characters, it is again crossed with other variety possessing genes for additional desirable attributes and selection continued. This stepwise improvement programme is not only slow but also limits the opportunities for desirable recombinations among linked genes due to rapid progress towards homozygosity. The traditional methods of breeding improved varieties have been very much successful in the past. But now superior germplasm needs to be built-up which carries superior genes for desirable traits, excellent than those, which have been utilized in the past. Moreover, improvement of qualitative traits or characters with high heritability is relatively easier, but further improvement of quantitative characters may be difficult unless germplasm is enriched with superior genes and gene combinations. Whether hybrids or varieties are the ultimate goal of the breeding programme the role of improved population is clear. Improved populations will provide basic source material for the development of both, and the level of production of varieties and hybrids will be a function of the level of production and genetic variability in the base population from which they are derived. An important point that needs to be remembered is that with each cycle of improvement of the base population, new varieties and hybrids with even better performance can be developed as shown in Figure 10.1.

According to Doggett (1972) traditional methods used to develop varieties, mainly of self-pollinating crops, have certain weaknesses, viz.,

(i) Because the inheritance of many economically important characters, such as yield, is governed by a large number of genes, with small, similar and cumulative effects, the size of population that needs to be grown in F_2 to select at least one superior combination, is prohibitively large. This is hardly executed.

(ii) The range of the parents from the entire germplasm that breeders of self-pollinated crops use is very small. Crossing is mostly confined to a few elite varieties of proven merit. In a crop like sorghum where several thousand varieties are available all over the world, only a tiny fraction of this is involved in the improvement programme. Several plus genes with desirable effects on quantitative traits are rarely infused in crossing programme.

(iii) Chances of crossing overs, and hence recombinations are very much limited in mainly self-pollinating crops. This is even more true when linkages are tight. Because opportunities for recombination are restricted, advantage of new combinations can be very limited.

Population improvement provides opportunities to utilize large number of varieties in random mating programme. Because hybridization continues with each cycle of selection, opportunities are more for breakage of linkages and formation of new combinations. Fixation of the type is delayed so long recurrent selection continues.

Maize is a naturally cross-fertilizing plant having separate male and female flower parts, and is ideally suited for random mating. Sorghum, on the other hand, is basically a self-pollinated crop that sometime out-crosses. It is not well adapted for random mating. However, male-sterility provides a favourable mechanism for out-crossing. After maize, sorghum is a crop, in which population improvement programme was initiated as early as 1963.

A reader may be wondering as to why population improvement in sorghum has been described in such details in this book. Sorghum is mostly a self-pollinated crop but with the use of genetic male-sterility various schemes of population improvement, proposed for maize, have to be utilized in sorghum with some modifications. This is possible in other self-pollinated crops also. Hence the population development and subsequently use of various schemes to improve the germplasm of sorghum have been described in some details.

Doggett and Jowett (1963, 1964) used cytoplasmic male-sterility to develop populations of sorghum and subsequently applied recurrent selection. Gilmore (1964) suggested methods of utilizing both cytoplasmic and genetic male-sterility for this purpose. According to Ross *et al.* (1971) first planned random mating population was synthesized by Webster in Nebraska. Webster (1965) carried out three cycles of selection in bulk population of sorghum using cytoplasmic-sterility and further set-up bulk population using Coes (ms_3) genetic male-sterility. Harvesting of only male-sterile heads ensures that all the seed from these heads is cross-pollinated. In 1965, Gardner and Nordquist initiated a programme to generate random mating populations of sorghum for quantitative genetic and recurrent selection studies.

There are several genetic male-steriles in sorghum, such as those possessing genes *al*, ms_1 (from India), ms_2, ms_3 Coes (from Nebraska, USA), ms_4 (from India), ms_5, ms_6, ms_7, Day etc. (Ross *et al.*, 1971), which for the most part are simply inherited and not affected by cytoplasm. Genetic male-sterile Coes, possessing ms_3ms_3 genes for male-sterility, has been used to develop sorghum populations. Two attributes make sterile line Coes more desirable as a source of male-sterility, viz.,

(i) Male-sterility is governed by a single pair of recessive genes, ms_3 ms_3 and hence the ratio of fertile to sterile plants (Ms_3 —: ms_3ms_3) in F_2 generation is 3: 1, with no modifying factors, and
(ii) Because the genetic male-sterility (ms_3) is not affected by the type of cytoplasm, it can be used to develop a population of maintainers (B lines) and restorers (R lines)

For developing a random mating population of sorghum, select superior lines. Suppose we want to develop a population of maintainers (i.e. those which maintain the sterility of cytoplasmic-genetic male-sterile line). Cross each of the selected *B* lines to Coes.

Two to three backcrosses are usually sufficient. There could be two approaches to develop a population. The F_1 seed of backcrosses is mixed together in equal amounts and F_2 generation grown out. In the F_2, male-steriles (which will be about 25 per cent) are tagged and harvested. The seed set on male-sterile will be due to random mating or out-crossing. The plants are grown at about 60,000 plants per hectare density. This process is continued for three generations. Alternatively, all possible single-crosses are made among the converted lines. Equal quantities of F_2 seed of diallel crosses is bulked and allowed to random mate. For hand made diallel between component lines male-sterile segregate in one line is pollinated by the pollen from fertile plant of another line. Only tagged, male-sterile heads, which have been pollinated, are harvested. It is important that male-sterile plants are tagged as soon as they emerge from boot leaf. It is very difficult to distinguish between a fertile head and a male-sterile head (if it has been adequately pollinated) at maturity because both have nearly complete seed set.

According to Ross (1973) one round of random mating may accomplish considerable mixing of germplasm. Nevertheless, when the primary factor of interest is yield, Ross (1973) prefers to have 3 rounds of random mating before starting the studies. A question may be asked, and is frequently asked, whether field random mating is really random? Actually it is not truly random mating. For instance, the plants in the proximity of a sterile plant have greater chance to pollinate it. Moreover, it will depend upon synchrony in flowering among fertile segregates and sterile segregates and the direction of wind current. Tagging the male-sterile plants, as soon as they are identified, and hand pollination with pollen collected randomly from several fertile plants may lead to considerable random mating among individuals.

Ross (1973) is of the view that about 1000 feet isolation for random mating population should be ensured. In order to minimize the genetic drift, population size for random mating should be adequately large. The population may be grown on at least one fourth of an acre.

10.4.1 Breeding Systems for Population Improvement

In sorghum, several population improvement schemes, suggested for maize have been utilized. These will be described briefly as used in sorghum. It will be clear to the reader from table 10.5 that hand pollination is required only in case of full-sib family testing and selection and reciprocal full-sib family testing and selection.

(a) Mass Selection

Readers know well that this is the oldest method of plant improvement used for self as well as cross-pollinated crops. This method is useful to sorghum for screening large number of plants for disease resistance, insect resistance, tolerance to a mineral element, grain quality attributes like protein and high yield. Gardner (1970) has shown that even in maize this method has been associated with large gains in yield as well as for other traits.

Mass selection in random mating population of sorghum involves selection of male-sterile plants before anthesis. These plants are tagged. Seed setting on these plants results from out-crossing. Thus one generation per cycle is required. But if male-fertile plant (*Ms ms*) is selected two generations per cycle will be required. Seeds of these plants shall be sown and male-sterile plants identified and tagged. Seeds from tagged male-sterile plants should be harvested.

Because linkages in sorghum appear to be closer, Ross *et al.* (1971) suggested treatment of seeds with mutagens like X-rays, gamma rays, fast neutrons or chemical mutagens. This may induce more recombinations and consequently lead to increased genetic variability.

Table 10.5: Population breeding methods in sorghum*

System of selection	*Generations per cycle*	*Hand pollination*	*Male identity*
Intra-population			
Mass selection	1 or 2	No	No
S_1 progeny testing and selection	3	No	No
Half-sib progeny test and selection	3	No	No
Full-sib progeny test and selection	3	Yes	No
Inter-population			
Reciprocal full-sib family testing and selection	3	Yes	Yes

*Ross, W.M., C.O. Gardner and P.T. Nordquist (1971)

(b) S_1 Progeny Testing

S_1 is simply maize breeders terminology for the first selfed generation. In sorghum it is the fertile plant in the random mating population. The S_1 progeny is then comparable to an F_2 because all fertile plants in random mating population are F_1s. The S_1 progenies are evaluated for yield and remnant S_1 seed of selected plants is grown in isolation for random mating. The experimental procedure actually involves blending of remnant seed of selected S_1s. Tag the sterile plants only and harvest seed from sterile plants. This will ensure considerable random mating among selected S_1s.

S_2 progenies could also be developed and evaluated. S_2 progeny selection and random mating utilizes 1.5 times the additive variance, existing in the population as compared to other systems (Eberhart, 1972). Nevertheless, this method of selection requires 4 generations per cycle.

(c) *Half-Sib Family Testing*

In random mating population of sorghum half-sib families are obtained very simply by allowing the male-sterile plants to be randomly pollinated. In order to avoid pollination of male-sterile plants by fertile plants in proximity, pollen is collected from random fertile plants and the male-sterile plants are hand pollinated which were tagged at flowering. Since several fertile plants pollinate each male-sterile, the progeny of a common male-sterile is half-sib. In this way several half-sibs are developed. According to Ross (1973) succeeding generation should be propagated from not less than 200 half-sibs (or S_1s) and usually the number should be twice than this.

The half-sib progeny families are evaluated and on the basis of performance data remnant seed of the half-sibs is used for recombination in isolation. This selection scheme also requires three generations per cycle.

(d) *Full-Sib Family Testing*

Making of full-sib family requires hand pollinations. Male-sterile plants are identified and bagged. If the tip of panicle has started blooming, the tip is cut off and bagged. When the stigmas are receptive each bagged male-sterile plant is pollinated by a selected fertile plant. One male-sterile plant should be pollinated by only one fertile plant. Pollinated plants are tagged but there is no need to maintain identity of fertile male parent because the F_1 seed (remnant) of superior full-sibs families shall be used for intermating.

Based on the performance of full-sib families, say for yield, remnant F_1 seed of superior families is composited and seed is sown in isolation to accomplish random mating. Full-sib family test and selection is very efficient in maize like S_1 selection (Sprague, 1966) and good gains are expected in sorghum. This system of selection requires three generations per cycle.

(e) Reciprocal Full-Sib Family Testing

Like maize, in sorghum also two genetically diverse populations are required for this selection scheme. Since cytoplasmic-genetic male-sterile lines are used as seed parents for the sorghum hybrids one of the two populations should be developed with *B* lines. Another population is developed with *R* lines. Remember, both the populations should be heterozygous for male-sterility gene i.e., Ms_3 ms_3. Populations *B* and *R* are sown in isolation. From population *B* a few fertile plants are chosen. Each plant from population *B* is used to pollinate a sterile plant from population *R*. Because one plant of population *B* is crossed to one plant from population *R*, the progeny of this mating is full-sib. In this way several full-sib matings should be made. Identity of male parent is maintained because remnant seed of superior males is used for intermating. In similar fashion a number of superior fertile plants from population *R* are selected and each fertile plant is used to pollinate only one male-sterile plant from population *B*. In this way several full-sib matings are accomplished, both in population *B* and *R*. In the next season, full-sib families from both the populations are evaluated in suitable experimental design at several locations. Even a single row plot will suffice. On the basis of evaluation trial superior full-sib families are identified. Remnant seed of superior male parents (fertile plants) is bulked in equal amounts and planted in isolation for random mating. Remember that composited seed of superior male-parents from population *B* and population *R* is to be sown in two separate and isolated fields for intermating. Three generations per cycle are required. This recurrent selection method will lead to evolution of elite *B* population and elite *R* population. Desirable plants from population *B* are selected and converted into male-sterile lines. *A* hybrid can be developed by crossing elite male-sterile with superior *R* line.

While developing populations, by using male-steriles, one has to be very cautious. A population developed by using a cytoplasmic-genetic male-sterile may succumb to a disease or an insect as observed for southern corn blight of maize and downy mildew of pearl millet. For example, sorghum populations use very extensively Coes male-sterile. If this male-sterile happens to be susceptible to an insect or a disease all the populations carrying ms_3 from Coes shall have a serious set back. A possible way to overcome this potential danger is to make hand-emasculation and pollination, which may be laborious.

Ross *et al.* (1971) have extended a valuable suggestion in connection with population improvement. "It should be stressed that population breeding is not terminal, but is only a means to an end. Inbreeding to produce homozygous lines, backcrossing male-sterility on to *B* lines and making of F_1 hybrids must be done".

Doggett (1972) developed eight separate populations of sorghum from a total of 158 varieties using genetic male-sterile gene ms_3, and composited for at least three generations before applying selection. Four selection schemes were applied to improve grain yield. Although selection was effective under all the four schemes, S_1 testing was most effective. Selections from populations out yielded the control, Serena. Effective progress was made from half-sib family testing, full-sib family testing and S_1 progeny testing in one cycle using population *NP3R*, by Ross (1973). However, S_1 progeny testing appeared more practical amongst the three selection schemes. Jinda *et al.* (1976) studied the effectiveness of selection schemes in the random mating grain sorghum population *NP3R*. This population was developed by Nordquist *et al.* (1973), which originated from elite *R* lines. After incorporating genetic male-sterile gene ms_3 into each of the 30 *R* lines, the 30 lines were crossed in a diallel fashion and the resulting F_1s were composited and subjected to 3 generations of random mating. Full-sib, half-sib and S_1 progenies were developed. A total of 196 families of each type were developed and evaluated. Jinda *et al.* (1976) also observed that S_1 family testing and selection was most promising selection scheme, particularly for improvement of yield and yield components. Results of population improvement in *NP3R* population of sorghum were summarized by Ross (1978). Significant increase in grain yield was observed in C_1 as compared to C_0 (base population) of full-sib, half-sib and S_1 family testing and selection (Table 10.6). However, S_1 progeny testing appeared more promising amongst the three schemes.

Table 10.6: Grain yield of sorghum population *NP* 3*R*. in 3 selections schemes*

Cycle	*Source*	*Grain yield (q/ha)*	*% of base*
C_0	Base	40.5	100.0
C_1	Half-sib	49.2	121.0
C_2	Half-sib	46.0	114.0
C_1	Full-sib	48.0	119.0
C_2	Full-sib	44.4	110.0
C_1	S_1	50.9	126.0
C_2	S_1	47.4	117.0
Hybrid check			
RS 626	–	47.2	
RS 671	–	51.6	
Entry *LSD* 5%	–	4.1	

* Ross 1978.

10.5 SYNTHETIC VARIETIES

Hybrid seeds are comparatively costlier than seeds of varieties. Moreover, every year farmers have to purchase seed of hybrid. Like varieties farmers can not retain part of the produce of hybrid crop as seed for next season. As pointed out earlier the objective of most of the breeders, engaged in improvement of cross-pollinated crops, is controlled utilization of heterosis, which is noticed in hybrids of certain genotypes. Although there are many crops, which do not possess the fortunate reproductive system, which enables easy selfing and crossing like maize, they do have male-sterility or self-incompatibility mechanism, which enables exploitation of the phenomenon of heterosis. Nevertheless, there are many crops in which production of F_1 seed on commercial scale every year is not possible. When this is true, synthetic varieties could be developed which can partly utilize the heterosis observed in certain crosses of crop. Hays and Garber (1919) were the first to advocate commercial utilization of synthetic varieties. Allard (1960) defined synthetic variety as "a variety that is maintained from open-pollinated seed, following its synthesis by hybridization in all combinations among a number of selected genotypes". Utilizing inbred lines, clones, mass selected populations or various other materials synthetic variety could be produced. The components of synthetics, e.g., inbred lines, clones, etc., are termed *Syn* 0. The F_1 generations resulting from random mating of component parts are designated as *Syn* 1 and F_2 as *Syn* 2. A synthetic variety is used commercially for a number of generations, usually *Syn* 2, *Syn* 3 and *Syn* 4. The component inbred lines or clones of the synthetic variety are maintained and synthetic varieties are reconstituted at regular intervals. The synthetic varieties are heterogeneous and may be comprised of homozygous and heterozygous individuals.

Reader will recall that a large random mating population (in case of cross-pollinated crops) attains equilibrium, in the absence of non-random mating, selection, migration and mutation. Hence, it is expected that the synthetic variety developed by random mating of *n* inbred lines should attain Hardy-Weinberg equilibrium. An important property of this equilibrium is that gene frequencies and consequently genotype frequencies do not change from generation to generation. Thus the mean and variability of the population do not change. According to Sewall Wright (1922) the expected performance of F_2 (i.e.*Syn* 2) of synthetic can be predicted as:

$$\hat{F}_2 = \bar{F}_1 - \frac{(\bar{F}_1 - \bar{P})}{n} \qquad \ldots(10.2)$$

where $\hat{F}_2$ is the estimated performance of F_2 (*Syn* 2) generation, $\bar{F}_1$ is the average performance of all possible single-crosses among the *n* inbred lines, $\bar{P}$ is average performance of all parental lines, and *n* is number of parental lines involved. This formula suggests that yield of F_2 and subsequent generations shall be $1/n$ times less than that portion of yield of F_1 that is attributed to heterosis i.e. $\left(\bar{F}_1 - P_1\right)$. The decline in yield of F_2 and subsequent generations as compared to F_1 could be accounted for departure from random mating that may be expected in synthetic. Departure from random mating could result due to differences in flowering behaviour of plants. Mating may take place between only those plants which show synchronous flowering. This assortive mating may lead to change of gene and genotype frequencies. Further, natural selection can also exert selection pressure. For example, a disease may cause preferential elimination of particular genotypes and exert a force similar to selection pressure. Inadequate size of population during multiplication of seed can also lead to genetic drift.

When such decline in performance is noticed, the synthetic variety can be synthesized afresh from selected components.

According to Allard (1960) performance of advanced generations of synthetic varieties depends upon three factors, *viz.*,

(i) The number of parental lines included,
(ii) The mean performance of these parental lines, and
(iii) The mean performance of all possible F_1s among the *n* lines.

The results of experiments by Neal (1935) showed decline in the performance of F_2 of a single-cross, three-way and double-cross, which was as per the expectations based on Wright's formula. Nevertheless, the yield of F_3 was same as F_2. Improving the yield of synthetic by increasing number of component lines appears to be the easier way. But it is difficult to find out additional lines, which will nick well with those which are already the components, in order to maintain high F_1 yields. Increasing the yield of synthetics is feasible via increasing the yielding ability of component lines. Once again population improvement involving recycling of selection and intermating among the selects shall enable development of high yielding inbreds.

Synthetic variety of pearl millet was developed at *ICRISAT*, Hyderabad (Govila, 1989). The parental materials employed to develop synthetic were crosses between Indian and African land races and elite inbreds. Several progenies were produced, which were screened in F_4 generation in downy mildew nursery and subjected to combining ability evaluation. Superior general combiners were intermated in isolation to generate a synthetic variety. Synthetic variety of pearl millet ICMS 7703 was developed and released in 1984. An experimental synthetic population of rapeseed (*Brassica campestris*), Syn 65, was developed at Hissar and released as 'Sangam' in 1974 (Rai, 1989). Likewise, 'Pusa Kalyani', a variety developed by synthetic breeding approach, of lotni brown sarson, was evolved. Singh (1989) quoted evolution of synthetic varieties of cotton in U.S.A. Synthetic varieties have been produced in cabbage (Pusa Synthetic) and cauliflower (Pusa early synthetic). Inbred lines are evaluated for their general combining ability through diallel-cross or top-cross and then superior inbred lines are intermated (Swarup, 1977). Synthetic varieties of carrot, radish, turnip, knolkhol can also be produced.

10.6 COMPOSITES

Beal in 1880 attempted to utilize varietal crosses of maize to increase yield. But because the magnitude of heterosis was much less varietal hybrids were abandoned. The first commercial use of advanced generations of varietal crosses of maize was made in India in 1966.

Dhawan (1963) outlined a population improvement methodology for developing commercial varieties called as "composites". The composites are advanced generations of varictal crosses. Component varieties are not evaluated for their general combining ability. Because random mating is expected in cross-pollinated crops, advanced generations of varietal crosses attain Hardy-Weinberg equilibrium, in the absence of agencies tending to change gene frequencies. Hence, after a few generations of random mating the mean and variance should not change. The initial synthesis of composites in India was done from high yielding inter-varietal crosses, which showed minimal depression in F_2 and subsequent generations. Random mated population of such selected crosses is a composite.

Experimental results of Mukherjee and Saha (1978) suggested that yield could be upgraded in maize population through population × population crosses (Table 10.7). Population × population crosses outyielded population × variety and variety × variety crosses. The magnitude of heterosis was also larger for population × population crosses. Thus heterosis could be maximized via population × population crosses. Inbreeding depression was also lower for population × population crosses. Composites could be developed from such a material.

Table 10.7: Mean performance of F_1s and advanced generations of maize*

	Grain yield (kg/ha)	*Per cent of best check*	*Mean heterosis*	*Inbreeding depression*
Parents				
Populations (average of 6 populations)	2793	105.4	—	—
Varieties (average of 6 varieties)	1620	54.4	—	—
Population × population	3256	122.9	47.59	59.90
Population × variety	2586	97.6	17.04	56.23
Variety × variety	2524	95.2	14.15	71.03
Advanced Generations				
$P \times P$	1307	49.3		
$P \times V$	1189	44.9		
$V \times V$	731	27.6		

*Mukherjee and Saha (1978)
Source: Mukherjee and Ahuja (1995)

Vikram developed from a cross of Basi and Eto Amarillo, was the first composite released in India in 1966. It yielded as good as double-cross hybrids. Subsequently, Kissan (J_1 × Coastal Tropical Flint), Ambar (B_1 (y), Jawahar (A_1 × Antigua G_1), Sona (J_1 × Cuba 11J), Vijay (A_6) were released in 1967. According to Rai (1995) for a successful composite breeding programme, in addition to high heterosis for yield, information regarding the magnitude of inbreeding depression in F_2 and subsequent generations is important.

Usually those crosses, which manifest larger inbreeding depression in F_2 for seed yield, indicate importance of non-additive gene action for yield. On the other hand, those F_2s, which show relatively less inbreeding depression suggest predominant role of additive genetic variance as a cause of variability (Jinks and Jones, 1958). Rai and Varshney (1963) are of the opinion that cross-combinations showing higher magnitude of heterosis for yield, low inbreeding depression in F_2 and subsequent generations with significantly high general mean than check would lead to development of superior composite.

Like the synthetics, composites are also heterogeneous populations made up of homozygotes and heterozygotes of various kinds. Provided large populations are grown in isolation, composites could be multiplied without much decline in yield. But their reconstruction is not possible. If composites are grown on larger area, in order to avoid random drift, farmers can retain part of the seed of the produce for planting in next season. At least 2000-3000 ears from central portion of field of maize may be retained for planting.

In maize, generally half-sib, full-sib, S_1 and bi-parental selection procedures were employed to develop populations. It was observed that highest yielding composites were obtained through S_1 selection followed by full-sib progeny selection (Mukherjee *et al.*, 1980). Using these methods a number of promising composites have been developed, e.g. Pusa Kundan, *EVA* 61-80, etc.

The objective of producing composites is to develop better source population and simultaneously produce commercial varieties.

Composites of toria (*Brassica campestris*) could be bred from advanced generations of inter-varietal hybrids of genetically diverse parents and/or their blends (Rai, 1989). A composite Ludhiana composite 1 was developed by following simple recurrent selection. It has been released in Punjab.

First successful attempt to develop composite varieties of pearl millet was made by *ICRISAT*, Hyderabad through recurrent selection

(Andrews *et al.*, 1985). The world composite random mating population was developed in Nigeria. It was constituted from derivatives of numerous crosses between diverse sources of pearl millet germplasm and Nigerian early maturing land races locally known as 'Gero millets'. Full-sib recurrent selection was employed to improve population. Composite WC-C-75 was developed from this material (Govila, 1989).

Because synthetic varieties and composites are heterogeneous, they are relatively phenotypically more stable in performance than varieties. This could be attributed to populational buffering.

Quality and uniformity of commercial produce is one of the desirable attributes of vegetables. If synthetic varieties or composites are developed in vegetables, a breeder should take special care that these are in no way inferior in quality and uniformity than the prevailing and predominant varieties.

References

Allard, R.W. (1960). *Principles of plant breeding*. John Wiley and Sons, Inc., NewYork.

Andrews, D.J.; S.C. Gupta and P. Singh (1985). Registration of WC-C-75 pearl millet. *Crop Sci.*, 25: 199-200.

Arunachalam, V. (1989). Genetic basis of plant breeding. In *Plant Breeding — Theory and Practice*. Ed. V.L. Chopra. Oxford and IBH Publishing Co.Pvt.Ltd., New Delhi, pp. 1-18.

Brim, C.A. and C.W. Stuber (1973). Application of genetic male-sterility in recurrent selection scheme in soybean. *Crop Sci.*, 13: 528-530.

Choo, T.M. and L.W. Kannenberg (1979). Relative efficiencies of population improvement methods in corn: A simulation study. *Crop Sci.*, 19: 179-185.

Compton, W.A. and R.E. Comstock (1976). More on modified ear-to-row selection. *Crop Sci.*, 16: 122.

Comstock, R.E.; H.F. Robinson and P.H. Harvey (1949). A breeding procedure designed to make maximum use of both general and specific combining ability. *Agron.J.*, 41: 360-367.

Devey, M.E. and W.A. Russell (1983). Evaluation of recurrent selection for stalk quality in a maize cultivar and effects of other agronomic traits. *Iowa State J.Res.* 58: 207-219.

Dhawan, N.L. (1963). Concepts, methodology and achievements in the genetic improvement of maize. Progress Report Coordinated Maize Breeding Scheme, ICAR, New Delhi.

Doggett, H. (1972). Recurrent selection in sorghum. *Heredity*, 28: 9-29.

Doggett, H. and D. Jowett (1963). Record of Research. *Ann.Rep.E.Afr.Agr. For Res.Org.*, 33-36.

Doggett, H. and D. Jowett (1964). Record of Research. *Ann.Rep.E.Afr.Agr. For Res.Org.*, 89-91.

Doggett, H. and S.A. Eberhart (1968). Recurrent selection in sorghum. *Crop Sci.*, 8: 119-121.

Dudley, J.W. (1977). Seventysix generations of selection for oil and protein percentage in maize, pp. 459-473. In "Proc. Int.Conf. Quant. Genet" Iowa State Uni., Ames.Ed.E.Pollak *et al.*, 16-21 Aug., 1976. Iowa State University Press, Ames.

Eberhart, S.A. (1969). Efficient plant breeding with recurrent selection. Genetics and plant improvement symposium of the XI International Botanical Congress. Seattle, Wagh.

Eberhart, S.A. (1972). Techniques and methods for more efficient population improvement in sorghum. In *Sorghum in seventies*, Eds. N.G.P. Rao and L.R. House, Oxford and IBH Publishing Co.Pvt.Ltd., New Delhi, pp. 197-213.

Empig, L.T.; C.O. Gardner and W.A. Compton (1972). Theoretical gains for different population improvement procedures. MP 26 (Revised) University of Nebraska, College of Agriculture, The Agricultural Experiment Station.

Gardner, C.O. (1961). An evaluation of effects of mass selection and seed irradiation with thermal neutrons on yield of corn. *Crop Sci.*, 1: 241-245.

Gardner, C.O. (1970). The role of mass selection and mutagenic treatment in modern corn breeding. In "Proc.20th Ann.Corn-Sorghum Res.Conf., Chicago, Ill. (1969), pp. 15-21.

Gardner, C.O. (1977). Quantitative genetic studies and population improvement in maize and sorghum pp.475-489. In "Proc. Int.Conf.Quant.Gent. Univ. of Iowa, Ames, 16-21 Aug., 1976. Iowa State University Press, Ames.

Gardner, C.O. (1978). Population improvement in maize. In *Maize breeding and genetics*. Ed. D.B. Walden, John Wiley and Sons, New York, pp. 207-228.

Gardner, C.O.; P.T. Nordquist and W.M. Ross (1971). New breeding system for sorghum improvement. *Nebr.Quar.*13: 4-5.

Gilmore, E.C. (1964). Suggested method of using reciprocal recurrent selection in some naturally self-pollinated species. *Crop Sci.*,4: 323-325.

Govila, O.P. (1989). Cross pollinated crops. Pearl millet. In *Plant breeding — theory and practice*. Ed. V.L. Chopra, Oxford and IBH Publishing Co.Pvt.Ltd., New Delhi, pp. 213-223.

Hallauer, A.R. (1967). Development of single-cross hybrids from two eared maize populations. *Crop Sci.*, 7: 192-195.

Hallauer, A.R. (1973). Hybrid development and population improvement in maize by reciprocal full-sib selection. *Egypt J.Genet. Cytol.*, 2: 84-101.

Hallauer, A.R. and J.B. Miranda (1981). *Quantitative genetics in maize breeding* Iowa State Uni. Press, Ames.

Hallauer, A.R. and S.A. Eberhart (1970). Reciprocal full-sib selection. *Crop Sci.*, 10: 315-336.

Hallauer, A.R.; W.A. Russcl and K.R. Lamkcy (1988). Corn breeding. In *Corn and corn improvement*, — Agronomy monograph No.18, 3rd Edition. ASA, CSSA-SSSA, 677, South Segoe Road, Madison, WI 53711, USA, pp. 463-564.

Hays, H.K. and R.J. Garber (1919). Synthetic production of high protein corn in relation to breeding. *J.Am.Soc.Agron.*, 11: 309-319.

Hopkins, C.G. (1899). Improvement in the chemical composition of corn kernel. *Agr.Exp.Sta.Bull.* 55: 205-240.

Hull, F.H. (1945). Recurrent selection and specific combining ability in corn. *J.Am.Soc.Agron.*, 37: 134-145.

Jenkins, M.K. (1935). The effect of inbreeding and of selection within inbred lines of maize upon hybrids made after successive generations of selfing. *Iowa State College J.Sci.*, 9: 429-450.

Jenkins, M.K. (1940). The segregation of genes affecting yield of grain in maize. *J.Am.Soc.Agron.*, 32: 55-63.

Jinda, Jan-orn, C.O. Gardner and W.M. Ross (1976). Genetic studies of the NP3R random mating grain sorghum populations. *Crop Sci.*, 16: 489-496.

Jinks, J.L. and R.M. Jones (1958). Estimation of components of heterosis. *Genetics*, 43: 223-234.

Jones, L.P.; W.A. Compton and C.O. Gardner (1971). Comparison of full-and half-sib reciprocal recurrent selection. *Theor.Appl.Genet.*, 41: 36-39.

Kojima, K. and T.M. Kelleher (1963). A comparison of pure bred and cross-bred selection schemes with two populations of *D. pseudoobscura*. *Genetics*, 48: 57-72.

Lantin, M.M. and A.R. Hallauer (1981). Response to reciprocal full-sib and mass selection in corn (*Zea mays*, L.) *Proc.Iowa Acad.Sci.*, 88 (4): 172-178.

Lonnquist, J.H. (1964). A modification of ear-to-row procedure for the improvement of maize populations. *Crop Sci.*, 4: 227-228.

Lonnquist, J.H. and N.E. Williams (1967). Development of maize hybrids through selection among full-sib families. *Crop Sci.*, 7: 369-370.

Mareck, J.H. and C.O. Gardner (1979). Responses to mass selection in maize and stability of resulting populations. *Crop Sci.*, 19: 779-783.

Martin, J.M. and A.R. Hallauer (1980). Seven cycles of reciprocal recurrent selection in BSSS and BSCB1 maize populations. *Crop Sci.*,20: 599-603.

Mather, K. (1943). Polygenic inheritance and natural selection. *Biol.Rev. Camb.Phil.Soc.*, 18: 32-64.

Moll, R.H. and W.D. Hanson (1984). Comparison of effects of intra-population vs. inter-population selection in maize. *Crop Sci.*, 24: 1047-1052.

Moll, R.H. and C.W. Stuber (1971). Comparison of alternative selection procedures initiated with two populations of maize (*Zea mays* L.). *Crop Sci.*, 11: 706-711.

Moreno-Gonzalez, J. and A.R. Hallauer (1982). Combined S_2 and cross- bred family selection in full-sib reciprocal recurrent selection. *Theor.Appl.Genet.*, 61: 353-358.

Mukherjee, B.K. (1989a). Concepts and methods: Breeding procedures for cross-pollinated crops. In *Plant Breeding — Theory and Practice*. Ed. V.L. Chopra, Oxford and IBH Publishing Co.Pvt.Ltd., New Delhi, pp. 173-195.

Mukherjee, B.K. (1989b). Cross-pollinated crops: maize In *Plant breeding — Theory and Practice*. Ed. V.L. Chopra. Oxford and IBH Publishing Co.Pvt.Ltd., New Delhi, pp. 199-212.

Mukherjee, B.K. and B.C. Saha (1978). Investigations on the nature of heterosis expressed in crosses of varieties and synthesized populations of maize. *Genetika*, 10: 277-285.

Mukherjee, B.K.; R.D. Singh; K.N. Agarwal and V.P. Ahuja (1980). Relative performance of maize population synthesized from various types of progenies. *Indian J.Agric.Sci.*, 50: 462-465.

Mukharjee, B.K. and V.P. Ahuja (1995). Upgrading yield levels of maize hybrids in India through conventional and non-conventional approaches. In *Genetic Research and Education — Current trends and the next fifty years*. Eds. B. Sharma *et al.*, Indian Society of Genetics and Plant Breeding, New Delhi, pp. 674-686.

Neal, N.P. (1935). The decrease in yielding capacity in advanced generations of hybrid corn. *Jour.Amer.Soc.Agron.* 27: 666-670.

Nordquist, P.T.; O.J.Webster; C.O. Gardner and W.M. Ross (1973). Registration of three sorghum germplasm random mating populations. *Crop. Sci.*, 13: 132.

Pandey, S.; A.O. Diallo; T.M.T., Islam and J. Deutsch (1986). Progress from selection in eight tropical maize populations using international testing. *Crop.Sci.*, 26: 879-884.

Penny, L.H.; G.H. Scott and W.D. Guthrie (1967). Recurrent selection for European corn borer resistance in maize. *Crop Sci.*, 7: 407-409.

Rachie, K.O. and C.O. Gardner (1978). Increasing efficiency of breeding partially out crossing grain legumes. In *Proceedings International Workshop on Grain Legumes*, ICRISAT, Hyderabad, 13-16 Jan., 1975, pp. 285-300.

Rai, B. (1989). Brassicas. In *Plant breeding — theory and practice*, Ed. V.L. Chopra, Oxford and IBH Publishing Co., Pvt.Ltd., New Delhi, pp. 159-170.

Rai, B. (1995). Extent of inbreeding depression and genetic basis for developing commercial composite varieties in Indian Rapeseed (*Brassica campestris* L.). In *Genetic Research and Education: Current trends and the next fifty years*, Eds. B. Sharma *et al.*, Indian Society of Genetics and Plant Breeding, New Delhi, pp. 902-906.

Rai, B. and S.K. Varshney (1983). Genetic basis of breeding better varieties of Indian rapeseed. Abstr. *International Genet.Congress*, New Delhi Part II, 593.

Robertson, A. (1960). A theory of limits in artificial selection. Proc.*Roy Soc.B.* 153: 234-249.

Ross, W.M. (1973). Use of population breeding in sorghum. Problems and progress. *Proc.Ann.Corn and Sorghum Ind.Res.Conf.Amer. Seed Trade Assn.* 28: 30-43.

Ross, W.M: (1978). Population breeding in sorghum. Phase II, 33rd *Annual Corn and Sorghum Research Conference*, pp. 153-166.

Ross, W.M.; C.O. Gardner and P.T. Nordquist (1971). Population breeding in sorghum. In "Grain sorghum producers Association, Grain Sorghum Research and Utilization Conference", 7th Biennial Program, March 2-4, 1971, Lubbock, Texas, pp. 93-98.

Singh, M.A.; A.S. Khera and B.S. Dhillon (1986). Direct and indirect correlated responses to recurrent full-sib selection for prolificacy in maize. *Crop Sci.*, 26: 275-278.

Sprague, G.F. (1955). Corn breeding. In *Corn and corn improvement.* Ed. Sprague, G.F. Academic Press, New York, pp. 221-292.

Sprague, G.F. (1966). Quantitative genetics in plant breeding. In *Plant breeding*, Ed. K.J.Frey, Iowa State University Press, Ames, Iowa, USA, pp. 315-354.

Shull, G.H. (1909). A pure line method of corn breeding. *Amer.Breed. Assoc. Rept.*, 5: 51-59.

Swarup, V. (1977). *Breeding procedures for cross-pollinated vegetable crops*. Indian Council of Agricultural Research, New Delhi, pp. 73.

Webster, O.J. (1965). Genetic studies in *Sorghum vulgare* (Pers.). *Crop Sci.*, 5: 207-210.

Wright, S. (1922). The effects of inbreeding and cross breeding on guinea pigs. *U.S.Dept.Agric.Bull.*, 1121.

11

Inbreeding Depression and Hybrid Vigour

Among the sexually reproducing crop species two mating systems exist, as was described in chapter 2, and the reader is also aware about this. It was shown in chapter 5 that by virtue of mating system the genetic composition of self-pollinated and cross-pollinated crops is different. Self-pollination leads to reduction of heterozygosity by half in each generation and ultimately leads to a population, which is heterogeneous, made up of different types of homozygous individuals. If during self-pollination selection is exercised, favouring a certain homozygous genotype, self-pollination will end up with a homozygous and homogeneous population. Random mating operating in cross-fertilizing crop species does not allow fixation of any genotype, but attains an equilibrium in which all the genotypes, that are expected are present, in the absence of migration, mutation, selection. Of course, the population is assumed to be larger.

It was also mentioned in chapter 5 that some F_1s between lines of self-pollinated crops are fully vigorous and may even display heterosis. What happens when cross-pollinated crops are self-fertilized? Studies on the effects of inbreeding (self-pollination is an extreme form of inbreeding while sib-mating is a milder form) of cross-pollinated crops revealed that it leads to plants with loss of vigour, productivity and so many deformities. However, as inbreeding advances, and if a species is able to tolerate inbreeding, individuals within a line become more uniform and resemble each other. The between line variability enhances. But when the inbred lines are crossed the F_1 between inbred lines, if they have 'nicked', appears much better than either parental lines. Let us see the effects of inbreeding a cross-pollinated crop species.

11.1 INBREEDING DEPRESSION

Inbreeding means mating together of individuals that are related to each other by ancestry. Of course the degree to which the two individuals are related depends upon the size of population. Thus in a small population the two individuals, even taken randomly, are more closely related to each other than the pairs mating in large random mating population. The essential consequence of two individuals having a common ancestor is that both of them may carry replicates of one of the genes present in their common ancestor and when they hybridize they may pass on these genes to their progeny. Thus, inbreds that is the individuals produced by inbreeding, may carry two genes at each locus that are replicates of genes present in the previous generation or their common ancestor. In this way inbreeding leads to homozygosity at all the loci and the progeny begins to resemble each other and becomes uniform in appearance. But the most striking observed consequence of inbreeding of a cross-pollinated crop species is the reduction in mean phenotypic value of the characters connected with reproductive capacity and physiological efficiency, a phenomenon called as inbreeding depression. Why should a cross-pollinated crop species, when inbred, manifest inbreeding depression? Why should continuous self-pollination or sib-mating an ordinarily cross-pollinated crop exhibit loss of vigour, reduction in productivity, reduced size, yield etc.?

Mutation is a recurrent process. Spontaneous mutations occur quite frequently (commonly accepted rates are 10^{-5}, Falconer, 1960). Mutations are mostly recessive (chapter 17) and deleterious. In self-pollinated crops, recessive deleterious mutations become homozygous soon after their origin and are eliminated promptly. However, in cross-pollinated crops the deleterious recessive mutations are not entirely purged. A considerable frequency of deleterious recessive mutants remain shielded in heterozygous state by their dominant favourable alleles. Upon inbreeding, the deleterious recessive genes become homozygous. Consequently, plants with deleterious effects such as plants that lack chlorophyll, with abnormal flowers, defective seeds, deficient root system, dwarf stature etc., appear. Lethals and sub-vitals also appear. In open-pollinated populations also such defective types, lethals and sub-vital types are seen but their frequency is very low. Further because genes controlling growth, productivity, vigour etc. mutate to recessive state, upon inbreeding homozygous recessive types exhibit adverse effects on growth, vigour, productivity etc. Inbreeding depression is thus reduction in the mean value of vigour, productivity and characters related to physiological efficiency. Inbreeding depression is not by itself a process of degeneration but is a consequence of Mendelian segregation. Magnitude of inbreeding depression, i.e., extent of reduction in mean phenotypic value of a character under consideration following inbreeding compared to open-pollinated population, depends upon the frequency of deleterious recessive genes sheltered in the population, which was inbred.

Several research workers have studied the effects of inbreeding in various cross-pollinated crop species. It may not be possible to quote all the results. A few examples will be presented so that reader gets a glimpse of the consequences of inbreeding in different crops.

Extensive research was carried out to learn about the effects of inbreeding in maize. Apparently the first inbreeding experiments with maize were those of Charles Darwin (1877). Because inbreeding was carried out only for two generations results were not much informative.

East at the Connecticut Agricultural Experiment Station and Shull at the Cold Spring Harbor started studies on effects of self-pollination in maize in 1905. Several others in USA, such as Collins from U.S.D.A., Wallace, Richey, Kyle and Holbert, initiated studies on self-fertilization of maize.

G.H. Shull (1910) and East and Hays (1912) summarized their results on effects of self-fertilization of maize. According to the observations made by them inbreeding of maize led to loss of vigour in all inbred lines. However, some inbred lines were much more vigorous than others, even though they did not differ in the degree of homozygosity. Progeny of every self-fertilized maize plant was inferior in size and productiveness as compared to the progeny of normally cross-bred plant derived from same source. The decline in size and vigour which accompanied inbreeding was greatest in first generation and became less and less with each succeeding generation until a stage reached when there was no more loss of vigour. Some strains were so lacking in vigour that they could not be propagated.

Despite considerable deterioration in maize, following inbreeding, this species is more tolerant of inbreeding. Alfalfa exhibits high frequency of morbid and lethal types upon inbreeding. Stewart (1934) was one of the earliest workers to report reduced vigour and viability among alfalfa progenies derived from self-pollination. There was marked loss of vigour in both vegetative and reproductive capacity in alfalfa after inbreeding (Tysdal *et al.*, 1942). Their results revealed that by seventh or eighth generations only about 26 to 30 per cent of the original vigour remained. Seed yield was reduced markedly.

Many studies have been conducted on the effects of inbreeding in cotton. Kearney (1923) studied the effect of controlled self-fertilization on Pima cotton. He found no harmful effect of continued self-pollination in this variety of cotton.

A high degree of inbreeding depression was reported for Brussels sprouts by Johnson (1966), broccoli (Watts, 1970) and Cabbage. A moderate level of inbreeding depression was noticed for radish, onion and beet. Cucurbits exhibit practically no ill effects of inbreeding. Muskmelon (*Cucumis melo*) is strictly cross-pollinated crop yet it does not suffer from inbreeding due to evolutionary rigors it had to go through (Scott, 1933). Carrot deteriorates drastically upon inbreeding (Allard, 1960).

11.2 QUANTITATIVE BASIS OF INBREEDING DEPRESSION

Let us assume that there are two alleles at a locus under consideration in the population, viz., *A* and *a*. Assume that the frequencies of genotypes *AA*, *Aa* and *aa* are $\frac{1}{4}$, $\frac{1}{2}$, and $\frac{1}{2}$, respectively. After one generation of self-pollination of these individuals the population will comprise of individuals *AA*, *Aa* and *aa* in proportions $\frac{3}{8}$, $\frac{1}{4}$ and $\frac{3}{8}$ respectively. One more generation of self-pollination will result in population comprising of individuals *AA*, *Aa* and *aa* in proportions $\frac{7}{16}$, $\frac{1}{8}$ and $\frac{7}{16}$ respectively. Thus with each generation of inbreeding heterozygosity declines by half as compared to the previous generation. Effects of inbreeding in random mating population can be made quantitative in terms of relative reduction in the frequency of heterozygotes. We can measure the amount of inbreeding by comparing the frequency of heterozygote in a population resulting from inbreeding with frequency of heterozygote in original random mating population. Assume that the frequencies of genotypes *AA*, *Aa* and *aa*, which belong to a random mating population, are p^2, $2pq$ and q^2, respectively. The frequency of *Aa* in original random mating population is denoted by symbol H_0 for instance and that of inbred population by H. The effect of inbreeding is given by expression $H_0 - H/H_0$ and is denoted in population genetics by F (Hartl, 1988). Thus $F = H_0 - H/H_0$. This is called as inbreeding coefficient. In biological terms F measures the fractional reduction in heterozygosity relative to that in random mating population. The frequency of heterozygosity of *Aa* was shown to be

$2pq$. Thus for this hypothetical population under consideration $H_0 = 2pq$, and F takes the form:

$$F = \frac{H_0 - H}{H_0}$$

$$= \frac{H_0}{H_0} - \frac{H}{H_0}$$

$$= 1 - \frac{H}{H_0},$$

Changing the sides, we get

$$\frac{H}{H_0} = 1 - F,$$

$\therefore$ $H = (1 - F)\, H_0$, and since $H_0 = 2pq$

$$H = (1 - F)\, 2pq \qquad \text{...(11.1)}$$

Thus frequency of heterozygote, Aa, reduces by a quantity $(1 - F)$, upon inbreeding.

The frequencies of homozygotes increase upon inbreeding. Let us show this for AA. The frequency of AA in random mating population was p^2, and the frequency of gene A is equal to p or $D + ½\, H$ (see equation 5.1, chapter 5). Thus $\hat{p} = D + ½\, H$. Since H denotes the frequency of heterozygote, after inbreeding the frequency of AA, i.e. D (or p^2), can be got as shown below:

$$\hat{p} = D + ½\, H$$

$\therefore$ $D = p - ½\, H$, where D is genotype frequency of AA. And because after inbreeding $H = 2pq\,(1 - F)$, substituting this expression in $p - ½H$, we get

$$D = p - pq + pqF, \text{ substituting } (1 - p) \text{ for } q$$

$$= p - p\,(1 - p) + pq\, F$$

$$= p - (p - p^2) + pq\, F$$

$$= p^2 + pq\, F \qquad \text{...(11.2)}$$

Similarly the genotype frequency of aa shall be $q^2 + pq\, F$ upon inbreeding a random mating population. The total genotype frequencies

Table 11.1 : Genotype frequencies, values of three genotypes after inbreeding.

Genotypes	*Genotype frequencies*		*Arbitrarily assigned genotypic values*
	Before inbreeding	*After inbreeding*	
AA	p^2	$p^2 + pqF$	$+a$
Aa	$2pq$	$2pq\,(1 - F)$	$+d$
aa	q^2	$q^2 + pq\,F$	$-a$

of homozygotes increase by an amount equal to the proportion by which heterozygote decreases. The genotype frequencies of three genotypes in a hypothetical population after inbreeding are shown in table 11.1. From this table mean of random mating population is obtained by adding the products of genotype frequency and corresponding genotypic value assigned to a genotype, over the three genotypes in the population. Division by the total of frequencies is not tobe done because

$p^2 + 2pq + q^2$ is unity (see chapter 5). Hence mean of initial population is (M_0):

$$p^2a + 2pqd - q^2a$$

On simplification it becomes

$$a\,(p - q) + 2pqd \qquad \text{...(11.3)}$$

In similar way the mean of inbred population can be got by using genotype frequencies given for 3 genotypes after inbreeding. Thus

$$M_F = a\,(p - q) + 2pqd\,(1 - F) \qquad \text{...(11.4)}$$

The mean genotypic value of a population with inbreeding coefficient F is thus:

$$= M_0 - 2pqdF \qquad \text{...(11.5)}$$

Expression 11.5 shows that inbreeding results in reduction of the mean observed in original random mating population. The reduction in mean, i.e. inbreeding depression, is equal to $-2pqdF$. Thus in the absence of dominance, i.e. $d = 0$, there will be no inbreeding depression. Inbreeding depression will also be influenced by gene frequency. Although the inbreeding depression was elucidated in quantitative terms for a single locus with two alleles, it holds true for several loci. With the assumption that effect of genes at various loci add their effects, the mean of a population upon inbreeding (M_F) shall be:

$$M_F = \Sigma a(\bar{p} - \bar{q}) + 2(\Sigma d\,\overline{pq})(1 - F)$$

The population mean of random mating population shall be

$$M_0 = \Sigma a(\bar{p} - \bar{q}) + 2\Sigma d\,\overline{pq}$$

Hence, reduction of mean upon inbreeding shall be

$$-2\,F\,\Sigma\,d\,\overline{pq} \qquad \text{...(11.6)}$$

This expression suggests that inbreeding depression shall be determined by degree of dominance, i.e. *d*, and gene frequencies, in addition to *F*, coefficient of inbreeding. The magnitude of *d* shall be larger if dominance at all the loci is in same direction, i.e. there is unidirectional dominance. Change of mean upon inbreeding shall be larger when $\overline{pq}$ is maximal. Obviously this implies that p = q = 0.5. When we say that a certain character manifests inbreeding depression, it refers to average change of mean value in a number of lines in the population.

Genetic consequences of inbreeding are:

(i) Increase in proportion of homozygotes accompanied by reduction in proportion of heterozygotes.
(ii) Recessive genes, which are sheltered by their favourable dominant genes, become homozygous, i.e. inbreeding uncovers the recessive genes.
(iii) Various homozygous genotypes are fixed.
(iv) Inbreeding depression, and finally
(v) Fitness is reduced (Proportionate contribution of offspring to the next generation is fitness).

Some scientists believe that the departure from breeding system normal for the species leads to loss of vigour. Hence, it is also believed the cross-pollinated populations upon inbreeding exhibit inbreeding depression and are not phenotypically stable. Heterozygotes are assumed to be phenotypically stable. The inbred lines manifest more variability in performance when grown in different locations and/or different years.

Inbreeding depression can be estimated as:

$$\frac{\bar{F}_1 - \bar{F}_2}{\bar{F}_1} \times 100$$

where $\bar{F}_1$ and $\bar{F}_2$ are means of F_1 and F_2 generations, respectively (Liang *et al.*, 1972).

11.3 HYBRID VIGOUR

The hybrid vigour observed when genetically diverse strains of a crop species are hybridized, has attracted the attention of man ever since the breeding of mule in Greece during pre-Christian era. The commercial exploitation of hybrid vigour started only after the rediscovery of Mendel's laws of inheritance and consequent understating of the theoretical basis of hybrid vigour. Hybrid vigour or heterosis may be defined as a manifestation of heterozygosity, expressed as increased vigour, size, fruitfulness and resistance to diseases, insects and/or climatic extremes (Briggs and Knowles, 1967). Hybrid vigour was reported as early as 1763 by Koelreuter in inter-specific crosses of several crops. Naudin (1865) noticed hybrid vigour in 24 inter-specific crosses in 11 different genera. Mendel (1865) also observed F_1s to be more vigorous. Beal (1880) published results of varietal hybrids of maize. He observed increased vigour in F_1. By crossing two varieties with contrasting characters he was able to obtain yield increase by 51 per cent over parents. Sanborne (1890) confirmed Beal's results and was able to produce hybrids yielding 31 per cent more than parents. During the investigations on the effects of inbreeding and cross-breeding with maize, Shull (1910) noticed that F_1 from a combination of plants belonging to certain self-pollinated/inbred families produced yield higher than the original crossbred stock. However, a cross between sibs within same inbred family exhibited little or no vigour and improvement over self-pollination in the same family. Shull (1914) coined the term "heterosis" for increased vigour exhibited by F_1s between inbred lines from different stocks in these words "To avoid the implication that all the genotypic differences, which stimulate cell division, growth and other physiological causes are Mendelian in inheritance, and to gain brevity in expression, the term heterosis is adopted instead of "stimulus of heterozygosis" "heterotic stimulation". Whaley (1944) pointed out that the term "heterosis" has been erroneously established as a synonym of hybrid vigour. He emphasized that the original definition of "heterosis" refers to developmental stimulation resulting by whatever mechanism, from union of gametes. Hybrid vigour manifests the effects of heterosis. Shull (1952) explained the heterosis concept as "interpretation of increased vigour, size, fruitfulness, speed of development, resistance to diseases and insect pests or climatic rigour of any kind, manifested by cross-bred organisms as compared to the corresponding inbreds, as a specific result of unlikeness in the constituents of the uniting gametes".

Hybrid vigour has been reported in almost all the crops being cultivated. Ashton (1946) summarized the evidence for heterosis in wheat, oats, barley, sorghum, rice, cotton, tobacco, tomato, eggplant and soybean. From his estimates occurrence of heterosis, manifested by F_1, relative to mid-parent is widespread in self-pollinated plants. In many, although not all the crops, F_1 performance better than the superior parent was observed for some characters.

Freeman (1919) was the first to report heterosis in wheat. Briggle (1963) reviewed the occurrence of hybrid vigour in wheat and rice. Heterosis in rice was recorded for the first time by Jones (1926). Subsequently several researchers studied occurrence of heterosis in rice (Ramiah, 1935; Kadam *et al.*, 1937). Cooner and Karper (1927) reported hybrid vigour in sorghum. In cotton, Mell (1894), Balls (1908), Durairaj (1968), and several others noticed heterosis. Heterosis was noticed in *Brassica juncea* by Labana and Badwal (1975). Deokar and Patil (1979), Karve and Deshmukh (1981) reported heterosis in safflower. Hybrid vigour was also reported in other oilseed crops like sesame (Pal, 1945; Dixit, 1976) and soybean (Gadag and Upadhyaya, 1995).

Heterosis is observed in vegetable crops also for yield of economically important attributes of plants e.g. swollen roots (e.g. carrot), modified stem (Knolkhol), tuber (e.g. potato), corm (e.g. Colocasia), hypocotyl (e.g. radish), leaf (e.g. spinach, fenugreek), flower (e.g. cauliflower), fruits and seed (garden pea). Heterosis for yield may be manifested in different ways. Higher yield of hybrids of cabbage, watermelon, onion may be attributable to larger size or weight of fruits, bulbs or heads. Early fruiting and larger fruit number may be responsible for heterosis in tomato, brinjal, okra. The size of heads or bulbs is uniform in case of hybrid cabbage and onion. Hybrids may exhibit better resistance to insect pests (e.g. onion), disease (e.g. tomato, cucumber), drought (watermelon). For more details on heterosis recorded for various characters of important vegetable crops reader is referred to Singh *et al.* (1998).

Hybrid vigour for tomato (*Lycopersicon lycopersicum* (Linn.) Karst.) was observed and reported for first time by Hedrick and Booth (1908). Hybrid vigour in tomato is manifested by increased number of fruits per plant (Hepler, 1929, Whaley, 1952; Shifriss, 1945). Hybrid tomatoes have improved fruit quality, fruits appear better and are resistant to cracking (Rick *et al.*, 1956). Hybrid vigour is manifested by enhanced resistance against diseases in tomato. Hybrids resistant to *Fusarium* wilt have been reported (Porte, 1932). Earliness in fruit setting and ripening is characteristic of hybrid tomatoes. Significant increase in yield has been reported in tomato by several researchers including Daskaloff (1937), Powers (1945) and Bhutani *et al.* (1973).

Nagai and Kida (1926) were probably the first to report hybrid vigour in brinjal (*Solanum melongena*). Viswanathan (1973) and Peter and Singh (1973) recorded hybrid vigour in brinjal. Pal and Singh (1946, 1949) observed that hybrid brinjals yield 48.8 to 56.6 per cent more than the better parent.

Not much work has been carried out to study heterosis in chillies (*Capsicum annum*). Deshpande (1933) reported occurrence of hybrid vigour in chillies. F_1s were heterotic for general vigour, maturity, plant height, thickness of fruits, total number of fruits produced per plant and consequently yield per plant. Attavar and Bhat (1973) noticed heterosis in bell pepper.

Increase in size of fruits, weight of fruits and number of fruits per plants was reflected in hybrid vigour of okra in studies carried out by Vijayaraghavan and Wariar (1946). Venkatramani (1952) observed that hybrids yielded about 5.4 to 14.5 per cent higher than better parent. Increase in total yield was attributed to production of more number of fruits per plant.

Hybrid onions were more vigorous than their parents. The weight of hybrid bulbs was more than either parent. Moreover bolting per cent observed for hybrids was much less than the male parent. Jones and Davis (1944), Jones and Clarke (1947) suggested crossing male-sterile lines to exploit hybrid vigour in onion. Numbers of bulbs per unit were more in case of hybrids. In addition the size, shape, colour and other attributes were superior in case of hybrid onions.

This is a glimpse of what can be known about hybrid vigour in vegetables. Research on vegetable crops and other crops has been extensively carried out to measure heterosis for various characters. In general hybrids are earlier, yield more, are uniform in phenotypic appearance and may be resistant to biotic and/or abiotic stresses.

Hybrid vigour, which is desirable in case of ornamental plants, is with regards to uniformity in vigour, flower size, colour, faster growth. Hybrids of petunia and Geranium grow faster (Vishnuswarup, 1995). Uniform height of plants, basal branching, simultaneous flowering and longer duration are some of the most desirable attributes expected from hybrids. Moreover, the flowers appear fresh and stay longer on the plants. This latter is possible by using male-sterility. Hybrids of marigold and zinnia are the examples.

Since market for cut flowers is developing faster, attention to hybrid breeding of ornamental plants is most essential. There is great demand in international markets for cut flowers. Major markets for Indian floricultural products are European countries such as Holland, Germany and U.K., Gulf countries and Japan. Cut flower exports in Japan have increased considerably during last few years. India earned from exports of floricultural products Rs.14.4 crores in 1991-92, which increased to 103 crores during 1998-99 (Raghava and Dadlani, 2000).

11.3.1 Various Definitions of Hybrid Vigour

In the literature enhanced yield, increase in vigour and speed of development, resistance to diseases and insect pests as well as to adverse conditions exhibited by the F_1 hybrid as compared to either parent has been expressed in different ways. A student of plant breeding should become familiar with various terms used in relation to hybrid vigour.

Heterozygosis

East and Hays (1912) coined this term to represent a situation where vigour decreases due to inbreeding the naturally cross-pollinated crop plants but the vigour increases upon crossing naturally self-pollinated crop plants. According to them both these observations result from one phenomenon, viz., heterozygosis (i.e. heterozygosity at a locus). Union of gametes with unlike genotypes is the cause of increased vigour, yield, etc.

Heterosis

This term was proposed by Shull (1914). It was suggested for shortness in explanation for the mechanism underlying the vigour observed for

F_1 between inbred lines. According to him vigour is due to 'stimulus of heterozygosis', 'heterozygotic stimulation'. Shull used this term to highlight the effect of the stimulus of heterozygosis upon cell division, growth and other physiological activities of an organism. But more precise definition of heterosis was suggested by Shull (1948). According this new definition heterosis refers to increased vigour, size, fruitfulness, speed of development, resistance to diseases and insect pests manifested by a cross-bred organism as compared to the corresponding inbred, as the specific result of unlikeliness in the genetic constitution of uniting parental gametes.

Hybrid vigour

Although vigour associated with F_1s between genetically diverse parents was recorded since pre-Mendelian era, it was Jones (1918) who used word hybrid vigour as a synonym of heterosis.

Heterobeltiosis

This term was coined by Bitzer *et al.* (1968) and Fonseca and Patterson (1968) for superior performance of F_1 as compared to its superior parent.

Euheterosis

It is a term proposed by Dobzhansky (1950) who classified hybrid vigour on the basis of fitness. It is also called "true heterosis". Hybrid vigour which confers higher fitness upon individuals in sexually reproducing and cross-fertilizing crop species is euheterosis. In such species higher level of adaptation under natural conditions is represented by this kind of heterosis. Heterosis is not merely a post-Mendelian phenomenon, which contributes only to attain improvement in crop productivity. It is instrumental in evolution acting both as a creative as well as a preservative force in nature. According to Powell (1971) heterosis provides a genetic system for maintaining flexibility in natural populations. Heterosis helps to generate superior individuals in population, which possess higher fitness and enable adaptation to adverse environments. Optimal degree of heterozygosity confers greater fitness upon individuals. Such individuals are favoured by natural selection during evolution. Euheterosis has been subdivided into mutational euheterosis and balanced euheterosis on the basis of origin of heterosis in natural and domesticated conditions.

Luxuriance

Dobzhansky (1950) suggested this term. From evolutionary point of view luxuriance is opposite to euheterosis, i.e. hybrids show poor fitness or adaptive value than parents. The term luxuriance is applied to special cases of hybrids between varieties or races or strains of self-pollinated or cross-pollinated crops, which manifest excessive growth as compared to parents but are inferior with respect to fitness and adaptation. Luxuriance is thought to be an evolutionary accident because the luxuriant plants can not survive in natural conditions. Luxuriant hybrids showing excessive vegetative growth and excessive size may be superior in man controlled environments. Hence luxuriance has also been called pseudoheterosis.

Positive and Negative Heterosis

Powers (1944) and Stern (1948) extended the concept of heterosis to include hybrid, which is inferior in performance compared to the parents as manifestation of heterosis. F_1 hybrids had relatively lower number of locules per fruit than either of the parents in tomato (Powers, 1944). He called this as 'negative heterosis' while superior performance of F_1 hybrid as compared to the parents as "positive heterosis". Some times negative heterosis may be a desirable attribute. For example, when F_1 hybrid flowers or matures earlier than the earlier parent. In other words the F_1 has faster rate of development than earlier parent.

Several other terms like adaptive heterosis, selective heterosis, labile and fixed heterosis and luxuriant heterosis are mentioned in the literature. For details on this aspect reader is directed to Mukherjee (1995).

11.4 GENETIC EXPLANATION OF HYBRID VIGOUR

Two principle hypotheses have been extended to explain genetic basis of heterosis and inbreeding depression. Although they may appear independent, they are not mutually exclusive. In reality both may be operating together to determine heterosis. These hypotheses are:

(a) The dominance hypothesis, and
(b) The overdominance hypothesis.

11.4.1 Dominance Hypothesis

This hypothesis assumes that growth and vigour are governed by a large number of genes, each with small and cumulative effect on the character. Because number of genes is large they are distributed on several chromosomes possessed by the organism in nucleus of each cell. In other words, several genes are linked. Further it is assumed that the genes controlling growth and vigour are dominant and favourable to the organism which possess these genes. Recessive genes are assumed to be deleterious or harmful. There is no non-allelic interaction.

As mentioned in chapter 5 and in section on inbreeding (11.2) individuals of cross-pollinated plant species carry a load of deleterious recessive genes shielded by the dominant alleles with favourable effect, in heterozygous state. Upon inbreeding the recessive genes are uncovered because they attain homozygosity. Consequently morbid, and lethal types appear. For example, plants which are deficient in chlorophyll, have defective flowers, defective seeds are seen in inbred population. Such plants also appear in open-pollinated population but the frequency of defective and deleterious types of plants is much larger in the inbred population. This is the result of inbreeding. Lines differ with respect to manifestation of deleterious effects. Some lines receive more favourable genes than others accounting for differences observed in the degree of inbreeding depression in different lines.

According to this hypothesis intercrossing of inbred lines leads to formation of hybrids in which deleterious recessive genes contributed by one parent are shielded by dominant favourable genes transmitted by other parent, as in the open-pollinated population. The performance of hybrid depends upon the genotypes of inbred lines, which are the parents. Some genotypes many complement each other to produce hybrids better than the average of open-pollinated variety while others may not 'nick' well. Nicely 'nicked' hybrids manifest heterosis. For example, let us assume that the genetic constitution of one inbred line is *AABBccddee* and that of the other is *aabbCCDDEE*. Suppose the effects of genes $A = B = C = D = E = 2$ units while effects of $a = b = c = d = e = 1$ unit on a character under consideration. Thus inbred *AABBccddee* measures 14 units while *aabbCCDDEE* measures 16 units. When these inbred lines are crossed, the genetic constitution of F_1 will be *AaBbCcDdEe*, and will measure 20 units, if degree of dominance is complete. Here the two inbred lines perfectly complemented each other. If some genes are common, the magnitude of heterosis will depend upon genetic constitutions of F_1. Hybrid vigour will be less if dominance is not complete.

Davenport (1908) was first to suggest that beneficial genes are dominant while recessive genes are deleterious. Keeble and Pellew (1910) supported Davenport's views to explain hybrid vigour in peas. They found that F_1 plants were taller than taller parent. Two dominant genes at separate loci were assumed to be involved, one of which governed the length of internodes and other the number of internodes. Bruce (1910) explained heterosis on the basis of combined action of favourable dominant or partially dominant genes.

Two objections were raised against this hypothesis, viz.,

(i) If heterosis is due to dominant genes with beneficial effects (and not due to heterozygosity *per se*) then it should be possible to develop true breeding lines with all dominant genes in homozygous condition and such lines should be as vigorous as heterozygous hybrid F_1, and

(ii) Because segregation is skewed in F_2 generation, when gene is completely dominant, segregation for heterotic characters should also be skewed rather than symmetrical, which is usually observed.

Jones (1917) explained inability to isolate dominant homozygous and true breeding lines as vigorous as F_1 on the grounds that growth is governed by many genes, which are located on different chromosomes carried by the organism. Moreover even a single chromosome will bear some dominant favourable and some recessive deleterious genes. In order to produce gametes carrying chromosomes with all dominant favourable genes, a series of precisely placed cross-overs between homologous chromosomes will have to take place. This will have to happen with all chromosomes, which carry genes influencing growth. And finally all the chromosomes carrying dominant favourable genes will have to go to same pole at anaphase so that they are included in same gamete. When such gametes unite, only then it will be possible to get a line possessing all dominant genes in homozygous condition. This is a rare possibility and hence chances of obtaining or producing a line homozygous for all dominant favourable genes are remote.

The other objection was against the observed symmetrical distribution for heterotic character. When a single locus with 2 alleles, viz., *A* and *a*, is considered, and *A* is completely dominant to *a*, in F_2 generation the phenotypic distribution is 3 dominant: 1 recessive. When 2 genes are involved the distribution is 15: 1 and so on. Hence the objection was, if dominance is responsible for heterotic traits, the distribution in F_2 should follow a pattern $(3/4 + 1/4)^n$. Jones again explained this discrepancy on the grounds that because dominant favourable genes are linked with recessive deleterious genes, distribution in F_2 is symmetrical. Further, Collins (1921) pointed out that even in the absence of linkage so long as the number genes involved is large skewed distribution will be difficult to detect.

These ideas were called as dominance hypothesis or dominance of linked genes hypothesis.

11.4.2 Overdominance Hypothesis

This alternative hypothesis was proposed independently by Shull (1908) and East (1908) and subsequently advocated by Stadler (1939) and Hull (1945). According to this hypothesis, proposed to explain the occurrence of hybrid vigour, heterozygosity at a locus is superior to homozygosity. The most vigorous individual is one, which carries largest number of heterozygous alleles. This hypothesis assumed that there is a physiological stimulus to development, which increases with diversity of uniting gametes. In other words, hybridity itself, i.e. the state of being heterozygous due to the union of gametes with diverse genic contents, has stimulating effect on physiological activities of an organism. In Mendelian terms, this implies that there are loci at which heterozygote is superior to either homozygote. Further, vigour increases in proportion to the amount of heterozygosity. This idea is known as single gene heterosis, superdominance, cumulative action of divergent alleles or stimulation of divergent alleles. The most commonly applied designation to this concept is overdominance and the hypothesis proposed is known as overdominance hypothesis of heterosis.

East (1936) presented Mendelian concept of interaction of alleles at a single locus to explain heterosis. He proposed that there may be several alleles at a locus with divergent functions, like say A_1, A_2, A_3. A heterozygous combination A_1A_2 or A_1A_3 or A_2A_3 will have greater effect than the homozygous condition, i.e. A_1A_1 or A_2A_2 or A_3A_3. The implication is that A_1 and A_2 perform different functions and the sum of their different products is superior than sum of the product of same gene in homozygous state. Moreover, the heterozygote becomes superior with diversity of uniting gametes, i.e. $A_1A_2 < A_1A_3 < A_2A_3$. According to this hypothesis no homozygote as superior as F_1 heterozygote can be isolated because the superiority is due to heterozygosity. This aspect can be explained with a hypothetical example. Let us assume a simple situation of a single locus with two alleles, viz., A_1 and A_2. Gene A_1 confers say 4 units on its carrier in homozygous state while A_2 confers 2 units. Hence the values of A_1A_1 and A_2A_2 would be 8 and 4 units, respectively, as shown in Figure 11.1. Because interaction between alleles at a locus confers superiority on the heterozygote, homozygote as superior as heterozygote can not be isolated.

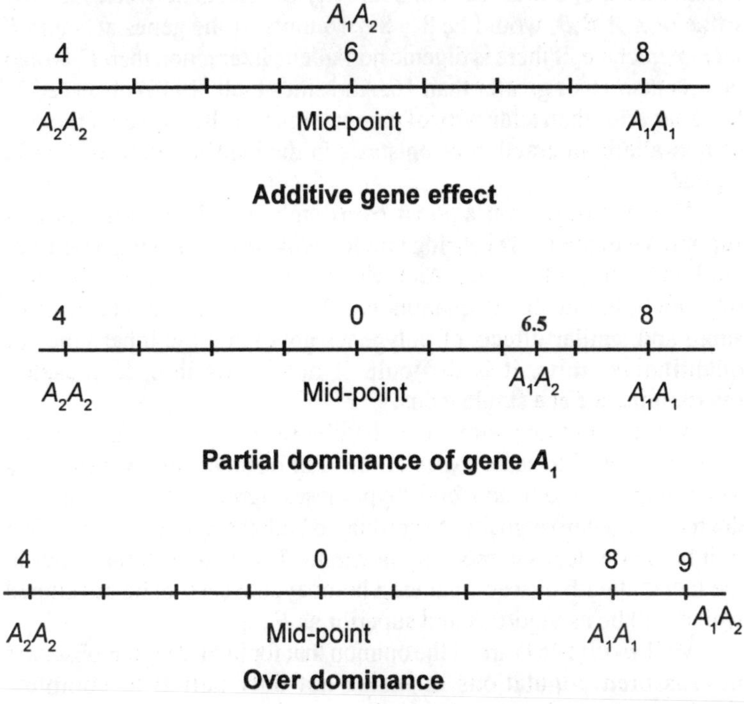

Fig. 11.1: Overdominance at a locus

At the time East and Shull independently propounded overdominance hypothesis there was no direct evidence supporting the superiority of heterozygote at a locus. Nevertheless, subsequently a few instances were noticed where a heterozygous condition at a given locus was superior to better homozygote (Karper, 1930; Quinby and Karper, 1946 in sorghum; Robertson, 1932 in barley).

The term overdominance was coined by Hull (1945) to denote heterozygote superiority over the homozygous parents. While working with maize he observed a number of F_1s which exceeded the sum of parental inbred lines with respect to yield. This could not be explained with dominant genes acting in a completely additive fashion. For example, two loci, each with two alleles, govern a trait, i.e. A_1, A_2 and B_1 and B_2. Let us assume that A_1 and B_1 are completely dominant. Assume that effects of $A_1 = B_1 = 4$ units and of $A_2 = B_2 = 2$ units. Then F_1 between $A_1A_1B_2B_2$ and $A_2A_2B_1B_1$ would be $A_1A_2B_1B_2$. With complete dominance at A and B loci but additivity of effects between loci, the value of $A_1A_2B_1B_2$ would be 8 + 8 = 16 units. If the genes at A and B loci interact, i.e. if there is digenic non-allelic interaction then F_1 would be less than 16 or greater than 16. And since Hull (1946) observed F_1 to be superior than additivity of parental inbreds, he expected the role of non-allelic interaction or epistasis in the manifestation of hybrid vigour.

A major objection against overdominance hypothesis is that supportive evidence favouring single locus heterosis has come from qualitative characters only. And what holds true for qualitative traits may not be applicable to quantitative characters. Actually because of small and similar effects of polygenes governing the inheritance of quantitative traits, it is difficult, if not impossible, to measure overdominance at a single locus.

Actually in many situations the dominance and overdominance hypotheses lead to same expectations. For instance, inbreeding leads to decline in vigour and both hypotheses agree that this is due to decrease of heterozygosity. According to both the hypotheses crossing of inbred lines leads to recovery in vigour. The main difference lies in the fact that with overdominance a homozygote can not be developed which will be as vigorous and superior as F_1.

Moll *et al.* (1964) are of the opinion that for hybrid vigour observed in crossbred populations of maize not only partial to complete dominance at the loci involved is important but also overdominance. The effect of repulsion phase linkages of favourable dominant genes mimic overdominance effects and can not be distinguished from effects of overdominance when recombination frequency is near zero, i.e., there is very tight linkage between genes concerned. Further, what needs consideration is that the overdominance does not tend to decline in magnitude with random mating, however, the magnitude of pseudodominance declines. Consequently not only the relative importance of dominance effects of genes versus overdominance has important implication regarding evolutionary significance of hybrid vigour and its practical utilization in plant breeding but also the rate at which pseudodominance diminishes with random mating.

11.5 BIOMETRICAL EXPLANATION OF HETEROSIS

Heterosis is explained as the superiority of F_1 as compared to the mean of two parents (i.e. called as mid-parent value) or better parent. Falconer (1960) has derived the biometrical expression of heterosis, estimated as the amount by which mean of F_1 exceeds mean of two parents.

Let us assume two random mating populations, each comprising of three genotypes, viz., A_1A_1, A_1A_2 and A_2A_2 (for simplicity of explanation we assume a single locus with two alleles). Let the genotype frequencies of A_1A_1, A_1A_2 and A_2A_2 in population 1 be p^2, $2pq$ and q^2, respectively. It is believed that the two populations, which are crossed to produce hybrid differ in gene frequency. The second population should have a different gene frequency than the first. Let us denote the gene frequencies of population one as p (for A_1) and q (for A_2). Remember that for random mating population $p^2 + 2pq + q^2 = 1$ and similarly $p + q = 1$. The difference between gene frequencies of populations 1 and 2 be denoted by y. The frequency of gene A_1 in population 2 is assumed to be less than in population 1 by amount y, i.e. is $p - y$. And gene frequency of A_2 is greater than in population 1 by an amount equal to y, i.e. $A_2 = q + y$. Using the values in table 11.2, let us calculate means of populations 1 and 2. To get the mean multiply value by corresponding genotype frequency and add the product over

Table 11.2: Genotypes, genotype frequencies and arbitrarily assigned values to the genotypes in population 1 and 2

Genotype	*Genotype frequency population 1*	*Arbitrary values population 1*	*Genotype frequency population 2*	*Arbitrary values population 2*
A_1A_1	p^2	$+a$	$(p-y)^2$	$+a$
A_1A_2	$2pq$	$+d$	$2(p-y)(q+y)$	$+d$
A_2A_2	q^2	$-a$	$(q+y)^2$	$-a$

genotypes. Thus $MP_1 = p^2a + 2pqd - q^2a$. Upon simplification this becomes $a(p-q)+2pqd$. Likewise, the mean of population 2 will be $a(p - q - 2y) + 2d[pq + y(p - q) - y^2]$. And the mid-parent value shall be the mean of MP_1 and MP_2, which is

$$M\overline{P} = a(p-q-y)+d[2pq+y(p-q)-y^2] \qquad \text{...(11.7)}$$

The population 1 produces two kinds of gametes, viz., A_1 and A_2 with frequencies p and q respectively (See chapter 5). Similarly population 2 produces gametes A_1 and A_2, with frequencies equal to $(p - y)$ and $(q + y)$, respectively. Random mating of the two populations could be described in terms of random union of gametes as shown below:

Gametes. Population 2	*Gametes. Population 1*	
	A_1 p	A_2 q
A_1 $(p-y)$	A_1A_1 $p(p-y)$	A_1A_2 $q(p-y)$
A_2 $(q+y)$	A_1A_2 $p(q+y)$	A_2A_2 $q(q+y)$

The mean of F_1 generation so produced will be sum of the products of value with genotype frequency over three genotypes. Thus

$$M\overline{F}_1 = a(p^2 - py) + d[2pq + y(p-q)] - a(q^2 + qy)$$

On simplification this becomes

$$a(p - q - y) + d[2pq + y(p - q)] \qquad \text{...(11.8)}$$

Heterosis measured as amount by which mean of F_1 exceeds mid-parental value is given by

$$M\overline{F}_1 - M\overline{P}$$

$$= dy^2 \qquad \text{...(11.9)}$$

The expression 11.9 shows that like the inbreeding depression, magnitude of heterosis, exhibited by the F_1 depends upon the degree of dominance (d) and also upon square of difference between gene frequencies. Heterosis will not occur if either dominance does not exist, i.e. $d = 0$, or the two populations do not differ with respect to gene frequencies, i.e. $y = 0$. When number of loci influencing the inheritance are considered, heterosis is expressed as Σdy^2, assuming absence of non-allelic interaction, and additivity of gene effects over loci. Heterosis will be larger in magnitude when the value of d is larger. It will be larger only when there is unidirectional dominance, i.e. dominance effects at each locus are in same direction. If the dominance effects at some loci are in positive direction while at others in negative direction, the sum of effects over all the loci will cause reduction in the value of Σd. Whenever the values of $+d$ and $-d$ (i.e. deviation of F_1 from mid-point in + and – direction) at different loci balance, sum of dominance deviations will be zero. Thus absence of heterosis should not lead to conclusion that there is no dominance at all the loci concerned.

Moll *et al.* (1962) observed that in maize the magnitude of heterosis increased with increase in genetic divergence of parent populations. Earlier work also suggested that crosses involving distantly related parents manifest more heterosis than crosses between more closely related parents. Nevertheless, there is an optimum degree of genetic divergence for maximum expression of heterosis in maize (Moll *et al.*, 1965). Heterosis increases with divergence within a restricted range of divergence but extremely divergent crosses resulted in decrease in magnitude of heterosis. Paterniani and Lonnquist (1963) observed that the F_1 performance of high × high crosses (with respect to grain yield) was higher than that of high × low or low × low crosses. Nevertheless, heterosis, showed exactly reverse pattern, i.e. high × high crosses exhibited lowest heterosis while low × low manifested highest magnitude.

Usually one is interested in heterosis expressed as superiority of $\bar{F}_1$ over the mean of better parent ($\bar{P}_1$). $M\bar{P}_1$ was shown to be $a(p-q) + 2pqd$ and mean of F_1 was $a(p-q-y) + d[2pq + y(p-q)]$. Hence heterosis over better parent is:

$$= -y\,[a + d(q-p)]$$

In this expression $[a + d(q-p)]$ is known as average effect of gene substitution (α), hence heterosis is

$$-y\,\alpha \qquad \text{...(11.10)}$$

Heterosis, therefore, depends upon average effect of gene and difference in gene frequency between the parent populations.

A reader can verify that the deviation of $\bar{F}_2$ from mid-parent equals half the deviation of $\bar{F}_1$ from $M\bar{P}$, i.e. ½ Σdy^2.

As a matter of fact the above discussion referred to a cross between populations. Actually for developing hybrids, a plant breeder uses true breeding varieties or inbred lines. Mather and Jinks (1971) described the method to calculate the theoretically expected value of heterosis, where heterosis is calculated as the amount by which mean of an F_1 exceeds better parent, i.e. $\overline{P}_1$. For simplicity to understand, let us assume a single locus with two alleles, i.e. A_1 and A_2. With respect to this locus there will be 3 genotypes, viz., A_1A_1, A_1A_2 and A_2A_2. Let us assign hypothetical genotypic values to A_1A_1 as $m + d$ and to A_2A_2 as $m - d$. Here $+d$ and $-d$ denote the extent of departure of A_1A_1 and A_2A_2, respectively from mid-point (m). Thus A_1A_1 is larger by $+d$ units from mid-point and A_2A_2 is smaller by $-d$ units. The departures of both A_1A_1 and A_2A_2 are identical in magnitude, except that A_1A_1 deviates in plus and A_2A_2 in minus direction. The difference between A_1A_1 and A_2A_2 is then $2d$ units and mid-point is equal to zero. When A_1A_1 and A_2A_2 are hybridized, the F_1 is A_1A_2 in genetic constitution. Because of dominance (intra-allelic interaction) the F_1 may deviate from mid-point in positive direction. But it may deviate in negative direction also. This departure of A_1A_2 from mid-point is designated as h. The mean expression of A_1A_2 can be therefore described as $m + h$. If A_1 gene is partially dominant A_1A_2 will assume value between m and $+d$, but if A_1 is completely dominant, A_1A_2 will be equal to A_1A_1, i.e. $h = d$. When there is overdominance h can be greater than $+d$ or smaller than $-d$. Likewise with partial dominance of A_2, A_1A_2 will lie between m and $-d$ but with complete dominance of gene A_2, A_1A_2 will be equal to A_2A_2, i.e. $m + h$ will be equal to $m - d$. When several loci, which govern the inheritance of a quantitative character are considered, then larger parent will possess genes A_1A_1, B_1B_1 and so on, and the phenotypic value is written as $m + [d]$. Where $[d]$ is the sum all the d increments, i.e. $da + db + dc$.... Likewise phenotypic value of smaller parent shall be $m - [d]$. $[d]$ or $[-d]$ take into consideration that $da \neq db \neq dc$ ..., and the departures of d increments at various loci may not be in same direction. The mean measurement of F_1 will be $m + [h]$. Here $[h]$ is sum of all the h increments. Remember that some h increments could be positive and some negative, and $[h]$ is balance of all the h increments. Moreover, magnitude of individual h increments may be different. In the present discussion simplified descriptions are given. For details reader may refer to Mather and Jinks (1971) and Dabholkar (1992). Absence of digenic non-allelic interaction is assumed to begin with. Departure of $\overline{F}_1$ from $\overline{P}_1$ if positive, is called positive heterosis. Thus positive heterosis would be:

$$\bar{F}_1 - \bar{P}_1$$
$$= m + [h] - m - [d]$$
$$= [h] - [d] \qquad ...(11.11)$$

For positive heterosis to occur $[h]$ must be larger in magnitude than $[d]$ and carry positive sign. $[d]$ will be smaller in magnitude when genes are dispersed between the parents. Sometimes breeders are interested in negative heterosis. For example, F_1 when earlier in maturity than early parent, heterosis will be negative. This is calculated as $\bar{F}_1 - \bar{P}_2$. Thus negative heterosis would be

$$(m + [h] - (m - [d])$$
$$= m + [h] - m + [d]$$
$$= [h] + [d] \qquad ...(11.12)$$

Thus negative heterosis will occur and occur with larger magnitude when $[h]$ is negative and greater than $[d]$.

Thus for heterosis to occur $\pm [h]$ should be larger than $[d]$ and one or both of the following conditions should be satisfied:

(i) h should be larger in magnitude than d at some or all the loci. This suggests that there should be overdominance at some or all the loci.

(ii) Genes should be dispersed between parents.

Heterosis with larger magnitude will occur when h increments at all loci are in desirable direction, i.e. unidirectional dominance.

When digenic non-allelic interaction is involved (see section 6.2.3) three parameters are to be included, viz., additive × additive interaction $[i]$, additive × dominance interaction $[j]$ and dominance × dominance interaction $[l]$. The mean measurements of P_1, P_2 and F_1 could be written as:

$$\bar{P}_1 = m + [d] + [i]$$

$$\bar{P}_2 = m - [d] + [i], \text{ and}$$

$$\bar{F}_1 = m + [h] + [l]$$

The positive heterosis would be

$$\bar{F}_1 - \bar{P}_1$$

$$= (m + [h] + [l]) - (m + [d] + [i])$$
$$= [h] + [l] - [d] - [i] \quad ...(11.13)$$

And similarly negative heterosis would be

$$[h] + [l] + [d] - [i]$$

For positive heterosis to occur, when digenic non-allelic interaction is involved, $[h] + [l]$ should be greater than $[d] + [i]$. In this case either $[h]$ or $[l]$ could be smaller or negative, but on the whole $([h] + [l])$ should be positive and larger than $([d] + [i])$. Further, genes should be dispersed between the two parents so that $[d]$ is smaller. But value of $[i]$ could be negative.

Jayasekara and Jinks (1976) and Pooni and Jinks (1981) provided evidence to show that in *Nicotiana rustica* heterosis for plant height was due to dispersion of genes between two parents.

11.6 COMPUTATION OF HETEROSIS

After growing the parents and F_1 hybrid in appropriately designed experiment, which is replicated, the mean measurements are statistically analysed. When significant differences are seen among the mean measurements heterosis is calculated. Heterosis may be calculated as deviation of $\bar{F}_1$ from better parent ($\bar{BP}$), known as heterobeltiosis, or as deviation of $\bar{F}_1$ from mid-parent, called as relative heterosis and departure of $\bar{F}_1$ from standard check (*st*), called standard heterosis. These estimates are got as:

(i) Heterobeltiosis

$$\hat{H} = \frac{\bar{F}_1 - \bar{BP}}{\bar{BP}} \times 100$$

Heterosis is estimated as per cent superiority. Significance of estimate of heterosis is tested by '*t*' test using standard error calculated as:

$$\sqrt{\frac{2Me}{r}}$$

where *Me* is error mean square and *r* is number of replications.

(ii) Relative Heterosis

$$\hat{H} = \frac{F_1 - \left(\frac{1}{2}\overline{P}_1 + \frac{1}{2}\overline{P}_2\right)}{\left(\frac{1}{2}\overline{P}_1 + \frac{1}{2}\overline{P}_2\right)} \times 100$$

The standard error is computed as:

$$\sqrt{\frac{3Me}{r}}$$

(iii) Standard Heterosis

$$\hat{H} = \frac{\overline{F}_1 - \overline{St}}{\overline{St}} \times 100$$

References

Allard, R.W. (1960). *Principles of plant breeding*. John Wiley and Sons, Inc., New York.

Ashton, T. (1946). The use of heterosis in the production of agricultural and horticultural crops. *Imper.Bur.Plant Breed. And Genetics*, School of Agric., Cambridge, England.

Attawar, M. and N.K. Bhat (1973). High yielding Indo-American capsicum-hybrid. *Lal Baugh*, 18 (12): 28-32.

Balls, W.L. (1908). Mendelian studies in Egyptian cottons. *J.agric.Sci.*, 2: 346-379.

Beal, W.J. (1880). *Indian Corn. Rept.Michigan State Board Agric.*, 19: 279-289.

Bhagchandani, P.M. and B. Choudhary (1970). Heterosis and inbreeding expression in carrot. *Progre.Hort.*, 2 (1): 65-73.

Bhutani, R.D.; Kalloo, G.; M.L.Pandita and G.P.Singh (1973). Studies on heterosis in tomato. *Haryana J.Hort.Sci.*, 2 (3-4): 156-163.

Bitzer, M.J.; S.Fonseca; Papathomasion, G. and F.L.Patterson (1968). 'Heterobeltiosis' (edited by Fonseca *et al.*, 1968).

Briggle, L.W. (1963). Heterosis in wheat — a review. *Crop Sci.*,3: 407-412.

Briggs, F.N. and P.F. Knowles (1967). *Introduction to plant breeding*. Reinhold Publishing Corp., New York.

Bruce, A.B. (1910). The Mendelian theory of heredity and augmentation of vigor. *Science*, 32: 627-628.

Collins, G.N. (1921). Dominance and vigor of first generation hybrids. *Amer.Nat.*, 55: 116-133.

Cooner, A.B. and R.E. Karper (1927). Hybrid vigour in sorghum. Texas Agricultural Experiment Station Bulletin No.359, College Station, Texas, U.S.A.

Dabholkar, A.R. (1992). *Elements of biometrical genetics*. Concept Publishing Co., New Delhi.

Darwin, Charles (1877). *In the effects of cross and self fertilization in the vegetable kingdom*. D.Appleton and Company, New York.

Daskaloff, C. (1937). Contribution to the study of heterosis in tomatoes in relation to production of heterotic varieties for practical purposes. *Gatenbauwiss*, 11: 129-143.

Davenport, C.B. (1908). Degeneration, albinism and inbreeding. *Science*, 28: 454-455.

Deokar, A.B. and F.B. Patil (1979). Heterosis in safflower. *Indian J.agric.Sci.*, 49: 82-86.

Deshpande, R.B. (1933). Studies in Indian chillis. III. The inheritance of some characters in *C.annum*, L. *Indian J.agirc.Sci.*, 3: 219-300.

Dixit, R.K. (1976). Heterosis and inbreeding depression in sesame. *Indian J.agric.Sci.*, 46: 514-517.

Dobzhansky, Th. (1950). Genetics of natural populations. XIX. Origin of heterosis through natural selection in populations of *Drosophila pseudoobscura*. *Genetics*, 35: 288-302.

Durairaj, M.S. (1968). Hybrid vigour in cotton improvement. *Madras agric.J.*, 55: 24-34.

East, E.M. (1908). Inbreeding in corn. *Rept.Connecticut Agric.Expt.Sta.*for 1907, pp. 419-428.

East, E.M. (1936). Heterosis. *Genetics*, 21: 375-397.

East, E.M. and H.K. Hays (1912). Heterozygosis in evolution and in plant breeding. U.S. Dept. *Agric.Bur.Plant Indust.Bull.*,243, p. 58.

Falconer, D.S. (1960). *Introduction to quantitative genetics*. Oliver and Boyd. *Edinburgh*, p. 362.

Fageria, M.S.; P.S. Arya and A.K. Choudhary (2000). *Vegetable crops. Breeding and seed production* Vol.I, Kalyani Publishers, Ludhiana.

Fonseca, S. and F.L. Patterson (1968). Hybrid vigour in seven parental diallel cross in common wheat (*Triticum aestivum* L.). *Crop Sci.*, 8: 85-88.

Freeman, G.F. (1919). Heredity of quantitative characters in wheat. *Genetics*, 4: 1-93.

Gadag, R.N. and H.D. Upadhayaya (1995). Heterosis in soybean (*Glycine max*. L.). *Indian J.Genet.*, 55: 308-314.

Hartl, D.L. (1988). *A primer of population genetics*. 2nd edition. Washington University School of Medicine, Sinauer Associates, Inc Sunderland, Massachusetts, U.S.A.

Hedrick, U.P. and N.O.Booth (1908). Medelian characters in tomatoes. *Proc.Amer.Soc.hort.Sci.*, 5: 19-24.

Hepler, J.R. (1929). Horticultural investigation at New Hampshire Station. *Bull.N.H.agric.Expt.Sta.*, 238: 12-27.

Hull, E.G. (1945). Recurrent selection and specific combining ability in corn. *J.Amer.Soc Agron.*, 37: 134-145.

Jayasekara, N.E.M. and J.L. Jinks (1976). Effects of gene dispersion and estimates of components of generation means and variances. *Heredity*, 36: 31-40.

Johnson, A.G. (1966). Inbreeding and production of commercial F_1 hybrid seed of Brussels sprouts. *Euphytica*, 15: 68.

Jones, D.F. (1917). Dominance of linked factors as a means of accounting for heterosis. *Genetics*, 2: 466-479.

Jones, D.F. (1918). The effects of inbreeding and crossbreeding upon development. *Connecticut Agric Expt.Sta.Bull.*, 207: 100p.

Jones, H.A. and A.E. Clarke (1947). *The story of hybrid onions*. Yb. U.S.D.A., 1943-47, pp. 320-326.

Jones, H.A. and G.N. Davis (1944). Inbreeding and heterosis and their relation to the development of new varieties of onions. *USDA Tech.Bull.*28:874.

Jones, J.W. (1926). Hybrid vigour in rice. *J.Amer.Soc.Agron.*, 18: 424-428.

Kadam, B.S.; G.C. Patil and V.K. Patankar (1937). Heterosis in rice. *Indian J.agric.Sci.*, 7: 118-126.

Karper, R.E. (1930). The effect of single gene upon development in heterozygote in sorghum. *Heredity*, 21: 187-192.

Karve, A.D. and A.K. Deshmukh (1981). Studies of F_1 hybrids of safflower (*Carthemus tinctorius* L.) *In Proc. First intern. Safflower conf.*, pp. 92-96. University of California, Davis, U.S.A.

Kearney, T.H. (1923). *USDA Tech.Bull.*1164

Keeble, F. and C. Pellew (1910). The mode of inheritance of stature and time of flowering in peas (*Pisum sativum*). *J.Genetics*, 1: 47-56.

Khanna-Chopra, R. and S.K. Sinha (1995). Physiological basis of homeostasis in heterotic hybrids of maize and sorghum. In "Hybrid research and development". Ed. Rai, M. and S. Mauria. *Indian Society of Seed Technology*, New Delhi. Proc. Of Seminar, 1993, pp. 217-229.

Koelreuter, J.G. (1766). Vorlau figen nachricht von einigen das geschlecht der pflanzen betreffenden versuchen und beobacnt Ungen., 266pp. Leipzig.

Labana, K.S.; S.S. Badal and B.D. Chaurasia (1975). Heterosis and combining ability analysis in *B.juncea*. *Crop Improv.*, 2: 46-51.

Liang, G.H.; C.R. Reddy and A.D. Dayton (1972). Heterosis, Inbreeding Depression and Heritability estimates in a systematic series of grain sorghum genotypes. *Crop Sci.*, 12: 409-411.

Mather, K. and J.L. Jinks (1971). *Biometrical genetics. A study of continuous variation.* Chapman and Hall Ltd., London.

Mell, P.H. (1894). Experiments in crossing for the purpose of improving the cotton fibre. *Albama* (U.S.) *Agri.Expt.Sta.Bull.*56.

Mendel, G. (1865). Versuche ubr Pflanzen-Hybriden. *Naturf.ver.* in Brunn, Verh, IV: 3-47.

Moll, R.H.; M.F. Lindsey and H.F. Robinson (1964). Estimates of genetic variances and levels of dominance in maize. *Genetics*, 49: 411-423.

Moll, R.H.; W.S. Salhuana and H.F. Robinson (1962). Heterosis and genetic diversity in variety crosses of maize. *Crop Sci.*, 2: 197-198.

Mukherjee, B.K. (1995). *The heterosis phenomenon.* Kalyani Publishers, Ludhiana, p. 142.

Naudin, C. (1865). Nouvelles recherches sur 1 hybridite dans les vegetaux. *Nouv.Arch.Mus.Hist.Nat.*Paris, 1: 25-174.

Pal, B.P. (1945). Studies in hybrid vigour. II. Notes on the manifestation of hybrid vigour in gram, sesamum, and chillis. *Indian J.Genet.*, 5: 106-121.

Pal, B.P. and H.B. Singh (1946). Studies in hybrid vigour. II Notes on the manifestation of hybrid vigour in the brinjal and bitter gourd. *Indian J.Genet.*, 6: 19-33.

Pal, B.P. and H.B. Singh (1949). Hybrid brinjals give increased yields. *Indian Fmg.*, 10: 378-380.

Paterniani, E. and J.H. Lonnquist (1963). Heterosis and inter racial crosses of corn (*Zea mays* L.). *Crop Sci.*, 3: 504-507.

Peter, K.V. and R.D. Singh (1973). Diallel analysis of economic traits in brinjal. *Indian J.agric.Sci.*, 43: 452-455.

Pooni, H.S. and J.L. Jinks (1981). The true nature of the non-allelic interactions in *Nicotiana rustica* revealed by association crosses. *Heredity*, 47: 253-258.

Porte, W.S. (1932). Tomato variety called Break O'Day succeeds in far scattered tests. Yb. U.S.D.A., pp. 373-376.

Powell, J.R. (1971). Genetic polymorphism in varied environments. *Science*, 174: 1035-1036.

Powers, L. (1944). An expansion of Jone's theory for explanation of heterosis. *Amer.Nat.*, 78: 275-280.

Powers, L. (1945). Relative yield of inbred lines and F_1 hybrids of tomato. *Bot.Gaz.*, 106: 247-268.

Quinby, J.R. and R.E. Karper (1946),. Heterosis in sorghum resulting from the heterozygous condition of a single gene that affects duration of growth. *Amer.J.Bot.*, 33: 716-721.

Ramiah, K. (1935). Rice genetics. *Proc.Ass.Eco.Biol.* Coimbatore, 3: 51-61.

Raghava, S.P.S. and N.K. Dadlani (2000). Revival of fortunes. In *The Hindu Survey of Indian Agriculture*. M/s Kasturi and Sons, Kasturi Buildings, Chennai, pp. 151-152.

Rick, C.M.; G.C. Baughn and B.J. Hall (1956). Three new hybrids of tomatoes. *Calif. Agric.*, 10: 4-5, 10.

Robertson, D.W. (1932). The effect of a lethal in the heterozygous condition on barley development. *Colorado Expt.Sta.Tech. Bull.*, 1: 12 pp.

Sanborn, J.W. (1890). *Indian corn Rept.Maine Dept. Agric.*, 33: 54-121.

Scott, G.W. (1933). Inbreeding with *Cucumis melo* L. *Proc.Amer.Soc. hort. Sci.*, 29: 185.

Shifriss, O. (1945). Hybrid tomato. 5th seedsman, 8: 15, 26, 36.

Shull, G.H. (1908). The composition of field of maize. *Rept. Amer. Breeders Assoc.*, 4: 296-301.

Shull, G.H. (1910). Hybridization methods in corn breeding. *Amer. Breeder's Mag.*, 1: 98-107.

Shull, G.H. (1914). Duplicate genes for capsule form in *Bursa bursa —pastoris.*, *Zeitscher. indukt Abstamm.*, 4, *Verebungst*, 12: 97-149.

Shull, G.H. (1948). What is heterosis? *Genetics*, 33: 439-446.

Shull, G.H. (1952). Beginnings of the heterosis concept. In *Heterosis*, ed. J.W.Gowen. Iowa State College Press, Amer., pp. 14-48.

Singh, H.B.; Vishnuswarup and B. Singh (1998). *Exploitation of hybrid vigour in vegetables*. Indian Council of Agricultural Research, New Delhi, pp.46.

Sinha, S.K. and R. Khanna (1975). Physiological, biological and genetic basis of heterosis. *Adv.Agron.*, 27: 123-174.

Stadler, L.J. (1942). Some observations on gene variability and spontaneous mutation. The Spragg Memorial lecture on plant breeding. Mich State College (Briggs and Knowles, 1967, quoted).

Stern, C. (1948). Negative heterosis and decreased effectiveness of alleles in heterozygotes. *Genetics*, 33: 215-219.

Stewart, G. (1934). Effect of inbreeding on variation in alfalfa. *J.Agric. Research*, 49: 669-694.

Tysdal, H.M.; T.A. Kisselbach and H.L. Westover (1942). Alfalfa breeding. *Nebraska, Agr.Exp.Sta. Research Bull.*, 124.

Venkatramani, K.S. (1952). A preliminary study of some inter-varietal crosses and hybrid vigour in *Hibiscus esculentus* (L.). *J.Madras Uni.*, 22: 183-200.

Vijayaraghavan, C. and U.A. Wariar (1946). Evolution of high yielding hybrid bhindi (*Hibiscius esculentus*). *Proc.33rd Indian Sci.Cong.* Abstr.Sect.10 (Abstr.30).

Vishnuswarup (1995). Heterosis breeding in flowerers. In *Hybrid research and development*, Eds. Rai, M. and S. Mauria. Indian Society of Seed Technology, New Delhi. Proceedings of seminar, 1993, pp. 183-187.

Viswanathan, T.V. (1973). Hybrid vigour in brinjal. *Proc,.Indian Acad.Sci.*(B), 77: 176-180.

Whaley, W.G. (1944). Heterosis. *Bot.Rev.*, 10: 461-498.

Whaley, W.G. (1952). Physiology of gene action in hybrids. In *Heterosis*, Ed. J.W.Gowen Iowa State College Press, Ames., pp. 98-113.

12

Hybrid Varieties

F_1 between genetically diverse parents used for commercial planting is popularly known as hybrid variety. Hybrid vigour is at its maximum in the F_1 generation and hence greatest emphasis is placed on the F_1 hybrids. Hybrid varieties can very well exploit the phenomenon of heterosis as compared to any other kind of varieties. Although in plants hybrid vigour was reported as early as 1763 by Koelreuter, its exploitation on commercial scale had to await Beal's suggestion. Beal in 1880 recommended the use of F_1 varietal crosses of maize for commercial cultivation. Hybrid seed was produced by detasseling of plants in alternate rows (employed as female or seed parent). Hybrids between heterozygous varieties of maize did not succeed because of low hybrid vigour. Further the hybrids between varieties could not be reproduced precisely. The genetic composition of individuals within varietal hybrids was not exactly same year after year. Hence they were not acceptable.

It may be interesting to note that F_1 hybrids of ornamental crops were known even before Shull (in 1911) who propounded the classical theory of hybrid vigour. In 1909, F_1 hybrid of Begonia (*Begonia semperflorens*) was released in Germany (Reimann-Philip, 1983, quoted by Vishnuswarup, 1995). It was after the Second World War and later that heterosis breeding attracted attention of flower breeders. In Japan, the first hybrid of Petunia was developed during world war second. From 1950 onwards, hybrid varieties were available in ornamental flowers.

Hybrids of several vegetable crops, viz., tomato, brinjal, okra, cabbage, cauliflower, carrot, radish, watermelon, muskmelon, cucumber, etc. are now commercially grown in India. Whereas the first hybrid of maize was available to the Indian farmers in 1961, hybrids of vegetables had to await till 1973 when the first hybrids of tomato and capsicum were marketed in India.

Much research on development of hybrid in crop plants was carried out on maize by virtue of its peculiar morphological reproductive structures. The male and female flowers are separated in space. The tassel represents staminate flowers while the ear or cob represents pistillate flowers. Moreover there is protandrous situation. This plant is therefore very well suited for studies on selfing and crossing. Hence discussion on various aspects of hybrid development will mostly refer to maize.

It was George Harrison Shull, who in 1909, for the first time suggested pure line method of hybrid maize development. According to this scheme hybrids of maize were to be developed by using inbred lines produced by continued self-fertilization of selected plants. However, the single-cross hybrids produced in this way were not successful because of some drawbacks, viz.,

(i) Inbred lines were not then available which upon crossing with other inbred lines could yield highly heterotic hybrids or yielding better than the existing best open-pollinated variety.

(ii) The hybrid seed was costly. This was because inbred lines were weaker and half to one third land had to be given to male parent in seed production plot. The male inbred parents were poor pollen producers. Finally,

(iii) The F_1 seed was often misshapen and poor in germination.

D.F. Jones in 1918 suggested production of double-cross hybrids so that seed production becomes economical, germination is better and seed not misshapen. Double-cross hybrid is produced by crossing four inbred lines, i.e. (A × B) (C × D). Here (A × B) and (C × D) denote single-cross between inbreds. First commercial production of double-cross hybrid of maize was started in 1921 in the United States of America. But it was only after 1944 that the maize hybrids occupied more than 80 per cent acreage under the hybrid.

Heterosis breeding in tomato was initiated as early as 1908. Commercial F_1 hybrids became available in Japan before 1925 (Kakizaki, 1930) beginning with brinjal followed by cucumber, tomato, and others. Methods of exploiting heterosis on field scale were first proposed using self-incompatibility system in cabbage (Pearson, 1933) and using cytoplasmic-genetic male-sterility in onion (Jones and Clarke, 1943). Today, hybrid varieties have almost entirely replaced open-pollinated varieties of vegetable crops in Japan (Seshadri, 1989).

All India Coordinated Maize Improvement Project started functioning in India in 1957. As a result of research work carried out in this project, four hybrids of maize were released for cultivation viz., Ganga 1, Ganga 101, Deccan and Ranjit in 1961. Subsequently several hybrids were developed and released. In 1964, first sorghum hybrid, based on cytoplasmic-genetic male-sterile line (CK 60A) as seed parent and yellow endosperm feterita IS 84 as pollen parent, was developed and released as CSH-1. HB-1 was the first pearl millet hybrid to be released in India in 1965. For production of this hybrid Tift 23A, a cytoplasmic-genetic male-sterile (CMS) line and BIL3B were employed as seed and pollen parents, respectively. In castor, a hybrid named Gujarat Castor Hybrid 3 (GCH-3) was released in 1968 (Gopani, 1969). For the first time at the global level, a manually produced cotton hybrid, viz., Hybrid 4 (H4) was developed in India and released for general cultivation in 1970. Sunflower hybrid BSH-1, with CMS 234A, a cytoplasmic-genetic male-sterile, as seed parent was released in 1980. Although a pigeonpea hybrid viz., ICPH 8, was developed by ICRISAT, Hyderabad and released in India, it was not successful. It was based on genetic male-sterile line ms Prabhat. Four hybrids of rice viz., APHR 1, APHR 2, MGR 1 and KRH 1, were released for cultivation in India in 1994. For production of rice hybrids two cytoplasmic genetic male-sterile lines developed at the International Rice Research Institute, Philippines, viz., IR 58025A and IR 62829A are being used as female parents.

Hybrids of vegetable crops heralded a new era in vegetable cultivation in India with the release of first hybrid of tomato (Karnataka) and capsicum (Bharat) in 1973. The Indo-American Hybrid Seeds Company, Bangalore, made these available to the Indian farmers. Several private companies and public sector organizations, as in other crops, are engaged in evolution of hybrid varieties of different vegetable crops. Pusa Hybrid 1, Pusa Hybrid 2, KT 4, Arka Vardan, Arka Vishal, Naveen, Rupali, Rashmi, Avinash 2, Larica, Ratna etc. are the tomato hybrids available to farmers of India. Pusa Hybrid 5, Azad hybrid, Pusa Hybrid 6, Arka Navneet, MBH 1, MBH 2, Sumex 9 and Sumex 19, HOE 404, HOE 414 are some brinjal hybrids. A few cabbage hybrids are Nath No.1, BSS 32. Pusa hybrid 2 is cauliflower hybrid available for cultivation. KT 1, Sel 2, Bharat, Early Bonty, Indira, Lario and HOE 801 are some capsicum hybrids. Hybrids are also available in muskmelon (Punjab hybrid 1, MH 10, Pusa Rasraj) and watermelon (Arka Jyoti and Arka Rajhans). This is not the complete list of hybrid varieties, there are many more. Hybrid varieties of pumpkin, chillies, cucumber, bottle gourd, carrot etc. have also been bred/are being marketed in India. Hybrids are not uniformly economical in all the vegetable crops to the same extent as in tomato. Hybrids of cabbage, cauliflower, okra, capsicum and chillies are profitable while the low value crops such as brinjal, gourds, melons and cucumber are moderately economical (Attavar, 2000).

Hybrids have been instrumental in boosting up not only the productivity of various crops but also responsible for improving quality and uniformity of economic produce of many crops. Hybrids are often more stable in performance over locations and years as compared to varieties.

12.1 STEPS INVOLVED IN HYBRID DEVELOPMENT

It has become quite apparent, from the studies carried out by 'hybrid corn markers', namely Richey, Jenkins, Hays, Shull, Jones and East, and other scientists engaged in research for development of hybrids that hybrid development of cross-pollinated crops essentially involves three steps, viz.,

(i) Selection of superior plants from heterozygous base population,
(ii) Inbreeding these plants (via selfing or sib-mating) for several generations to produce homozygous inbred lines, and finally,
(iii) Cross the selected inbred lines to produce hybrid.

In general, for producing hybrid seed emasculation of flowers of line to be used as seed parent and its pollination by collecting pollen from the pollen parent are the essential steps. In case of maize, detasseling is done in female rows, and because of cross-pollinated nature of the crop, hybrid seed is produced in the ears of detasseled rows. Emasculation is carried out in cotton whereas cytoplasmic-genetic male-sterile lines are used as seed parents for production of hybrid seeds of sorghum, pearl millet, sunflower and rice. In case of vegetable crops four approaches are adopted for hybrid seed production (Singh *et al.*, 1998).

(i) Emasculation and pollination, e.g. tomato, brinjal and okra.

(ii) Use of self-incompatible lines as seed parent, e.g. cole crops.
(iii) Use of male-sterility, e.g. onion, and
(iv) Use of variation in sex expression, as in cucurbits.

In case of dioecious crops, like spinach, male plants are removed from rows to be used as seed or female parent. In monoecious crops, such as cucurbits, male flowers are removed from plants of female rows.

It is important to recognize that the three steps involved in the production of hybrids, viz., selection of superior plants from open-pollinated population, inbreeding these plants to homozygosity and utilization of inbreds for hybrid production, merely reproduce the same population genotype that characterizes the material from which the inbred lines are derived, provided (*a*) individuals to be inbred are selected at random from base population, (*b*) there is no selection during inbreeding and (*c*) inbred lines so developed are crossed at random. In other words, if an open pollinated variety is inbred to fixation without selection of any kind and the inbred lines so developed are crossed at random, the cycle merely reproduces an open-pollinated variety. Thus selection plays an important role in the evolution of superior hybrids. Selection while choosing individuals to be inbred from open-pollinated population and during the course of inbreeding is not much effective for gain in productivity that is expected (Allard, 1960). Choice of inbred lines to be crossed to produce hybrid has significant contribution to superior performance of hybrid.

For initiating a successful hybrid programme following prerequisites should be satisfied :

(i) Substantial magnitude of non-additive gene effects should be responsible for inheritance of yield and other important traits.
(ii) Sufficient genetic diversity should be present in the populations being handled.
(iii) For economizing the hybrid seed production either male-sterility or self-incompatibility or some suitable flower morphology is desirable.

Basically there are three kinds of hybrids in vogue in different crops. These are:

Single-cross

It is the first generation of cross between two inbred lines ($A \times B$) used for commercial planting. Single-cross hybrids are uniform with respect to characters like plant height, days to maturity, fruit size and colour etc.

Three-Way Cross

The first generation of cross between two inbred lines ($A \times B$) is crossed to another inbred line (C).

In the Indian maize improvement programme the hybrid programme was slightly modified in sixties to develop double top-cross hybrids. Crossing an open-pollinated variety to a single-cross leads to development of double top-cross hybrid. In the three-way cross an inbred line is employed as third parent. Ganga 5, Ganga Safed 2 are, for example, double top-cross hybrids of maize released in 1989.

Double-Cross

Double-cross hybrid is developed by crossing two single-crosses, i.e. ($A \times B$) ($C \times D$). First generation of cross of two single-crosses is used for commercial planting. Four inbred lines are involved in double-cross. Four maize hybrids, Ranjit, Deccan, Ganga 1 and Ganga 101, were double-cross hybrids. Odland and Noll (1950) outlined a programme for double-cross hybrid production of cabbage. Although double-cross and three-way cross hybrids are economically viable for vegetables they lack uniformity (Kalloo, 1988). Hence, in vegetable crops single-cross hybrids are preferred.

Distinction is made between conventional and non-conventional hybrids in maize. Conventional hybrids are those, which utilize only inbred parents for their synthesis. The constituent lines may have varying levels of inbreeding, ranging from partial inbreeding to completely fixed lines (Vasal and Srinivasan, 1995). There may be modified conventional maize hybrids, viz., modified single-cross [($A \times A'$) $\times B$], double modified single-cross [($A \times A'$) $\times$ ($B \times B'$)], and modified three-way cross [($A \times B$) ($C \times C'$)] hybrids, where prime is used to indicate sister lines. Sister line crosses are used to improve yield and vigour of inbred parents with least adverse effect on performance and uniformity of commercial hybrids (Vasal *et al.*, 1995).

The hybrids, which involve at least one non-inbred progenitor, are called non-conventional hybrids (Vasal, 1987). There is a range of non-inbred parents that can be crossed with each other to produce a number of different kinds of non-conventional hybrids, e.g., broad genetic base gene pools, broad genetic base synthetics, experimental populations, natural populations, etc. but the main type of non-conventional hybrids are inter-varietal, inter-family, top-cross (variety × inbred), double top-cross (single cross × variety) hybrids.

At CIMMYT, Mexico, hybrid maize programme was started in 1985. The CIMMYT caters the requirements of developing countries, most of which do not have well developed breeding programmes and seed industry is non-existent. In order to cater the needs of such countries and limit the exploitation of heterosis, CIMMYT decided to lay more emphasis initially on the use of non-inbred parents in hybrid variety development. And hence the origin of new terminologies, viz., conventional and non-conventional hybrids (Vasal, 1986, 1987).

The merits of conventional hybrids are:

(i) The purity of parental lines. Hence genetically similar F_1 seed of hybrid can be produced year after year.
(ii) Hybrids are appealing to the grower because of their uniform appearance for most of the characters and better phenotype.

The time span required for developing and identifying superior double-cross or three-way cross of maize is about six to eight years, depending upon the level of inbreeding of the lines involved and stages at which top-cross evaluation can be accomplished. Double-cross and three- way cross hybrid development involves two stage testing, viz., one for evaluation of single-crosses and another to test the actual three-way or double-cross hybrids. For development of single-cross hybrids the time span could be relatively less.

The merits of non-conventional hybrids are:

(i) Short time required to identify hybrid combinations.
(ii) Reduced cost of seed and hence more affordable to farmers.
(iii) Less vulnerability to epidemics due to their built-in genetic variability, and
(iv) More yield than open-pollinated varieties.

But it will be realized that it is difficult to maintain the parents in genetically pure state because they are variable. The non-conventional hybrids have, of course, yield disadvantage as compared to the conventional hybrids but they have their own place (Vasal and Srinivasan, 1995).

Experience of the maize breeders of India in development of composites revealed that continuous selfing of plants selected from open-pollinated population is not essential for derivation of hybrids and exploitation of heterosis. One of the inbred parents of hybrid could be replaced by an heterozygous parent (Mukherjee and Ahuja, 1995). As a result non-conventional double-top-cross hybrids were produced and released in India for general cultivation.

12.2 INBRED LINES

An inbred line is a 'pure line' developed by continued self-pollination (and/or sib-mating) of plants selected on the basis of their superiority from open-pollinated population. Selection undoubtedly accompanies selfing. Selfing continues till apparently homozygous plants are obtained. Purpose of inbreeding, obviously, is to fix the characters in homozygous condition in order that the line may be maintained in genetically pure form without any change. Consequently hybrid can be produced of a given genotype year after year. Inbred lines are an essential prerequisite for developing hybrids in case of cross-pollinated crops. This may not be essential in case of self-pollinated crops. Most of the varieties are homozygous at most, if not all, the loci.

In case of cross-pollinated vegetable crops such as onion, cabbage, cauliflower, radish, turnip, carrot and beet it is necessary to produce inbred lines. However, in cucurbits, e.g. muskmelon, watermelon, cucumber, squash, pumpkin, in which varieties can be maintained in pure form by inbreeding without loss of vigour, it is not necessary to produce inbred lines. In these crops homozygous varieties can be straight away used for hybrid seed production. In self-pollinated vegetable crops homozygous varieties are directly used to produce hybrid.

12.2.1 Source Population

In case of maize the source population for isolating superior plants for developing inbreds can be open-pollinated varieties, composites, synthetics, germplasm complexes or any other heterozygous population. This would be true for other cross-pollinated crops also. Elite varieties in other crops like sorghum, cotton, rice, tomatoes, brinjals and chillies etc. could form the source population. An essential requirement for a base population is that it should be rich in superior gene contents, should exhibit high performance for yield and other important traits over locations and years and should possess genes for resistance to prevailing diseases and insect pests. The base population should have desirable genes for quality attributes of the economically important produce of plant, e.g. lint in case of cotton, fruits in case of tomato, brinjal, okra etc., grains/kernels in cereals, pulses, oil in case of oilseed crops and so on. In other words, the base population should be a reservoir of superior genes for several desirable traits.

12.2.2 Developing Inbred Lines

Self-fertilization is almost always preferred for development of inbred lines but depending upon inbreeding depression sib-mating may be carried out. Plants to be selfed are selected on the basis of general vigour, standing ability, plant height, days required for maturity, disease and insect resistance etc. Selection of vigorous plants during the course of inbreeding neither delays the attainment of homozygosity nor does it adversely influence the general combining ability. Selection is practiced at harvesting time as well for characters such as length of ear of maize, colour of kernel or fruit, size of fruit, shape of fruit, shape of head/curd etc. Undesirable plants are out right rejected. Seeds from selected plants are used to plant progeny row in the next season. Usually progeny of 20-30 plants of each selection is planted in single hills spaced at distance recommended for the crop. In the standard method of inbred development selection is practiced both between and within progeny rows and only the best plants in best rows are selected for further inbreeding. With continuation of inbreeding there is a considerable decline in vigour, in case of majority of cross-pollinated crops, but there is increased uniformity of plants within each row. In maize and several other crops, several strains are lost, during the course of inbreeding, because of deficiencies and lack of appeal. After about five to six generations of self-pollination every plant within a row is like others but differences among lines are conspicuous. At this state self-pollination is often discontinued and lines are propagated by sib-mating. Sibs are brothers-sisters, with both parents common in case of full-sibs and any one parent common in case of half-sibs. In self-pollinated crops all the plants within a progeny row originate from common parents and hence are full-sibs. In maize and other cross-pollinated crops progeny of self-pollinated plant also results in full-sibs. In case of maize, pollen from all the plants (from tassel) within a row are collected and then placed on silks of ears of plants within that row. This accomplishes sib-mating or sibbing.

Jones and Singleton (1934) proposed a modification of this procedure used for development of inbred lines of maize. The method is known as "*single hill method*". In this method progeny of each selected plant, during inbreeding, is represented by a single three plant hill. Thus in each inbreeding stage the selected plant is represented by a single hill of 3 plants. This method permits growing of a large number of plants originating from a number of selections made initially. However, this scheme does not have scope for within family selection.

Another approach advocated to attain homozygosity in a few steps is the use of haploids. This method has been tried only in maize. Einset (1943) suggested use of doubled haploids to produce inbred lines. Chase (1952) observed occasional androgenic haploids in maize.

Self-pollination and/or sib-mating is also practiced in vegetable crops to produce inbred lines. Most of the Cole crops, such as cabbage, cauliflower, etc. are self-incompatible. Majority of plants in the population are, therefore, heterozygous with respect to self-incompatibility (s) locus. For development of inbred lines in these crops good plants are chosen at heading stage (Odland and Noll, 1950). These plants serve as foundation stock. These plants are self-pollinated by adopting bud-pollination (I_0). The progeny of bud-pollinated plants constitutes I_1 generation. Hand-pollination is involved for maintenance and increase of inbred lines through bud-pollination. Buds due to open about a day or two days later and of convenient size to work with, when pollinated with mature pollen (collected from previously bagged flowers) results in good seed set. Selfed progeny of each plant is grown. About 10 or more plants are randomly selected from the progeny and crosses are made *inter se*. The pod and seed setting in various combinations indicates the genetic constitution of plants with respect to self-incompatibility alleles. Starting with original plant, say with genetic constitution S_1S_2, the inbred progeny would be S_1S_1, or S_1S_2 or S_2S_2. These plants will be selfed through bud-pollination. In this way at the end of inbreeding S_1S_1 and S_2S_2 inbred lines can be obtained.

In cabbage, cauliflower, onion, radish and carrot greater uniformity of phenotype is expected for hybrids than other crops like cucurbits. The degree to which parents should be inbred depends upon the extent of uniformity expected for hybrid and inbreeding depression. Onion exhibits pronounced inbreeding depression and hence 2-3 generations of inbreeding will suffice. Inbreeding can be practiced for 3-4 generations in cabbage and cauliflower and much longer in cucurbits (Singh *et al.*, 1998).

12.3 EVALUATION OF INBRED LINES

Hybrid between two selected inbred lines is expected to exhibit at least some increase in vigour and yield as compared to its inbred parents. But what a breeder aims is that the hybrid should be significantly superior than the open-pollinated variety or even the hybrid being grown. For this purpose identification of inbred lines, which are expected to transmit superior genes to their progeny is imperative. Kisselbach observed a general correlation between the yields of inbred lines themselves and their single cross hybrids. Subsequently the results of research work by Hays and Johnson (1939) revealed that visual selection can yield inbred lines with superior combining value.

Superiority of inbred lines in crosses depends upon their ability to transmit desirable genes to their progeny. A number of inbreds lines are usually developed by a breeder initiating a hybrid programme because very few turn out to be desirable parents of hybrids. It is relatively easier to develop inbred lines but it is very difficult to develop and identify superior inbred lines which when crossed with another inbred will produce a superior hybrid combination. As mentioned in chapter 11, nicking of genes between two inbreds leads to development of superior hybrid.

Concept of combining ability has relevance to this aspect. Combining ability is the relative ability of a biotype to transmit desirable performance to its crosses. Sprague and Tatum (1942) delineated the importance of general combining ability while choosing a superior inbred line. Average performance of a strain or an inbred line in series of crosses is its general combining ability (*gca*). An inbred line identified as a good general combiner tends to transmit superior genes to its progeny when crossed to a heterozygous tester parent. Griffing (1956) attributed variance due to general combining ability to additive genetic variance and digenic non-allelic interaction variance of additive × additive kind. It is, therefore, fixable kind of variance.

In the early days all possible crosses were made among *n* inbred lines and average performance of a given inbred line involved in $n(n-1)/2$ crosses was estimated in order to pick up superior inbred line. Thus with 10 inbreds, $n(n-1)/2$ =, 45 crosses are to be made (ignoring reciprocal crosses) for evaluating the inbred lines. The number of crosses to be made would be enormous as the number of inbreds increase. For instance, with 100 inbred lines, 4950 single-crosses would have to be effected. In 1927, Davis suggested a method called top-cross test, for evaluation of general combining ability of inbred lines. Only *n* crosses would be required for evaluating general combining ability of *n* inbred lines rather than $n(n-1)/2$. An open-pollinated variety with broad genetic base would be an appropriate tester. All the inbred lines are crossed to the tester and the progeny is evaluated over locations and years. Jenkins and Brunson (1932) provided evidence to show the usefulness of top-cross test for evaluating general combining ability of inbred lines. They compared the rankings of *n* inbred lines obtained by two procedures, viz., the top-cross test and average performance of an inbred line in all possible single-crosses with other inbreds being considered. They observed that those inbred lines, which produced low yields in top-crosses, were found to produce low yielding single-crosses. On the other hand, high yielding single-crosses resulted from inbred lines identified as superior in top-cross. On the basis of these observations Jenkins and Brunson (1932) concluded that it was safe to discard the lower yielding half of the lines in top-crosses without serious risk of losing promising material. It is now generally accepted that top-cross is a satisfactory way of evaluating the inbred lines for their general combining ability, especially when the material so developed is tested over locations and years. Sprague (1939) described procedure to be followed for making top-crosses.

The choice of tester to be employed differs with the objectives. For example, if the inbred lines are to be isolated for resistance to a specific disease and disease resistance is monogenic recessive, then the tester should carry recessive genes in homozygous condition otherwise superior lines can not be identified. Tester with dominant gene will not permit correct identification of disease resistant inbreds as shown below:

Genotypes of tester	*Genotypes of inbreds* AA	aa	
AA	AA susceptible	Aa susceptible	Genotype of progeny disease reaction
aa	Aa susceptible	aa resistant	

Desirable tester is one that provides maximum information about the inbred lines. It should be simple to use. No single tester fulfils all the requirements for all the circumstances.

Combining ability of inbred lines or parental varieties of vegetable crops can be evaluated through top-cross, single-cross, diallel-cross and polycross (Singh *et al.*, 1998). Top-cross can be used for crops like beet, diallel and polycross for crops such as cabbage, cauliflower, radish and carrot, diallel for onion, tomato, cucurbits, brinjal, capsicum and okra.

In maize, the common procedure to test the general combining ability is by adopting top-cross test. When the inbred lines have attained considerable degree of homozygosity they are subjected to the top-cross test. About four to five generations of inbreeding is sufficient in maize to attain sufficient homozygosity. This is certainly time consuming. Moreover after four to five generations of inbreeding if the top-cross reveals that majority of the inbred lines developed are poor general combiners, it is wastage of money, efforts and time. A breeder has to start the inbred development programme again. Jenkins (1935) was the first who proposed early testing of plants to be inbred. According to the early testing scheme original selections from the base population (S_0) or the first generation selfed plants (S_1) are outcrossed to a broad based tester (heterogeneous) and the resulting progeny is evaluated for yield and general performance. Lines showing poor general combining ability are rejected. Rest of the lines are continued by selfing/sibbing and developed into inbred lines. This approach is based on the assumption that S_0 or S_1 plants differ in general combining ability and these differences could be detected by the top-cross test, despite the problem of sampling posed by the heterozygous nature of S_0 or S_1 plants. Jenkins (1935) believed that inbred lines acquire their individuality as parents of top-cross very early in the inbreeding process and tend to remain fairly stable for combining ability thereafter. Scientists supporting the concept of early testing provided adequate evidence. An objection against this concept is that the performance of inbred lines in crosses may change with the degree of homozygosity attained with inbreeding.

Polycross test is employed to evaluate the combining ability of forage grasses and some vegetable crops such as cabbage. Polycross test was independently proposed by Tysdal *et al.* (1942) in United States of America and Wellensiek (1952) in Holland. The term polycross was proposed by Tysdal *et al.* (1942) to designate "the progeny from seed of the line that was subjected to out crossing with other selected lines growing in the same nursery". The test essentially comprises of a polycross nursery, where crosses are made, and the polycross progeny test where progenies obtained from seed of polycross nursery are evaluated. About 100-300 inbred lines are established and evaluated for two to three years. Best 25 to 50 selected lines or clones are subject to polycross test. Selected lines or clones are planted in four to ten replications in the polycross nursery. Planting is carried out in such a manner which will encourage random intermating. Seeds from all replicates of each inbred line or clone is composited separately which is known as polycross seed. Polycross test comprises of replicated trial of these progenies. One or more check may be included. Better parental lines or clones are then selected and used in hybrid combinations or for production of synthetic variety.

It is assumed that in polycross nursery random pollination takes place and seed produced on a line or a clone in polycross nursery should be same as composite of all single-crosses of same line or clone. Random pollination may be adversely influenced by:

(i) The volume of pollen produced,
(ii) The time of pollen shedding,
(iii) Differential cross-compatibilities, and
(iv) Level of selfing.

Provided the pollen travels for only a short distance random mating will be adversely influenced.

Comparison of the polycross test with other types of tests revealed that polycross test provides essentially same information about general combining ability as others.

12.4 SPECIFIC COMBINING ABILITY OF CROSSES

The top-cross test or the polycross test permits a breeder to identify inbred lines which have ability to transmit superior genes to their progeny. A number of superior inbred lines can thus be isolated. However, hybrid between two (or 3 or 4) specific inbred lines (or varieties in case of self-pollinated crops) can be superior. With additive gene effects the hybrid performance will be equal to sum of effects of genes transmitted to the progeny by the two parents. However, with intra-allelic and inter-allelic interaction the hybrid will deviate from sum of the effects of genes contributed by the two parents. Even when the genes are dispersed between two parents and with partial to complete dominance, hybrid will exhibit excellent performance as compared to the two parents. Which two (or more) parents when crossed exceed the sum of effects contributed by two parents is an important aspect of hybrid breeding designated by Sprague and Tatum (1942) as specific combining ability. The term specific combining ability, according to them is used to designate those cases in which certain combinations do relatively better or worse than would be expected on the basis of average performance of lines involved. This assumes that the inbred lines confer general combining ability effects upon the hybrid and the expected performance, on the basis of additive gene effects would be the sum of general combining ability effects of two parents i.e. $g_i + g_j$. The mean performance of hybrid between the ith and j^{th} line would therefore be:

$$x_{ij} = \mu + g_i + g_j$$

Where μ is the overall mean performance of population of inbreds, g_i and g_j denote the general combining ability effects of ith and j^{th} parent, respectively. However, when the genes contributed by the two parents to the F_1 interact (intra-allelic and/or inter-allelic) the mean performance of hybrid would be linearly written as:

$$x_{ij} = \mu + g_i + g_j + s_{ij} \qquad \text{...(12.2)}$$

where s_{ij} denotes the specific combining ability effect. S_{ij} could assume positive or negative sign, i.e. the mean performance could be enhanced or reduced, than expected on the basis of additivity of effects, by an amount equal to s_{ij} whenever the genes from two parents interact. Griffing (1956) ascribed variance attributable to specific combining ability effects to dominance variance and digenic non-allelic interaction of additive × dominance and dominance × dominance kind. Hence this variance is not fixable.

With n inbred lines in stock, the double-crosses that are possible, taking 4 inbreds at a time, would be given by $3n! / [(4!)(n-4)!]$. For example, with 10 inbred lines the number of double-crosses that are possible, excluding reciprocals, would be 630. Identification of a superior double-cross combination is a cumbersome job as the number of inbred lines increase. When a breeder has 10 inbred lines with him, he has to produce 630 possible double-crosses and grow them over locations and years in replicated trials to evaluate them. After identifying good general combining lines on the basis of top-cross or other designs appropriate for the crop, there should be a method which will identify a superior double- cross combination without having to actually produce all possible double- cross hybrids and test them. Jenkins (1934) correlated the yields of 42 double-crosses with those expected on the basis of 4 prediction methods, viz.:

(i) Average yield of all possible 6 single-crosses involving the 4 inbred lines (*A*, *B*, *C* and *D*), i.e. average of ($A \times B$), ($A \times C$), ($A \times D$), ($B \times C$), ($B \times D$) and ($C \times D$). This mean was used to predict performance of ($A \times B$) ($C \times D$).

(ii) Average yield of four single-crosses, excluding the two single-crosses making up the double-cross. For instance with 4 inbreds (*A*, *B*, *C* and *D*), mean performance of ($A \times C$), ($A \times D$), ($B \times C$) and ($B \times D$) could be used to predict mean performance of ($A \times B$) ($C \times D$).

(iii) Mean yielding ability of every inbred line in all possible single-crosses with 10 other inbreds is obtained, and lastly

(iv) Mean performance of top-crosses of the four inbreds lines that are involved in a double-cross is used to predict performance of double-cross. For example, with 4 inbreds the mean performance of ($A \times B$) ($C \times D$) is predicted from:

$$\frac{(\text{tester} \times \text{A}) + (\text{tester} \times \text{B}) + (\text{tester} \times \text{C}) + (\text{tester} \times \text{D}).}{4}$$

Investigations revealed that most accurate estimate of the yield of a double-cross can be predicted from mean yields of four non-parental single-crosses, i.e. method (*ii*). Further investigations revealed that crosses between inbred lines from different sources are more productive than crosses between inbreds from same population. For example, when inbreds *A* and *B* have been derived from one source population and inbreds *C* and *D* from another population, most productive double-cross would be ($A \times B$) ($C \times D$). Evidence for this concept was provided by Eckhardt and Bryan (1940).

12.5 HYBRID VARIETIES IN DIFFERENT CROPS

Hybrid varieties of several crops, such as maize, sorghum, pearl millet, cotton, rice, sunflower and several vegetable crops (e.g. tomato, brinjal, chillies, cabbage, cauliflower, carrots, watermelons, muskmelons, etc.) are being produced and marketed in India. Hybrid seed production of some of the crops will be briefly described. This will give reader an idea about the diverse ways by which hybrids can be produced. Details of the procedures are not given.

12.5.1 Cereals

Double-cross, double top-cross and single-cross hybrids of maize have been released in India. Deccan, Ranjit are double-cross hybrids, Ganga 5 and Ganga Safed 2 are double top-cross hybrids and Pusa Early Hybrid Makka (*CM* 135 × *CM* 136) and Pusa Early Hybrid Makka 2 (*CM* 137 × *CM* 138) are the examples of single-cross hybrids released by All India Coordinated Maize Improvement Project. Private seed companies have also produced several maize hybrids.

In the United States of America double-cross hybrids of maize were the predominant types of hybrids in cultivation. Double-cross hybrids were in vogue initially because of the problems with hybrid seed production of single-crosses. One of the most significant changes in maize industry in recent years is production of single-cross hybrids by seed companies of USA and acceptance of these hybrids by the farmers (Eberhart, 1969). Seed of single-cross hybrid can now be produced at a reasonable cost from the now available superior inbred lines. Theoretically double-cross hybrids are expected to be more stable in performance when grown over several environments as compared to the single-cross hybrids because they are heterogeneous. This has been designated as genetic homeostasis by Lerner (1958) and populational buffering by Allard and Bradshaw (1964).

On the other hand, some single-crosses may be as stable as most of the double-cross hybrids (Eberhart, 1969), which is attributed to developmental homeostasis (Lerner, 1958) or individual buffering (Allard and Bradshaw, 1964). Some single-cross hybrids of maize, which yield more than highest yielding double-crosses and as phenotypically stable as double-cross hybrids have been identified (Eberhart and Russell, 1969).

The inbred parents of hybrids are multiplied in isolation (not less than 400 m from any maize with same kernel colour and texture and 600 m from maize with different kernel colour and texture). For production of single-cross the female and male parents are planted in field in a ratio of 4 : 2 in isolation. Plants in female rows are detasseled when tassel is well out of the leaf sheath but before the anthers start shedding pollen. Double-cross hybrid seed is produced by planting single-cross employed as seed parent alternately with another single-cross used as pollen parent The planting ratio of female to male rows is 6 : 2. Female rows are detasseled at appropriate time. Seed set in ear heads of female rows is hybrid seed.

In United States of America cytoplasmic-genetic male-sterile lines are used as female parents for production of hybrid seed. Three cytoplasmic sources are recognized, viz., T-cytoplasm (Texas cytoplasm), C-cytoplasm (Charrua cytoplasm) and S-cytoplasm (Beckett, 1971). The male-sterility results from interaction of nuclear genes with factors in cytoplasm. Genes causing male-sterility are recessive. Fertility restoration of lines with T-cytoplasm is determined by two dominant nuclear genes, viz., RF_1, RF_2 (Duvick, 1965), while the fertility of C-cytoplasm is restored by single gene, RF_4 (Pour *et al.*, 1981). Fertility of S-cytoplasm is restored by a single dominant gene RF_3 (Laughnan and Gabay, 1975). Before 1970, about 85 per cent hybrids of USA were based on T-cytoplasm. In 1970 epidemic caused by *Cochliobolus heterostrophus* (Southern corn leaf blight) devasted the maize fields in USA. Maize hybrids based on other cytoplasms were not affected.

The Indian Council of Agricultural Research (I.C.A.R.), New Delhi in collaboration with the Rockefeller Foundation of U.S.A. launched an Accelerated Sorghum Improvement Programme in 1961 to develop hybrids of sorghum suitable for Indian conditions. Combine Kafir 60A (CK 60A), a cytoplasmic-genetic male-sterile line, was introduced from U.S.A. and straight away used as seed parent to produce hybrids. This male-sterility results from interaction of recessive nuclear (ms ms) genes with heritable factors in mitochondria present in cytoplasm. The nuclear genes are from Kafir and cytoplasm is from Milo types of sorghum, which is classified as A_1 cytoplasm (Stephens and Holland, 1954). Alternate cytoplasms causing male-sterility are also known in sorghum, viz., A_2 (Schertz, 1977), A_3 isolated from an Indian variety Nilwa (Quinby, 1980) and A4 from local material being grown in Nigeria (Worstell *et al.*, 1984). Except the A_1 cytoplasm others are not presently commercially utilized. Because only one cytoplasm is used for development of hybrids of sorghum in the country utilization of germplasm is restricted because only those lines can be used as pollinators (or restorers), which carry gene for restoration of A_1 cytoplasm. If other cytoplasms could be used larger germplasm will be utilized. Moreover use of only one cytoplasm may lead to catastrophy if disease epidemic appears to which the A_1 cytoplasm is susceptible. It has happened in maize and pearl millet.

The cytoplasmic-genetic male-sterile line can not produce seed when grown in isolation because it does not produce pollen (or viable pollen, if at all some is produced). Maintainer lines have been developed, which are isogenic, i.e. possess exactly same genotype like its male-sterile counter-part, but possess a fertile cytoplasm. Hence it produces viable pollen. It is called *B* line or maintainer line and the male-sterile line is called as the *A* line. The *A* line is multiplied in an isolated field (400 m isolation from sorghum and Johnson grass). The *A* and *B* lines are planted alternately in a ratio of 2 rows: 1 row (Figure 12.1). The lines, which restore the fertility of cytoplasmic-genetic male-sterile (CMS) line with A_1 cytoplasm have dominant gene (*Ms Ms*) but may have fertile or sterile cytoplasm. For hybrid seed production 4 rows of male-sterile (*A*) line and 2 rows of restorer line (*R* line) are planted alternately in entire seed plot. Indian sorghum breeders have evolved several male-sterile lines, such as 2219*A*, 2077*A*, 296*A*, *AKMS* 14*A*, *IMS*9*A*, etc. with A_1 cytoplasm. New restorers have also been developed which led to evolution of hybrids like Coordinated Sorghum Hybrid 5 (*CSH*-5), 9, 13, 14 and 18.

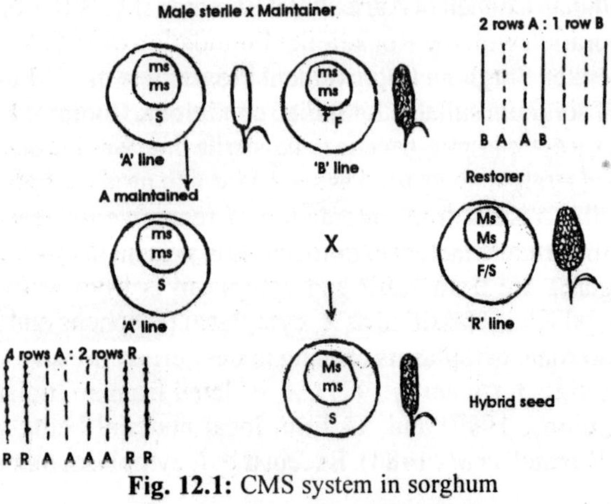

Fig. 12.1: CMS system in sorghum

Pearl millet is a highly cross-pollinated crop. It manifests protogynous nature. In 1965, first pearl millet hybrid, viz., *HB* 1, was released in India. Cytoplasmic-genetic male-sterile line Tift 23*A* developed by Burton (1965) was made available. This male-sterility results from interaction of recessive nuclear genes (ms_1 ms_1) with heritable factors in cytoplasm. Cytoplasmic factors associated with male-sterility have been shown to be encoded by the mitochondrial *DNA* (Lonsdale, 1987). Restriction Fragment Length Polymorphism (*RFLP*) of the mitochondrial *DNA* (*mt DNA*) is rapid method to classify the male-sterile cytoplasm (Pring *et al*., 1982; Bailey-Serres *et al*., 1986).

The restorer lines carry genes Ms_1Ms_1 and fertile or sterile cytoplasm. During 1970 an epidemic of downy mildew (causal organism *Sclerospora graminicola*) destroyed fields of *HB*1 in Haryana. Subsequently epidemic spread to other parts of the country. It was found that Tift 23*A* was susceptible parent. This was ascribed to its cytoplasm because Tift 23*B* was not susceptible. Cytoplasmic uniformity of hybrids being grown in vast area of the country makes them vulnerable to diseases and insect pest epidemics. Disease susceptibility in maize (Ullstrup, 1972; Pring and Lonsdale, 1985) and tobacco (Durbin and Uchytil, 1977) is associated with cytoplasmic genome variants. Sujata *et al*. (1994) and Chhabra *et al*. (1998) attempted to classify various cytoplasms responsible for male-sterility.

In order to develop male-sterile lines resistant to downy mildew efforts were made to induce mutations through irradiation of Tift 23B seeds (Govila, 1989). MS 5071A was found to possess some degree of resistance. Breeding efforts were also started at the Punjab Agricultural University and the Indian Agricultural Research Institute, New Delhi.

Multiplication of the male-sterile (*A* line) is carried out by planting *A* and *B* lines in a ratio of 4 rows: 2 rows in isolation. An isolation of 1000 m has to be maintained. For hybrid seed production the *A* and *R* lines are planted alternately in a ratio 6 rows to 2 rows, respectively, in isolation.

Hybrids of rice released in India for general cultivation in 1994 utilized cytoplasmic-genetic male-sterile lines developed at the International Rice Research Institute (*IRRI*), Philippines, viz., *IR* 58025*A* and *IR* 62829*A*. The cytoplasm responsible for the male-sterility is 'Wild Abortive' (*WA*). In 1973, cytoplasmic-genetic male-sterile rice lines were developed in China (Yuan, 1996). Recessive gene *rf* in homozygous condition when present in *WA* cytoplasmic background results in male-sterility. The maintainer line (*B* line) carries fertile cytoplasm but is isogenic with its corresponding *A* line. The fertility is restored by *Rf Rf* genes (Ahmed, 1996). A search for other cytoplasms in the germplasm, which result in male-sterility after interaction with nuclear genes, revealed that cytoplasms of *CMS DA*, *IP*, *DI*, *BT*, *KR*, *HL*, *Oryza perennis*, *ARC*, *GAM* and *TN* are useful in this respect (Mishra, 2002), however, only *WA*, *BT*, *DI* and *IP* are being utilized because their male-sterility is stable in all the environments and fertility is easily restorable.

Rice breeders designate the cytoplasmic-genetic male-sterility system as three line system since *A*, *B* and *R* lines are involved in hybrid seed production. Use of a specific cytoplasm restricts the use of germplasm to only those lines which can restore the fertility in hybrid. In rice it is expected that only about one third germplasm is useful with *WA* cytoplasm. Moreover, unitary source of sterility inducing cytoplasm is prone to the risk of susceptibility to a disease or an insect-pest (Ahmed, 1996). Sexuality in higher plants is influenced by nuclear as well as cytoplasmic hereditary factors, chemicals, environment and their interactions (Kaul, 1988). Among the environmental factors, temperature is the predominant factor influencing male-sterile gene action followed by photoperiod (Kaul, 1988). Thermo-and/or photosensitive male-sterile genes have been reported in some self-pollinated crops such as tomato (Rick, 1948; Rick and Boynton, 1967), barley (Ahokas and Hocket, 1977), wheat (Jan, 1974) and rice. As an alternative two-line approach has been suggested. The two-line approach utilizes either the environmental sensitive genic male-sterility (*EGMS*) or chemically induced male-sterility.

The genic male-sterility, which changes fertility with the alteration in certain environmental factors and *vice versa*, is known as environmental sensitive genic male-sterility. Scientist in China (Sun *et al.*, 1989) and Japan (Maruyama *et al.*, 1991) identified thermo-sensitive rice lines and Shi (1981) discovered photoperiod-sensitive rice plant called *NONGKEN* 58S in *japonica* variety *NONGKEN* 58. New sources of photoperiod-sensitive genic male-sterility (PGMS) and temperature-sensitive genic male-sterility (*TGMS*) were also isolated (Lu and Wang, 1986). The *TGMS* lines have total pollen sterility between 30 to 32°C but are partial to normal pollen fertile between 24 and 27°C. In this way the requirement of maintainer line is eliminated. Hybrid can be produced with only two lines, viz., *TGMS* and restorer.

International Rice Research Institute (IRRI), Philippines, is working in collaboration with research workers of Japan to develop *TGMS indica* rice. The Asian cultivated rices (*Oryza sativa*) have been broadly divided into two groups, viz., *indica* and *japonica*. These groups exhibit certain morphological, physiological and genetical differences. A third group *javanica*, has now been recognized among Indonesian types. In India, *indica* types of rice is grown. And hence the necessity to develop *indica* type *TGMS* lines. *A japonica* rice *TGMS* mutant (*PL* 12) was introduced from Japan and its single recessive gene responsible for thermo-sensitive male-sterility is being incorporated into elite *indica* rices. *TGMS* rice lines have been indigenously identified in India, e.g. *ATG* 1, *ATG* 4, *ATG* 8 (Siddiq, 1996). *ATG* 8 is a reverse *TGMS* in that it is sterile below 24°C and fertile above 28°C.

Most of the photoperiod sensitive genic male-sterile (*PGMS*) lines remain male-sterile under long day (more than 13.75 hours) condition and revert back to fertility under short day condition (Mishra, 2002).

Photoperiod sensitive genic male-sterile (*PGMS*) system in rice, discovered in Hubei province of China, showed complete to high sterility when the line flowered before September 1 or 2 (day length 14 hours), however, male-sterile plants flowering after this date (day length less than 13.75 hours) showed various degrees of spikelet fertility ranging from 10 to 40 per cent (Lu and Wang, 1988). The sterility of sterile stage was stable but the degree of fertility was not stable in fertile stage. The sterility of *PGMS* line was controlled by a single recessive gene pair and some modifiers. *PGMS* system is already in use to develop F_1 rice hybrids in China (Virmani *et al*., 1990).

In India, *TGMS* male-sterility system can be profitably utilized for hybrid rice seed production. During the sterile phase the *EGMS* line can be used as a seed parent for hybrid seed production, whereas at the fertile phase it can be multiplied. Since there are wide temperature differences in the country, *TGMS* lines are being utilized for two line hybrid programme.

The planting ratio for multiplication of cytoplasmic-genetic male-sterile lines is 2 rows of *B* to 10 rows of *A*. However, for hybrid seed production 12 rows of *A* are alternately planted to 2 rows of restorer line. Application of GA_3 at panicle emergence leads to higher seed yields.

12.5.2 Cotton

India is the only country, which cultivates cotton hybrids on commercial scale. Research to develop cotton hybrid was being conducted as early as 1952 in India (Patel and Patel, 1952). Although hybrid number 1, 2 and 3 were developed at Surat, they were not promising. First cotton hybrid with extra long staple length, viz., hybrid 4 (*H*4) was released for commercial cultivation in1970 (Patel, 1971). Hybrid 4 involved two *hirsutum* lines, namely Gujarat 67 and American nectariless. Subsequently several *hirsutum* × *hirsutum* hybrids, such as *JKHy*1, *NHH*4, Savita, *MECH* 4 etc. and *hirsutum* × *barbadense* hybrids (Varalaxmi, *DCH* 32, *JKHy* 11 etc.) were developed and released. These hybrids were developed by hand emasculation of the seed parent and hand pollination of emasculated flower. Because the pollen grains of cotton are oily and heavy they have to be physically transported to the desired place. Hybrids produced mechanically, i.e. hand emasculation and hand pollination, are called by the cotton breeders as conventional hybrids. A few conventional hybrids exhibited 200 per cent heterosis (Patil, 2002). Hybrids of desi cotton (diploid) such as *DH* 7 and *DH* 9 were released in 1985 and 1988, respectively. Both of these are *arboreum* × *herbaceum* hybrids.

For exploitation of heterosis in cotton through hybrids, parents have to be carefully chosen as in other crops. Good general combining ability is mandatory for male as well as female parents. Apart from this the female parent should have good boll weight because higher boll weight contributes to more hybrid yield (Patil, 2002). Further the seed index should be good. If 100 seed weight is more the hybrid seed yield will be more. More seeds per boll is another important requirement. Obviously this will result in more hybrid seeds per boll. Lastly, emasculation should be easier. The anther column should be loose for its removal in one stroke. Conventional method of hybrid seed production involves Doak (1934) method of hand emasculation or its modification (Mehta and Patel, 1983).

A flower bud, which is about to open the next day is emasculated during afternoon or evening by removing the anther column. Such emasculated flowers are enclosed inside a red tissue paper bag or with a twine thread for identification of bud to be pollinated the next day and to identify crossed boll at the time of harvesting, respectively. Next day morning, between 8 to 12 noon, when the stigma is receptive, male flowers (of the male parent) are collected in a tray and pollen from one flower of male variety is used to pollinate 4 to 5 emasculated flowers. Uncrossed flowers should be removed.

For hybrid seed production the female and male parents are sown in contiguous fields in a ratio of 4: 1 (Singh, 1989).

Male-sterility is also employed in cotton for hybrid seed production/ hybrid development. Genetic and cytoplasmic-genetic sterility is available in cotton.

Eleven loci have been identified to condition genetic male-sterility in cotton, 10 of them are in *G. hirsutum* and one in *G. barbadense*. The Genetic Male-Sterility (*GMS*) system involving genes $ms_5\ ms_5\ ms_6\ ms_6$ found in Gregg male-sterile is a good source of male-sterility. It is stable and is being utilized in India, Pakistan and USA. All the *hirsutum* genotypes carry $Ms_5\ Ms_5$ or $Ms_6\ Ms_6$ or both genes for restoration. Any *Gossypium hirsutum* line can be converted into a GMS line by repeated back-crossing with alternate selfing and selection (Patil, 2002). Maintenance of GMS can be accomplished by sib-mating of male-sterile plants ($ms_5\ ms_5\ ms_6\ ms_6$) with male-fertile plants ($ms_5\ ms_5\ Ms_6\ Ms_6$ or $Ms_5\ Ms_5\ ms_6\ ms_6$). The seed production plot of *GMS* will comprise of fifty per cent male-sterile and 50 per cent male-fertile plants. The latter have essentially to be rouged during flowering and before pollination. Hybrid Saguna (*CPH* 2) was developed using Gregg genetic male-sterile (Srinivasan and Gururajan, 1978).

Interaction between the genes of *G. hirsutum* and factors in cytoplasm of different sources viz., *G. arboreum, G. anomalum, G. harknessii, G. longicalyx, G. aridum, G. trilobum* was found to produce male-sterility (Meyer, 1973; 1975; Stewart, 1992; 1996). However, *G. harknessii* (D_2 CMS), *G. trilobium* (D_8 CMS) and *G. aridum* (D_4 CMS) were found to be more stable male-sterile cytoplasms and are available for hybrid seed production (Patil, 2002). Most of the *G. hirsutum* lines are maintainers (*B*). Any B line can be converted to male-sterile by repeated back-crossing to a CMS line and selection. Restoration of CMS with *G. harknessii* cytoplasm (D_2) involves monogenic dominant action with one enhancer gene (Sheetz and Weaver, 1980; Stewart, 1992).

Both the tetraploid and diploid cotton hybrids based on cytoplasmic-genetic male-sterile lines have been evolved in India. According to Bhale (1998, personal communication) and Bhale and Bhat (1990) the CMS based cotton hybrids, including those with common genomes, yield 8 to 12 per cent less than conventional and GMS based hybrids. This aspect needs thorough investigation. Is reduction in yield due to cytoplasmic/ maternal effects or the male-sterile line during the conversion of a line could not be isogenic to the original line at least with respect to genes for yield and yield components?

12.5.3 Oilseeds

Castor is normally monoecious with pistillate flowers on the upper portion of the raceme and staminate flower are borne on lower portion of same raceme. Maize is also monoecious but staminate flowers are borne in tassel, which is at the top of stem while ears bear female flowers, which are borne at lower positions on the stem. The staminate and pistillate flowers are not present on same inflorescence as in castor. The proportion of pistillate and staminate flowers on same raceme manifest wide variation both within and between genotypes. In all the currently cultivated castor varieties, which are monoecious, the percentage of pistillate flowers along the raceme axis is usually maximum on the first raceme and declines thereafter progressively in subsequent orders, i.e. secondary, tertiary, etc., There is proportionate increase in the number of staminate flowers in subsequently formed racemes. This variation in proportions of staminate and pistillate flowers within line or within a plant is attributed to plant age, environmental conditions etc. At more than 32°C daily temperatures the spike tends to show more male flowers on different sequential spikes. Femaleness is generally stronger in younger plants grown on soil with higher fertility levels and lower temperatures in a particular time. Classen and Hoffman (1950) described variation in pistillate and staminate flowers on the racemes. For hybrid seed production 90 to 100 per cent pistillate character is used. This character is governed by a recessive pair of genes (Shiffris, 1956).

Three different types of pistillate mechanisms have been noticed in castor, viz., *N*, *S* and *NES* types. In the *N* type a recessive sex-switching gene (*F*) governs femaleness. Sib-mating can maintain the system. The *S* type mechanism is based on sex-reversals of variable types, viz., early and late. In sex reversals plants which are females to start with revert to monoecism later in one or more sequential order branches. If female plant reverts to monoecism in secondary or tertiary order it is known as early revertant. If the reversion takes place in quaternary branches or subsequent to it, it is late reversion. Among the late revertants more than ninety per cent are pistillate, and thus suitable for hybrid seed production. The female parent VP1 of castor hybrid GAUCH 1 is based on typical S type mechanism, characterized by variable degrees of reversions to monoecy any time after 5 to 10 sequential order racemes besides possessing environmental sensitive gene(s) for interspersed staminate flowers.

The *NES* mechanism comprises of gene(s) sensitive to environment for interspersed staminate flowers besides being homozygous for sex switching gene (*F*).

In United States of America a 100 per cent pistillate line was developed. This line, i.e. *TSP* 10*R* (Texas Pistillate 10) was released in USA in 1962 (Ramachandram and Prasad, 1995). It was received in India from Texas in 1965. Based on *TSP* 10*R* an indigenous pistillate line of castor was developed at Vijapur, which was designated as *VP*1. Gujarat Castor Hybrid 3 based on *TSP* 10*R* was released in 1968 (Gopani, 1969). Subsequently *GAUCH* 1 (*VP* 1 × *VP* 9) was released in 1973. A number of pistillate lines have now been evolved.

Although 177 per cent heterosis for yield and 50-80 per cent for oil output was reported for safflower, commercial utilization of heterosis was very limited due to lack of an efficient system to economize and ease the hybrid seed production. Genetic male-sterility in safflower was discovered by Heaton and Knowles (1980). After this discovery the research on hybrid safflower development was pursued seriously. A single pair of recessive gene governs the inheritance of genetic male-sterility discovered by Heaton and Knowles (1980). Singh (1997) developed a dwarf genetic male-sterile line, *DMS*. A heterozygous line with respect to sterility locus maintains the genetic male-sterile line. Heterozygous line produces two kinds of pollen, viz., half with gene for sterility and rest half carrying dominant gene for fertility. Plants which are fertilized by pollen carrying dominant gene set seed. Linkage of gene for dwarf stature with gene for male-sterility makes identification of male-sterile plants easy. At the age of 30 to 40 days after germination male-sterile plants are about 5 to 10 cm dwarfer than their normal sibs. This helps rouging of fertile plants.

Hybrid *DSH* 129 (*MS* 9 (0) × *A*1) was released in 1997. Subsequently several hybrids were developed.

Large scale cultivation of sunflower began in India in 1972 with the introduction of varieties of Soviet Union origin. These were Armavirski 3497 (*EC* 68415) and Peredovik (*EC* 68414).

Cytoplasmic-genetic male-sterility was discovered in sunflower by Leclercq (1969) in the progeny of a cross between *Helianthis petiolaris* and the cultivated sunflower, *H. annuus* (cv. Armavirkii 9345). Cytoplasmic factors of *Helianthus petiolaris* have been responsible for interaction with nuclear gene(s) of *H.annuus* for male-sterility. The male-sterility discovered by Leclercq is stable. Genes for fertility restoration were found in wild species by Kinman (1970).

Subsequently Leclercq (1971) and Enns *et al.* (1970) also reported gene for fertility restoration. A single dominant gene is said to restore fertility of male-sterile plants but there is controversy over the number of genes. First sunflower hybrid based on cytoplasmic-genetic male-sterile line, *CMS* 234*A*, was released in 1980 (Seetharam, 1980). The hybrid was *BSH* 1 (*CMS* 234*A* × *RH* 274). Subsequently several hybrids have been developed and released. The presently produced sunflower hybrids are single-cross hybrids, however, some private organizations are marketing three-way hybrids.

At Bangalore centre of the All India Coordinated Sunflower Improvement Project two new sources of cytoplasms responsible for male-sterility, have been identified. These are *CMS* I (*Helianthus annuus* spp. *lenticularis*) and *CMS PF* (*H.petiolaris* spp. *flallax* × *H.annuus*). These two sources have been distinguished in terms of maintenance and restoration behaviour (Virupakshappa *et al.*, 1991).

12.5.4 Vegetable Crops

For production of hybrid seed, where male-sterility or other genetic mechanism of emasculation is not available, hand emasculation and hand pollination of emasculated flowers is the only recourse available. Hand emasculation and hand pollination may be economically viable where:

(i) Hand emasculation of a single flower and subsequently hand pollination results in production of adequate number of hybrid seeds. For example, in tomato, brinjal, chilli, okra, tobacco etc. a single flower after being emasculated and pollinated with desired pollen yields enough seeds. On the other hand, only a single seed is obtained per spikelet in case of wheat, rice, sorghum etc.

(ii) In those crops which manifest considerable heterosis (as compared to better parent) and require relatively less seed to plant a unit area. Vegetable crops require few grams of seed to plant an hectare as against few kilogrammes in cereals, pulses, oilseeds.

(iii) In those crops where high yields compensate for higher cost of small quantity of seed required to plant a unit area. Other means have to be sought for those crops where because of difficulties hand emasculation can not be conveniently carried out. Male-sterile lines, self-incompatibility and different forms of sex expression are some genetic tools, which could be employed to economize hybrid seed production.

Various techniques, which could be utilized for production of hybrids in vegetables, were classified by Singh *et al.* (1998) into five categories.

(i) *Hand emasculation and hand pollination*: This is practiced in self-pollinated and often cross-pollinated crops with hermaphrodite flowers e.g. tomato, brinjal, pepper, okra.

(ii) *Hand pollination without emasculation*: This is practiced in monoecious crops like cucurbits. Pistillate lines are used as seed parents and hence only pollination has to be done. In crops like tomato, brinjal and capsicum where male-sterility is available, this technique is useful.

(iii) *Emasculation and wind pollination*: This technique is possible with dioecious vegetables like spinach (*Spinacea oleracea*) and asparagus (*Asparagus officinalis*). Actually in these crops male and female flowers are borne on different plants. But some intersexes appear, such as male intersex and female intersex (with terminal androecious flowers). The male plants appearing in female rows are pulled out and similarly female plants are pulled out from male rows. Monoecious plants are often found in spinach, which have been used for hybrid seed production (Thompson, 1955). In monoecious plants both male and female flowers are borne in same cluster in varying proportions.

(iv) *Emasculation and use of insect pollination*: This technique is also possible in cucurbits. Male flowers are removed from lines to be used as females. These are insect pollinated. For example, in Canada hybrid seed of cucumber is produced by planting male parent alternately with 2 rows of female parent (Walkof and Nuttal, 1955).

(v) *Free insect or wind pollination without emasculation:* This technique is useful in cross-pollinated vegetable crops, such as onion and carrot, where emasculation is very tedious, or is comparatively easy but pollination results in very few seeds, such as cabbage, radish and other cole crops. Here advantage is taken of multiple allelism for self-incompatibility.

All the above techniques are used in cereals, oilseeds and pulses, except the use of self-incompatibility and differences in sex expression.

(a) Tomato, Brinjal, Okra and Pepper

All these vegetable plants have hermaphrodite flowers and can be conveniently hand emasculated and hand pollinated.

In case of tomato the variety to be used as pollen parent is usually planted 2-3 weeks ahead of the female variety, which ensures adequate supply of pollen. For producing hybrid seed the ratio in which male and female parents are planted is 1: 5 unit areas (Chhabra, 1992). Flowers of brinjal are conspicuous as compared to those of tomato and pepper. Hence, hand emasculation is fairly easy. Hand emasculation is carried out in tomato and pepper.

For emasculation, flowers buds, which are expected to open the following day are selected. Anthers are removed with the help of needle and forceps. Emasculated buds are enclosed inside paper bags. Care should be exercised during emasculation that anthers do not burst open.

Flower buds are hygroscopic in the morning hours, hence flowers for collection of pollen are plucked from the male plants in the afternoon and pollen is collected in the next morning (Kalloo, 1998). Pollen is usually collected by tapping the anthers on watch glass and pollination is carried out by dipping the protruding stigma in pollen mass. In case of brinjals, buds could be pollinated a day prior to flowering and removing the petals immediately thereafter to prevent the visit of insects.

In okra, Joshi *et al.* (1958) investigated various possibilities of hybrid seed production and found that scraping the anthers soon after the opening of flowers of female parent, followed by hand pollination without bagging resulted in 85 per cent hybrid seed. Singh *et al.* (1998) are of the view that seed yield could be increased upto 90 per cent by bagging the pollinated flower buds.

Because male-sterile lines of tomato and brinjal are available these could be used to produce hybrid seed. The male-sterile line and pollen parent could be planted in alternate rows. Currence and Jenkins (1942) reported that upto 4 per cent natural out-crossing could take place in tomato.

A chilli hybrid *CH* 1 was produced at Punjab Agricultural University by employing a genetic male-sterile line *MS* 12. The pollen parent was *LLS*. *MS*12 was developed by utilizing male-sterile genes from the source line imported from France (Kalloo, 1995).

(b) Cabbage

Hybrids of cabbage are characterized by uniform, compact and heavy heads. In Brassicas the F_1 hybrids are produced by hand pollination or by free insect pollination with the help of self-incompatible lines (Sampson, 1966). Use of male-sterility and marker has also been suggested. Hand pollination is not economical for commercial seed production (Swarup, 1977).

Self-incompatibility in cole crops is attributed to S factors. For production of hybrids self-incompatible but cross-compatible lines with good general combining ability are selected. Self-incompatibility utilized for hybrid production in cole crops is based on 'oppositional factor hypothesis' (Singh *et al.*, 1998). Pollen grain carrying a specific S allele, which is present in stylar tissues of female line, will not germinate on stigma or will be unable to accomplish fertilization. Only when pollen grains carry different genes, they will germinate and effect fertilization. There are about 50 alleles at S locus.

Odland and Noll (1950) suggested a scheme for hybrid production in cabbage. Two inbred lines, which are self-incompatible but are cross-compatible are developed, such as S_1S_1 and S_2S_2 possessing genes with respect to self-incompatibility locus. Two other inbred lines possessing still other self-incompatibility alleles are developed from another strain, say S_3S_3 and S_4S_4. S_1S_1 and S_2S_2 from one source are planted in alternate rows in isolation (1000 m). Similarly S_3S_3 and S_4S_4 carrying inbred lines are sown alternately. This will lead to development of two single-crosses, viz. (S_1S_2) and (S_3S_4). Interplanting the two single-crosses will produce a double-cross hybrid. As mentioned in section 12.2.2 inbred lines of cabbage are developed by adopting bud pollination (because of self- incompatibility). Hand pollination has to be practiced for maintenance and increase of inbred lines by bud pollination. Synchronous flowering of the inbred lines is the most desirable requirement for seed production.

Single-crosses, employing self-incompatibility mechanism, have been produced in Brussels sprouts (Johnson, 1966). F_1 hybrids of cabbage (Single-cross) have been released in United States of America (York *et al.*, 1957).

Although male-sterility has been reported in several cole crops, like cabbage, Brussels sprouts, cauliflower, etc. it has not been used for production of hybrid seeds because of its instability varying with environment.

Gibberellic acid (*GA* 4/7) in isopropyl alcohol was found useful as gametocide. It induced complete male-sterility when sprayed on cabbage, cauliflower and Brussels sprouts.

(c) Cucumber (Cucumis Sativus)

There are two types of cucumber, viz., slicing cucumber and pickling cucumber. Latter are grown in India. Hybrid cucumbers are early and prolific because they fruit at every flowering node. They are uniform with respect to fruit size and shape, attractive in colour and flesh texture.

Cucumbers are actually monoecious but a large number of gynoecious lines have been identified. Peterson and Weigle (1958) reported that Korean variety 'Shogoin' produces gynoecious plants. Shogoin was crossed and back-crossed to a pickling variety Wisconsin *SMR* 18, to develop a gynomonoecious line that produces a high percentage of gynoecious plants (Peterson, 1960). Percentage of gynoecious plants in different gynomonoecious lines is variable. Peterson and Weigle (1958) suggested use of gynomonoecious lines for hybrid seed production of cucumber.

The gynomonoecious lines and monoecious lines to be used as seed and pollen parents, respectively, are planted alternately. At the 10 node stage monoecious and other intermediate sex types of plants are rouged out from the gynomonoecious lines. Hybrid seed formed on gynoecious plants from gynomonoecious rows is collected. Rouging should be accomplished before the pistillate flowers open on gynoecious plants.

Peterson and Anhder (1960) and Peterson (1960) suggested the use of homozygous gynoecious lines of cucumber for hybrid seed production. Two gynoecious lines, which exhibit good specific combining ability, are planted alternately. The gynoecious line to be used as male parent is sprayed with GA_3/GA_7 @ 2000 ppm at first true leaf stage and spray repeated after 14 days. This induced about 10 male flower formations per 100 nodes. Tolla and Peterson (1979) suggested spray of silver nitrate at 2 to 3 true leaf stage to induce formation of male flowers. Gynoecious plants can be maintained by selfing. For this male flower formation is induced in gynoecious lines. Remember gynoecious plants bear only pistillate flowers. Some silver compounds like silver nitrate (50 to 100 ppm) or silver thiosulphate at 2 to 3 leaf stage has been found to successfully induce formation of male flowers in gynoecious lines of cucumber (Nijs and Visser, 1979).

Recently ethrel (2-choroethylphosphonic acid) has been reported to be effective for modification of sex expression in cucurbits, particularly cucumber. It increases early femaleness in gynomonoecious and gynoecious lines (Robinson *et al.*, 1968; Tronickova, 1979). Most effective concentration is 250 to 500 ppm. The spraying should be done between first to third true leaf stages.

Pusa Sanyog (Japanese gynoecious line × Green Long Naples), a cucumber hybrid, has been identified by the All India Coordinated Vegetable Improvement Project (*AICVIP*) and recommended for cultivation.

(d) Muskmelon

Although potentialities for exploitation of hybrid vigour were demonstrated in muskmelon by Munger (1942), commercial development of hybrid lagged.

The lines to be used as female and male parents are planted alternately. Due to monoecious nature of flowering the female lines will bear male flowers, which may be removed before anthesis or pollen production. Pollen from male flowers of male line is placed on stigma of female flowers of female line to produce hybrid seed. A single pollination produces large number of seeds.

In muskmelon, gynoecious lines like *MSU-IG* are being used for hybrid seed production. Silver nitrate (100 ppm) can be used to induce perfect flowers in gynoecious lines for maintenance.

Genetic male-sterility has been utilized recently to produce hybrid seed. Punjab Hybrid 1 was developed by using genetic male-sterility (*MS*1 × Hara Madhu). Another hybrid, 'Pusa Rasraj' was developed at the Indian Agricultural Research Institute, New Delhi by crossing male-sterile line *M*3 with Durgapura Madhu (Fageria *et al.*, 2001).

(e) Onion

The concept of hybrid onions ows its origin to the classic discovery of male-sterility in variety Italian Red in 1925 by Dr. H.A.Jones. He observed a bulb (13-53) in variety Italian Red, which produced sterile plants. Vegetative propagation, through bulbils, of this bulb provided male- sterile lines (Jones and Emsweller, 1936). The male-sterility of bulb 13-53 was due to interaction of recessive nuclear gene (ms) with heritable factors in cytoplasm (Jones and Clarke, 1943). For maintenance/multiplication of the male-sterile line an isogenic line was developed with the difference that it had a fertile cytoplasm. With respect to male-sterility the constitution of male-sterile line could be written as *S* msms. The line, which maintains the male-sterile, has same genotype as the male-sterile line (*A* line) but fertile cytoplasm and hence it produces viable pollen grains. The constitution of the maintainer line (*B* line) is then *N ms ms*, where *N* denotes normal cytoplasm. Because the pollen grains of *B* line bear identical genes with *A* line, upon fertilization of egg cell of *A* line zygote is produced with same genetic constitution (nuclear as well as extra nuclear) as *A*. Utilization of the male-sterile lines for production of hybrids was suggested by Jones and Davis (1944). For production of hybrids restorer lines (*R* lines) are identified from existing varieties and/or germplasm, which carry nuclear genes *MsMs*. The male-sterile condition is thus monogenic recessive. Genes restoring fertility are monogenic dominant. The cytoplasm of *R* line could be either sterile or fertile. In the presence of even a single Ms gene the cytoplasmic factor is not effective and hence normal pollen is produced by the F_1. The constitution of F_1 would be *S Ms ms*.

Banga and Petiet (1958) proposed a scheme to convert a variety of breeder's choice to male-sterile line. While choosing a male-sterile line or converting a variety to male-sterility a breeder should keep in mind that the keeping quality of bulbs or storage quality should be good and should be suitable to Indian conditions with respect to their reaction to photoperiod and temperatures. The lines imported from USA, Japan or Holland may not meet these requirements and hence breeder should be cautious.

For producing hybrid seed the planting ratio of *A* to *R* lines would be 8 to 12: 1 to 2. The *A* and *R* lines are planted alternately, e.g. 8 rows of *A* alternating with 1 row of *R* and so on.

Schweisguth (1973) discovered another cytoplasm inducing male-sterility in onion (*T cytoplasm*) after interaction with three nuclear genes. Among the three gene pairs involved, gene *a* is independent while *b* and *c* are complementary.

Van Bennekom and Van der Meer (1978) observed gametocidal action of GA_3 and GA_7 in Jumbo and Durga varieties of onion. The lines were completely male-sterile after application of these gametocidal hormones. Use can be made of the gametocides for hybrid seed production. An advantage of using this technique is that a breeder is not limited to use only those varieties as restorers, which can restore fertility of male-sterile line. Nor is he restricted to use male-sterile to produce hybrid seed.

Advantage of hybrid onion over the open-pollinated varieties is uniform maturity, uniform but large size and weight of bulb.

Hybrid varieties of some of these crops are being grown in India by the farmers. But this is not the complete list of crops of which hybrids are produced and marketed. An attempt has been made here to give one example of the techniques by which hybrids are produced. But there could be other ways also.

References

Ahmed, M.I. (1996). Outlines of heterosis breeding programme in rice. In *Hybrid Rice Technology*, Ed. Ahmed, H.I. *et al.* Directorate of Rice Research, Hyderabad, pp. 55-65.

Ahokas, H. and E.A. Hocket (1977). Male-sterile mutants of barley. IV. Different fertility levels of msg 9 ci (cv. Vantage) an ecoclinal response. *Barley Genet.Newsl.*, 7:10-11.

Allard, R.W. (1960). *Principles of plant breeding*. John Wiley and Sons, New York.

Allard, R.W., and A.D. Bradshaw (1964). Implications of genotype-environment interaction in applied plant breeding. *Crop Sci.*, 4: 503-508.

Attawar, M.M. (2000). Tool for productivity gains. In *The Hindu survey of Indian Agriculture*. Publ. Kasturi and Sons Ltd., National Press, Kasturi Buildings, Chennai 600 002, pp. 145-149,

Bailey-Serres, J.; L.K. Dixon; A.D. Liddell and C.J. Leaver (1986). Nuclear-mitochondrial interactions in cytoplasmic male-sterile sorghum. *Theor.Appl.Genet.*, 73: 252-260.

Banga, O. and J.Petiet (1958). Breeding male-sterile lines of Dutch onion varieties as a preliminary to the breeding of hybrid varieties. *Euphytica*, 7: 21-30.

Beckett, J.B. (1971). Classification of male-sterile cytoplasms in maize (*Zea mays*). *Crop Sci.*, 11: 724-726.

Bhale, N.L. and M.G. Bhat (1990). Investigations on exploitation of heterosis in cotton (*Gossypium hirsutum* L.) using male-sterility. *Indian J.Genet.*, 50: 37-44.

Burton, G.W. (1965). Cytoplasmic male-sterility in pearl millet (*Pennisetum glaucum* (L.). *Agron.J.* 40: 230.

Chase, S.S. (1952). Production of homozygous diploids of maize from monoploids. *Agron.J.* 44: 263-267.

Chhabra, A.; K.N. Rai; I.S. Khairwal; S. Sivaramakrishnan and C.T. Hash (1998). Mitochondrial DNA-RFLP analysis distinguishes new CMS sources in pearl-millet (*Pennisetum glaucum* (L.) R.Br.). *J.Plant Biochemistry and Biotechnology*, 7: 85-92.

Chhabra, N. (1992). The tomato. *Indian Horticulture*, April-June, p.11-20.

Classen, E. and A. Hoffman (1950). The inheritance of pistillate character in castor and its possible utilization in production of hybrid seed. *Agron.J.* 42: 79-82.

Currence, T.M. and J.M. Jenkins (1942). Natural crossing in tomatoes as related to distance and direction. *Proc.Am.Soc.hort.Sci.*, 41: 273-276.

Doak, C.C. (1934). A new technique in cotton hybridizing, suggested change in existing methods of emasculation and bagging cotton flowers. *J.Hered.*, 25: 201-204.

Durbin, R.D. and T.F. Uchytil (1977). Cytoplasmic inheritance of coupling factor 1 sub units. *Biochem.Genet.*, 15: 1143-1146.

Duvick, D.N. (1965). Cytoplasmic pollen sterility in corn. *Adv.Genet.*, 13: 1-56.

Eberhart, S.A. (1969). Yield stability of single-cross genotypes. Proceedings of *24th Corn and Sorghum Research Conference*, 24: 22-35.

Eberhart, S.A. and W.A. Russell (1969). Yield and stability for a 10 diallel of single-cross and double-cross maize hybrids. *Crop Sci.*, 9: 357-361.

Eckhardt, R.C. and A.A. Bryan (1940). Effect of method of combining the four inbred lines of a double-cross of maize upon the yield and variability of the resulting hybrid. *Jour.Amer.Soc. Agron.*, 32: 347-353.

Einset, J. (1943). Chromosomal length in relation to transmission frequency of maize trisomics. *Genetics*, 28: 349-364.

Enns, H.; D.G. Dornell and W.O. Chubb (1970). Sunflower research — a progress report. *Proc. 4th Intern. Sunflower Conf. Memphis* T.N. pp. 162-168. *Intern. Sunflower Assoc.* Toowoomba, Australia.

Fageria, M.S.; P.S. Arya and A.K. Choudhary (2001). *Vegetable crops. Breeding and seed production.* Vol.I Kalyani Publishers, Ludhiana, pp. 104-123.

Gopani, D.D. (1969). And now Gujarat Castor Hybrid 3. *Indian Farmers Digest*, 11: 11-14.

Govila, O.P. (1989). Cross-pollinated crops: Pearl millet. In *Plant breeding — theory and practice*. Ed. V.L. Chopra. Oxford and IBH Publishing Co.Pvt.Ltd., New Delhi, pp. 213-223.

Griffing, B. (1956). A generalized treatment of the use of diallel crosses in quantitative inheritance. *Heredity*, 10: 31-50.

Hays, H.K.and I.J. Johnson (1939). The breeding of improved selfed lines of corn. *J.Am. Soc.Agron.*, 31: 710-724.

Heaton, T.C. and P.F. Knowles (1980). Registration of UC 148 and UC 149, male-sterile safflower germplasm. *Crop Sci.*, 20: 554.

Jan, C.C. (1974). Genetic male-sterility in wheat (*Triticum aestivum*): Expression, stability, inheritance and practical use. Ph.D. thesis Univ.California, Davis, USA.

Jenkins, M.T. (1934). Methods of estimating the performance of double-crosses in corn. *J.Amer.Soc.Agron.*, 26: 199-204.

Jenkins, M.T. (1935). The effect of inbreeding and selection within inbred lines of maize upon the hybrids made after successive generations of selfing. *Iowa State Coll.Jour.Sci.*, 3: 429-450.

Jenkins, M.T. and A.M. Brunson (1932). A method of testing inbred lines of maize in crossbred combinations. *J.Amer.Soc.Agron.*, 24: 523-530.

Johnson, A.G. (1966). Inbreeding and production of commercial F_1 hybrid seed of Brussels sprouts. *Euphytica*, 15: 58-79.

Jones, D.F. (1918). The effects of inbreeding and cross-breeding upon development. *Conn. Agric.Expt.Sta.Bull.*, 207: 100.

Jones, D.F. and W.R. Singleton (1934). Crossed sweet corn. *Conn.Agric. Expt. Sta.Bull.*, 361: 489-536.

Jones, H.A. and A.E. Clarke (1943). Inheritance of male-sterility in onion and the production of hybrid seeds. *Proc. Amer.Soc.Hort.Sci.*, 43: 189-194.

Jones, H.A. and G.N. Davis (1944). Inbreeding and heterosis and their relation to the development of new varieties of onions. *U.S. Dept. Agric.Tech.Bull.*, 874: 1-28.

Jones, H.A. and S.L. Emsweller (1936). A male-sterile onion. *Proc.Amer. Soc. Hort.Sci.*, 34: 582-585.

Joshi, B.S.; H.B. Singh and P.S. Gupta (1958). Studies in hybrid vigour III. Bhindi (*Abelmoschus esculentus* (L.) Moench). *Indian J.Genet.*, 18: 1-12.

Kakizaki, Y. (1930). Breeding crossed eggplants. *Japan J.Hered.*, 21: 253-258.

Kalloo, G. (1988). *Vegetable breeding* Vol.I. C.R.C. Press, Florida, USA.

Kalloo, G. (1995). Heterosis breeding in vegetable crops: Present status and future prospects. In *Hybrid research and development*. Proc. of Seminar Indian Society of Seed Technology, New Delhi, 1993. Ed. M.Rai and S. Mauria, pp. 165-172.

Kaul, M.L.H. (1988). *Male-sterility in higher plants*. Springer-Verlag. Berlin-Heidelberg: 1005.

Kinman, M.L. (1970). New developments in the U.S.D.A. and State experiment station sunflower breeding programme. *Proc. 4th Intern. Sunflower Association*, Toowoomba, Australia.

Laughnan, J.R. and S.J. Gabay (1975). In "Genetics and biogenesis of mitochondria and chloroplasts". Eds. Birky, W.W.; P.S. Perlman Jr. and T.J. Byers (Ohio State Uni. Press, Columbus, O.H.), pp. 330-349.

Leclercq, P. (1969). Une sterilite male cytoplasmique chezle tournesol. *Ann. Amelior.Plant*, 19(2): 99-106.

Lerner, I.M. (1958). *Genetic basis of selection*. John Wiley and Sons, New York.

Lonsdale, D. (1987). Cytoplasmic male-sterility: a molecular perspective. *Plant Physiol. Biochem.*, 25: 256-271.

Lu, Xing Gui and JiLin Wang (1988). Fertility transformation and genetic behaviour of Hubei photoperiod sensitive genic male-sterile rice. In *Hybrid Rice*, International Rice Research Institute, Manila, Philippines, 129-138.

Lu, Y.G. and J. Wang (1986). Fertility transformation and genetic behaviour of Hubei photoperiod sensitive genic male-sterile rice. In *Hybrid rice*, IRRI, Philippines, pp. 129-138.

Maruyama, K.; H.Araki and H. Kato (1991). Thermo-sensitive genetic male-sterility induction by irradiation. In *Rice Genetics*. International Rice Research Institute, Manila, Philippines, pp. 227-232.

Mehta, N.P. and C.T. Patel (1983). Hybrid seed production technique for exploitation of heterosis in cotton. *Indian Farming*, 23-31.

Meyer, V.G. (1973). A study of reciprocal hybrids between upland cotton (*Gossypium hirsutum* L.) and experimental lines with cytoplasm from 7 other species. *Crop Sci.*, 14: 658-663.

Mishra, R. (2002). Hybrid rice breeding for increasing yield. In *Genetic improvement of field crops*. Eds. Singh, C.B. and D.Khare, Scientific Publishers (India), Jodhpur, pp. 44-59.

Mukherjee, B.K. (1989). Cross-pollinated crops: Maize. In *Plant breeding — theory and practice*. Ed. V.L. Chopra. Oxford and IBH Publishing Co.Pvt.Ltd., New Delhi, pp. 199-212.

Mukherjee, B.K. and V.P. Ahuja (1995). Upgrading yield levels of maize hybrids in India through conventional and non-conventional approaches. In *Genetic research and education: Current trends and next fifty years*. Ed. B.Sharma *et al*. Indian Society of Genetics and Plant Breeding, New Delhi, pp. 674-686.

Munger, H.M. (1942). The possible utilization of first generation muskmelon hybrids and improvement method of hybridization. *Proc. Amer. Soc. Hort. Sci.*, 40: 405-410.

Nijs, A.P.M. and D.L. Den Visser (1979). Silver compounds inducing male flowers in gynoecious cucumbers. *Report Cucurbit Genet. Cooperative*, 2: 14-15.

Odland, M.L.and C.J. Noll (1950). The utilization of cross-compatibility and self-incompatibility in production of F_1 hybrid cabbage. *Proc. Amer. Soc. Hort. Sci.*, 55: 391-402.

Patel, C.T. (1971). Hybrid 4 — a new hope towards self sufficiency in cotton in India. *Cot.Dev.*, 1: 1-6.

Patel, G.B. and C.T. Patel (1952). Long staple cotton from hybrid seed crosses between *hirsutum* and *barbadense* species. *Indian Cot.Grow.Rev.*, 6: 205-210.

Patil, S.A. (2002). Hybrid vigour in cotton — conventional and MS based approaches. In *Souvenir XI National seed seminar on quality seed to enhance agricultural productivity*. Jan.18-20, 2002. Indian Society of Seed Technology, New Delhi and University of Agricultural Sciences, Dharwar, pp. 79-87

Pearson, O.H. (1933). Breeding plants of the cabbage group. *California Agric.Exp.Sta.Bull.*, 532: 3-22.

Peterson, C.E. (1960). A gynoecious line of cucumber. *Quart.Bull. Mich. agric. Exp.Sta.*, 43: 40-42.

Peterson, C.E. and L. Anhder (1960). Induction of staminate flowers on gynoecious cucumbers with gibberellin A_3. *Science*, 131: 1673-1674.

Peterson, C.E. and J.L. Wiegle (1958). A new method for producing hybrid cucumber seed. *Quart.Bull.Mich.agric.Exp.Sta.*, 40: 960-965.

Pring, D.R.; M.F. Conde and K.F. Schertz (1982). Organelle genome diversity in sorghum: Male-sterile cytoplasms. *Crop Sci.*, 22: 414-421.

Pring, D.R. and D.M. Lonsdale (1985). Molecular biology of higher plant mitochondrial DNA. *Int.Rev.Cytol.*, 97: 1

Pour, A.R.; V.E. Gracen and H.L. Everett (1981). Genetics of fertility restoration in the C group of cytoplasmic male-sterility in maize. *Genetics*, 98: 379-388.

Quinby, J.R.(1980). Interaction of genes and cytoplasm in male-sterility in sorghum. In "Loden, H.D. and D. Wilkinson (eds.) American Seed Trade Association. Proc. 35th *Annu.Corn. Sorghum Res.Conf.*, Chicago, III, pp. 175-184.

Ramachandram, M. and M.V.R. Prasad (1995). Hybrid castor research and development in India. In *Hybrid research and development*. Proc. of Seminar Indian Society of Seed Technology, New Delhi, 1993. Ed. M.Rai and S. Mauria, pp. 93-103.

Rick, C.M. (1948). Genetics and development of nine male-sterile tomato mutants. *Hilgardia*, 18: 599-633.

Rick, C.M. and J.E. Boynton (1967). A temperature sensitive male-sterile mutant of the tomato. *Am.J.Bot.*, 54: 601-611.

Robinson, R.W.; S. Shannon and M.D. Guardia (1968). Regulation of sex expression in cucumber. *Biol.Sci.*, 99: 141-142.

Sampson, S.R. (1966). Linkage of genetic male-sterility with seedling marker and its use in producing F_1 hybrid seed of *Brassica oleracea* (Cabbage, broccoli, Kale etc.). *Canadian J.Pl.Sci.*, 46: 703.

Schertz, K.F. (1977). Registration of A2 Tx 2753 and BTx 2753 sorghum germplasm. (Reg.No.GP 30 and 31). *Crop Sci.*, 17: 983.

Schweisguth, B. (1973). Study of new type of male-sterility in onion. *Ann.Amel.Plant*, 13: 221.

Seetharam, A.; K.Giriraj and P. Kusumakumari (1980). Phenotypic stability of seed yield in sunflower hybrids. *Indian J.Genet.*, 40: 102-104.

Seshadri, V.S. (1989). Vegetable crops. In *Plant breeding — theory and practice*.

Ed. V.L. Chopra. Oxford and IBH Publishing Co. Pvt. Ltd., New Delhi, pp. 391-408.

Sheetz, R.H. and Jr.J.B.Weaver (1980). Inheritance of fertility enhancer factor from Pima cotton when transferred into upland cotton with *Gosspium harknessii*, Brandegee, cytoplasm. *Crop Sci.*, 20: 272-275.

Shiffris, O. (1956). Sex instability in *Ricinus*. *Genetics*, 41: 265-280.

Shi, M.S. (1981). Preliminary report of breeding and utilization of late *indica* natural double purpose line. *J.Hubei Agric.Sci.*, 1-3.

Singh, Har Bhajan; Vishnu Swarup and B. Singh (1998). *Exploitation of hybrid vigour in vegetables*. Reprinted May, 1998. Indian Council of Agricultural Research, New Delhi.

Singh, M. (1989). Often cross-pollinated crops: cotton. In *Plant breeding — theory and practice*. Ed. V.L. Chopra. Oxford and IBH Publishing Co.Pvt.Ltd., New Delhi, pp. 237-249.

Sprague, G.F. (1939). An estimation of number of top-crossed plants required for adequate representation of corn variety: *J.Amer.Soc.Agron.*, 31: 11-16.

Sprague, G.F. and L.A. Tatum (1942). General vs. specific combining ability in single-crosses of corn. *J.Amer.Soc.Agron.*, 34: 923-932.

Srinivasan, K.and K.N. Gururajan (1978). Transference of male-sterility to some improved cotton varieties. *ISCI Jr.* 3 (1): 17-20.

Stephens, J.C. and R.F. Holland (1954). Cytoplasmic male-sterility for hybrid sorghum seed production. *Agron.J.*, 46: 20-23.

Stewart, J. McD (1992). A new cytoplasmic male-sterile and restorer in cotton. In Proceedings *The 1992 cotton research meeting*. Arkansas Cotton Research/Extension/Production and Marketing group, Arkansas Agricultural Experiment Station, University of Arkansas, Special Report, 158 USA.

Stewart, J. McD (1996). Inheritance of fertility restoration in D_8 cytoplasmic male-sterile line. In proceedings *The 1996 cotton research meeting and summaries of cotton research in progress*: Arkansas Agricultural Experiment Station, University of Arkansas, Special Report, 178 USA.

Sujata, V.S.; Sivaramakrishnan; K.N. Rai and K.Seetha (1994). A new source of cytoplasmic male-sterility in pearl millet: RFLP analysis of mitochondrial DNA. *Genome*, 37: 482-486.

Sun, Z.X.; Xiong, Z.M.; S.K. Min and H.M. Si (1989). Identification of the temperature-sensitive male-sterile rice. *Chinese J. Rice Sci.*, 3(2): 49-55.

Swarup, Vishnu (1977). *Breeding procedures for cross-pollinated vegetable crops*. Indian Council of Agricultural Research, New Delhi.

Swarup, Vishnu (1995). Heterosis breeding in flowers. In *Hybrid research and developments*. Proc. Seminar held in 1993. Indian Society of Seed Technology, New Delhi, pp. 183-187.

Thompson, A.E. (1955). The extent of hybrid vigour in spinach. *Proc.Am. Soc. Hort.Sci.* 67: 440-444.

Thompson, R.F. (1959). Breeding better kales. *Agriculture* (London), 65: 487-491.

Tolla, G.E. and C.E. Peterson (1979). Comparison of gibberellins and silver nitrate for induction of staminate flowers in gynoecious cucumber line. *Hort.Sci.*, 14: 542-544.

Tronickova, E. (1979). The effect of ethrel on sex expression in the maternal components in the production of hybrid seed pickling cuçumber. *Genetica slechteni*, 15: 267-272.

Tysdal, H.M.; J.A. Kissellbach and H.L. Westover (1942). Alfalfa breeding. *Nebraska Agric.Exp.Sta.Bull.*, 124: 1-46.

Ullstrup, A.J. (1972). The impacts of the southern corn leaf blight epidemics of 1970-71.*Ann.Rev.Phytopathology*, 10: 37-50.

Van Bennekom, H. and Q.P. Van der Meer (1978). Use of gibberrellic acid as a gametocide in onions, III Zaabelahgen.

Vasal, S.K. (1986). Approaches and methodology in the development of QPM hybrids. *Anais do* XV *Congresso National de Milho e Sorgo*. Brasilia, EMBRAPA-CNPMS, Documentos, 5 pp. 419-430.

Vasal, S.K. (1987). *Development of quality protein maize hybrids*. In "K.S. Gill, A.S. Khera and M.M. Verma (ed.) Proc.First Sym. On crop improvement Vol.I, pp. 57-75". Crop Improvement Society of India, PAU, Ludhiana, India.

Vasal, S.K.; B.S. Dhillon and G. Srinivasan (1995). Changing scenario of hybrid maize breeding and research strategies to develop two parent hybrids. In *Hybrid research and development*, Proc.Seminar held 1993. Indian Society of Seed Technology, New Delhi, eds. M. Rai and S. Mauria, pp. 19-36.

Virmani, S.S.; H.P. Moon and P.D.Aurin (1990). Heterosis in intervarietal rice hybrids grown under tropical and temperate environments. *Agron.Abst.Am.Soc.Agron.*San Antonio. Texas, 114.

Virupakshappa, K.; A. Seetharam; Jayarame Gowda and R.L. Ravikumar (1991). Maintainer and restorer behaviour of some sunflower lines on new cytoplasmic male-sterile sources. *J.Oilseeds Res.*, 8: 195-198.

Walkof, C.and W.W. Nuttal (1952). Hybrid vegetables for short season gardens. Pub. 947. *Can.Dep.Agric*, p.7.

Wellensiek, S.J. (1952). The theoretical basis of polycross test. *Euphytica*, 1: 15-19.

Worstell, J.V.; H.J. Kidd and K.F. Schertz (1984). Relationship among male-sterility inducing cytoplasms of sorghum. *Crop Sci.*, 24: 186-189.

York, J.L.; H.M. Munger and D.M. Wallace (1957). Experimental release of the cabbage inbred, NY 52: 153 *Mimeogr. Pub.Dep. Pl. Breed.* Cornell University, Itheca, USA.

Yuan, L.P. (1996). Hybrid rice in China. In *Hybrid rice technology*, compiled and edited by Ahmed, M.I. *et al.* Directorate of Rice Research, Hyderabad, pp. 51-54.

13

Resistance of Plants to Fungal Diseases

Despite the control measures employed, considerable amount of country's crop production is lost each year through the activities of diseases. Annual crop losses due to insects, diseases and weeds have been estimated to be about rupees fifty thousand crores (Jayaraj, 2002). The man and the pests have always competed for food and fibre. The pests and diseases are as old as the agriculture itself. Several methods of controlling diseases and insect-pests are employed by man, *viz.*, avoidance (comprising of quarantine, good crop husbandry, crop rotation, use of clean seed and propagating material, hygiene, control of vectors, etc.), direct control measures (which include use of chemicals for control of pests), biological control (use of hyperparasites and predators) and the use of resistant varieties.

After the arrival of green revolution in India several high yielding varieties (*HYVs*) and/or hybrids of various crops were bred in India and made available to the farmers. The farmers extensively cultivated the high yielding types. Adoption of the high yielding varieties prompted high nutrient application. This led to enhanced cropping intensity and change in cropping pattern. The adoption of *HYVs* probably led to enhanced build-up of newer and more destructive diseases. For example, sugary disease of sorghum caused by *Sphacelia sorghi* became a serious disease after the release of hybrids. The sorghum hybrids have cytoplasmic-genetic male-sterile lines as seed parent. Male-sterile lines are more susceptible to ergot than normal lines, even their B counterparts. Similarly, hybrids were more susceptible to grain molds. They are caught in rains, but the indigenous varieties flower in about 90 days and hence usually escape infection by mold pathogens. Only when late rains are received grain molds develop on indigenous varieties.

The imported hybrids of okra are not tolerant to Yellow Vein Mosaic virus. Similarly imported hybrids of tomato are not tolerant to native diseases like leaf curl virus, local races of bacterial wilt and other diseases (Attavar, 2000).

Fig. 13.1: Yellow vein mosaic of okra

In the past, disease causing pathogens have been responsible for epiphytotics (unarrested spread of disease). One of the best known examples of catastrophic epiphytotic of plant disease caused by a fungus is the late blight of potato (causal organism: *Phytophthora infestans*) which occurred in 1845 and 1846 in Western Europe. Other disease epiphytotics caused by fungi were the rusts of cereals (causal organism: *Puccinia* spp.) in many parts of the world, brown spot of rice (causal organism: *Helminthosporium oryzae*) in Bengal and coffee rust (*Hemileia vastatrix*) in Sri Lanka.

The disease, referred to in this chapter, is the disorder of crop plants caused by microorganisms such as fungi, mycoplasma, bacteria and viruses.

Plant resistance is one of the most effective and ideal approach to combat diseases and insect pests. Use of resistant varieties in vegetables is very much desirable. It is because the residual toxic effect of pesticides in the vegetable produce is an health hazard. Fruits of vegetables like okra, brinjal, tomato, etc. are plucked at different times. Farmers are usually not particular about the spray schedule. Consequently, vegetables reach the consumer just after a few days of spray of fungicides. Moreover, for cosmetic purposes also vegetables should be resistant to diseases and insect pests.

Excellent examples are on record, which support the value of disease and insect resistant varieties. In United States of America, Hessian fly (*Mayetiola destructor*) was controlled by evolving hessian fly resistant wheat varieties. Some other examples are the control of brown plant hopper (*Nilaparvata lugens*) of rice in Asia and suppression of rust epidemics by growing rust resistant varieties of wheat throughout Europe, North America, Australia and Asia. Disease resistant varieties ensure greater yield stability and guarantee food security. A classic example is that of a rice variety IR 36, evolved by the International Rice Research Institute, Philippines and released for general cultivation. This variety is resistant to four diseases, viz., blast (caused by *Pyricularia grisea*), bacterial blight (caused by *Xanthomonas oryzae* pv *oryzae*), grassy stunt virus and tungro virus, and five insect pests (brown plant hopper, green leaf hopper, yellow stem borer, stripped stem borer and gall midge). This variety was grown annually on more than 10 million hectares of land between 1978 and 1984. Its cultivation alone yielded an additional income of one billion dollars to rice growers and processors (Khush, 1995). Although breeding for insect and disease resistance is very important objective of a plant breeder, he should not forget that the resistant varieties shall not be acceptable to both the farmers and consumers unless they are high yielding and superior in quality. Thus breeding for resistance is one of the objectives for a breeder. The host plant, the disease causing organisms and insect-pests are biological entities. And hence, as will be shown later, all these exhibit variability. Plants differ in their ability to withstand disease (and/or insects). The fungi etc. manifest variability for virulence and insects show differences in their preferences to attack plants.

Among the disease causing organisms fungi are predominant and hence following sections will occupy more discussion about these pathogens.

13.1 RESISTANCE OF PLANTS TO FUNGAL DISEASES

It was long known that cultivated varieties of various crop plants differ in their ability to avoid diseases. Biffen (1905) was the first to report on the inheritance of resistance to disease. He studied the inheritance of resistance to yellow rust (causal organism — *Puccinia striiformis*). At the University of Cambridge, Biffen crossed two susceptible varieties of wheat, viz., Michigan Bronze and Red King, with a yellow rust resistant variety Rivet. In the F_2 generation of both the crosses he observed seventyfive per cent population susceptible to yellow rust while twentyfive per cent was resistant. He concluded that rust resistance was monogenic in inheritance and susceptibility was dominant.

The plant pathologists expressed doubt on this conclusion in the light of the fact that varieties resistant in one locality succumbed to the disease in another locality and that resistance varied from season to season even in the same locality. However, the position was cleared by the report that within a species of the pathogen there are distinct races, which although morphologically alike, differ from each other in their ability to attack only certain varieties within the host species and not others (Eriksson, 1894; Ward, 1902; Barrus, 1911; Stakman and Piemeisel, 1917). Further it was found that genes occurred in the host species, which specifically conditioned resistance to one or a group of physiological races of the pathogen.

The concept/definition of resistance to diseases varies with the vision of person addressing the issue. Actually host response or reaction is a continuous attribute and varies between the extremes of resistance (perhaps no symptoms) and total susceptibility (probably death of plants). The point of distinction between resistance and susceptibility is not clear. Resistance is a relative state whereby one genotype (or population) displays less disease and/or experiences lower losses in agricultural productivity than other genotype or population (Mc Intosh, 1995). According to Russell (1978) resistance is any inherited characteristic of host plant, which lessens the effect of parasitism. The resistant plants are thus expected to be relatively less damaged by the parasite as compared to susceptible ones. Simmonds (1983) is of the view that resistance is a state of less disease with immunity (no disease) representing one extreme. Ability of a host to hinder the pathogen or disease causing agent is resistance, in the opinion of Robinson (1969).

Many plants are non-hosts to most of the parasites and hence are immune to them. The non-host immunity is absolute and there is no disease or pest attack whatsoever. Such immune plants are never susceptible to pests under any circumstances whereas resistant plants of a host species can be rendered susceptible by different treatments. Resistance can be quantified as high, moderate and low, because there may be a complete gradation from extreme resistance to extreme susceptibility.

13.1.1 Types of Resistance

The prime objective of a plant breeder is to evolve a variety, which suffers least damage from a disease. Resistance to fungal diseases may comprise of several kinds (Russell, 1978), viz., disease escape, resistance to disease itself and tolerance. Resistance to a pathogen, as will be discussed in the next section, could be a pre-existing or responsive mechanism.

13.1.1 (a) Disease Escape

Plants may have several mechanisms where by they escape or avoid direct contact with the pathogen and hence infection. For example, *Claviceps purpurea* causes a serious disease of inflorescence of cereals, the ergot. It does not affect those varieties of wheat and barley in which flowers remain closed until pollination has occurred. In such varieties the spores of pathogen have no opportunity to enter the flowers and infect the stigma at its susceptible stage. Closed flowering habit is an example of disease escape. Futrell and Webster (1966) showed that the unfertilized florets of cytoplasmic-genetic male-sterile plants of sorghum, e.g. *CK* 60*A*, were more susceptible to the ergot infection. The disease was not serious on its fertile counterpart, namely *CK* 60*B*. Ramakrishnan (1948) reported that sorghum florets could become infected with ergot only between the time flowers open and the time of fertilization.

Glume blotch of wheat is caused by *Septoria nodorum*. The spores of *Septoria* are usually splash dispersed from leaf to leaf in the crop by the rain drops. Wheat varieties, which have tall plants with longer internodes are less affected by this disease, because splash dispersal of spores is less effective as compared to varieties with short straws (Sharp *et al.*, 1972). Upper leaves of tall strawed varieties are, therefore, free from *Septoria* under conditions of natural spread of disease in the field, although they are susceptible to *Septoria* when artificially inoculated.

13.1.1 (b) Resistance to a Pathogen

Resistance to a fungal pathogen refers to any inherited attribute of a host plant, which restricts the establishment or growth of that pathogen. In case of disease escape, there is no cell-to-cell contact between the host and pathogen.

Resistance could be classified on the basis of mode of inheritance. Monogenic inheritance is governed by a single gene pair. The gene has a large measurable effect on the trait. It is relatively easier to handle a monogenic resistance during breeding, especially when it is dominant to susceptibility. Oligogenic resistance is controlled by a few genes with large measurable influence on the character. Monogenic resistance is often race-specific and fails to control the race of a pathogen effectively which arises *de novo*. Resistance could be polygenic *i.e.* governed by a relatively large number of genes. It may involve mechanisms such as heavy lignification of cell walls, which is the end result of several distinct chemical reactions. Resistance that is governed by many genes is likely to be more durable and less race-specific as compared to resistance controlled by only a few genes.

Heritable factors in cytoplasm could also determine the resistance. For example, maize lines possessing *T* cytoplasm are susceptible to southern corn leaf blight. The fungus, *Cochliobolus heterostrophus* produces a toxin in plants with *T* cytoplasm. This toxin is known to adversely affect the membranes of mitochondria of host cells, causing them to swell irreversibly (Miller and Koeppee, 1971). The maize varieties possessing *S* and *C* cytoplasms are not susceptible to this fungus. Narayanan et al. (1990) reported that cytoplasms of *Gossypium harknessii*, *G. arboreum*, *G. anomalam*, etc. show relatively higher levels of resistance/tolerance to boll worms.

Van der Plank (1963) proposed two main types of resistance, *viz.*, vertical and horizontal. When a variety is more resistant to some races of pathogen than to others, the resistance is designated as vertical (*VR*). On the other hand, when resistance is evenly spread against all the races of a pathogen, the resistance is known as horizontal (*HR*). Horizontal resistance is regarded as effective against all the variants of a particular parasite and vertical resistance is effective against certain variants only. Vertical resistance is governed by major genes. Genes responsible for horizontal resistance involve some mechanism, which is beyond the capacity of a pathogen to overcome. It is believed that polygenic inheritance is involved for horizontal resistance. Robinson (1976) is of the opinion that not all the horizontal resistance is polygenic and not all the oligogenic resistance is vertical. Since vertical resistance is against some races of a pathogen, such a resistance exercises a selection pressure on the non-pathogenic races and the resistance may break down because of appearance of new race. Contrary to the belief that vertical resistance is less durable than the horizontal resistance; actual situation could be different in different cases. In many situations resistance governed by a single gene pair has remained effective for long periods. Ellingboe (1981), however, maintained that horizontal resistance is the resistance that has not yet shown to be vertical. According to him when experiment is of cursory nature, resistance is frequently considered to be horizontal in nature. When experiments are more precise and analytical resistance seems to be of a vertical nature.

Horizontal and non-race specific resistance have been used as synonyms. However, these terms should be used with caution since it is practically impossible to test the varieties or the breeding material against all possible genetic variants of a pathogen. Resistance which is thought of as non-race specific breaks down some times when a variety is attacked by resistance breaking variant of a pathogen. Hence, the term durable has been proposed in place of non-race specific resistance (Johnson and Law, 1975). Durability of resistance, nevertheless, does not imply that it is effective against all the variants of a pathogen. It only suggests that resistance is imparting effective control over the pathogenesis for many years and it is still effective. Transient resistance, on the contrary, is a term used to describe resistance which has given effective control of a pathogen for a relatively shorter period because resistance breaking variants of a pathogen can develop.

Resistance can be distinguished according to the manifestation at various growth stages of plants. In case of wheat, scientists talk of seedling and adult-plant resistance. This resistance was early recognized and it had been shown that in some varieties adult plants become resistant after being susceptible in seedling stage. In 1927, Goulden *et al.* demonstrated in H-44- this adult plant or mature plant resistance was simply inherited. The nature of this resistance was studied by Popp (1933). He reported that adult plant resistance became effective during 4th leaf stage with resistance being increased with further development. Adult plant resistance is the resistance that can not be easily detected at the seedling stage but is best expressed in post-seedling stage. On the other hand, rust resistance that expresses at the seedling stage, is usually effective throughout the life of the plant although there are exceptions. Occasionally seedling resistance is not effective in adult plants (Knott, 1995). Adult-plant resistance is regarded as more durable. Wheat scientists regard the resistance as general, which is effective against all pathotypes or races of a pathogen. Slow rusting is a type of general resistance where the disease progresses at a relatively retarded rate resulting in intermediate to low disease levels against all the pathotypes of a pathogen (Rajaram *et al.*, 1995), compared to a fast rusting susceptible checks. On the basis of mechanisms operating at different stages to impart resistance against a pathogen, the resistance could be classified as 'active' and 'passive' (Russell, 1978). The 'active' or 'responsive' or 'reactive' resistance refers to the resistance mechanism of the host plant, which occurs in response to the attack by a parasite. Production of phytoalexins by plants upon inoculation by some pathogens is an example of active resistance. The term phytoalexin (phyto = plant, alexin = warding off compound) was defined by Muller and Borger (1940) as substances that are formed when living plant tissues are invaded by fungal parasites. Pisatin and phaseolin are two phytoalexins whose induction has been mostly studied. Gossypol is a prominent pigment found in the glands distributed throughout leaves, stems, root cortices and floral parts of most cotton varieties (Adams *et al.*, 1960). Gossypol was identified as phytoalexin toxic to fourteen forms of *Fusarium oxysporum* (Davis, 1964). Active resistance also includes formation of cork layer against common scab as in potato, formation of abscission layer against *Xanthomonas* as in peach, and deposition of gum as in rice blast disease. Other mechanisms are swelling of cell walls of the invaded cells as in cucumber against scale, and change in the texture of cytoplasm of cells of certain plants (Kumar and Phul, 1977). Many plants produce certain specific metabolites after they come in contact with the pathogen. These interfere with the normal growth and development of the pathogen. In another mechanism certain compounds are deposited on cell wall at or near the point of attempted penetration by the pathogen. Some of the compounds deposited in this way comprise of callose (Hardwick *et al.*, 1971; Heath, 1971), lignin (Hijwegen,

1963; Friend, 1973), etc. The 'passive' resistance involves mechanism of host plants, which is already present before the attack. This is also known as 'pre-existing' mechanism. These mechanisms could be physical or physiological. Many plants produce certain metabolites either before or after they have come in contact with the pathogen. These interfere with the normal growth and development of the parasite. Resistant varieties of tomato, sugar beet and gram produce fungitoxic substances on their leaves so that the spores can not germinate. In onion, red-scale varieties are resistant to smudge because they have protocatechuic acid and catechol, which are toxic to spores. Also, certain substances produced in normal metabolic pathways of many resistant plants are toxic to their parasites. Potato resistant to potato scab has more chlorogenic acid than the susceptible varieties. Pre-infectional inhibitors in the vegetable crops include catechol, protocatechuic acid, phenols, tarpens, flavonoids and tomatine. Tomatine and solanine are inhibitors found in tomato. Some prohibitors are toxic to specific pathogens and hence result in lessend disease incidence. Such compounds are sinigrin, gossypol, α tomatine and solanine, etc. Cotton varieties having no or low levels of an alkaloid, gossypol, are more vulnerable to their pathogens than those having high gossypol.

Some studies have shown increased production of certain fungitoxic substances in resistant varieties following attack by the pathogen. Aglucones and peroxidases are such substances produced in resistant varieties of wheat and maize. Productions of phytoalexins have also been reported in peas, beans, sweet potato, soybean, carrot and many other crops.

Some plants are extra sensitive to the attack of certain pathogens. As soon as pathogen establishes itself slightly in the host, the tissues immediately surrounding the pathogen die, cutting off its nutrition and the pathogen is not able to multiply further. This behaviour of plants is called hypersensitive reaction and occurs in a variety of crops. Hypersensitivity is a kind of resistance reaction shown by a plant. Stakman (1954) defined hypersensitivity as the rapid death of infected cells, which restricts the spread of obligate pathogen. Obligate pathogens parasitize only living host plant tissues. Hypersensitive reaction has recently been described as "programmed cell death".

Some plant pathogens are unable to colonize host cells, which have heavily lignified walls. Probably these act as physical barriers. It could be a physiological barrier also because the cell walls may not be easily degradable by the enzymes produced by the pathogenic fungi. Lignified walls may contain compounds such as ferulic acid, which are fungitoxic (Hartley *et al.*, 1976).

Field resistance is a term that is commonly used to describe the resistance, which gives an effective control of a parasite under natural conditions of field. Genetic basis of different resistant stocks may be different and can be characterized by laboratory or a glasshouse test under controlled conditions.

13.1.1 (c) Tolerance

The plants that are diseased to the same extent as the other plants but are not damaged as much by the infection may be designated as tolerant. As a matter of fact, the development of a pathogen is not adversely affected by the tolerant variety. Some plant breeders are not in favour of developing tolerant varieties because such varieties provide a source of infection to other less tolerant varieties. The tolerant varieties manifest more sporulation of pathogen compared to the resistant ones.

Tolerance can be expressed in many ways. For instance, the tolerant variety may exhibit relatively less growth rate of root or shoot and may have relatively less effect on yield of economic produce. A plant breeder is interested in the effect of disease on the agriculturally important end products. If a breeder is evolving a variety of forage crop, he will be mainly concerned with the yield of fodder, and yield of seed will be of secondary importance.

An excellent example of tolerance or recovery resistance is that found in sorghum against shoot fly (*Atherigona soccata*). The fly lays eggs on the underside of leaves of seedlings aged seven to eight days old. The eggs hatch in two to three days and the maggots crawl down into the plant between outer leaves and the inner tightly furled leaves. Then they bore horizontally into the young shoot and growing point. While feeding, the maggot damages the growing point, causing the inner leaf whorl to turn yellow and rot at base resulting in typical 'dead heart' symptom. The plants of variety Namatare from East Africa and Serena also from Africa show very good recovery resistance. The plants recover after damage by the maggot by developing tillers from the base of plant, which bear ear-heads. Under severe conditions even the tillers are damaged by the shoot-fly but over 70 per cent of the attacked plants recover and give good yield. The heritability of recovery resistance is very high and it is easy to use this trait in breeding programme (Doggett *et al.*, 1969; Starks *et al.*, 1969).

13.2 THE FUNGAL PATHOGEN

Because a large number of plant diseases are incited by fungal pathogens, the following discussion will be mostly confined to fungi. Although there are more than one lakh different species of fungi, only 200 of these are serious pathogens of crop plants. Fungi reproduce asexually, vegetatively and some are known to reproduce sexually also. Conidia are asexual spores borne on spore bearing hyphae called conidiophores. Sexual spores are variously known as oospores, zygospores, ascospores and basidiospores, depending upon the manner of their formation and the class of fungi to which they belong.

Spores germinate to give rise to one or more fine filaments called the germ tubes. Germ tubes grow and give rise to branches, each of which is called hyphae. The collective term for a group of hyphae is mycelium. Mycelium is the vegetative part of the fungus. Some of the hyphae rise vertically from the surface of plant part, *e.g.*, leaf, and conidia are budded off from them. A fungal spore germinates when it comes in contact with surface of an appropriate host plant, provided environmental conditions are favourable. Important requirements for germination of spores are high humidity of surrounding air or a film of free water on the substrate, appropriate light and temperature. There are several other factors, which favour or discourage spore germination. For example, germination of a spore can be delayed or inhibited by compounds, which are present on the surface of some plant parts.

The fungi are commonly considered as haploid organisms, however, nuclear fusion of the gametes may give rise to a short lived diploid stage. The sexual reproduction starts with the fusion of sex cells, each of which may contain one or more haploid nuclei. The fusion of gametes is called plasmogamy. The nuclei of fusing gametes may not fuse immediately. The nuclei of two gametes become associated in pairs, which may be one or more. The pairs of associated nuclei are known as dikaryon. In dikaryon the genetic material of two associated nuclei are separated from each other by nuclear membranes. The dikaryotic phase may persist more or less indefinitely as a result of mitotic cell divisions (known as conjugate division). Ultimately the associated nuclei fuse to form a diploid nucleus. This fusion of associated nuclei is called caryogamy. Thus the male and female nuclei do not fuse immediately after the union of protoplasts from the male and female thalli. Raper (1954) distinguished seven basic types of life cycles in fungi based on the duration of haploid, dikaryotic and diploid phases

(Life cycle, as you know is the changes taking place between production of gametes by one generation and production of gametes by the next generation).

During the sexual reproduction of fungi only the compatible nuclei fuse to form the diploid phase. Fungi can be broadly divided into two groups with respect to mating system. Where the haploid cells fusing to form a dikaryon are from same mycelium, they are called as homothallic and when the two haploid cells are from different mycelia, called as plus and minus, they are heterothallic. In the homothallic fungi sexual fusion occurs between cells produced on any two hyphae, even between the cells of same hyphae. On the other hand, sexual fusion occurs only between the cells from thalli of different mating groups in heterothallic fungi. Heterothallism was discovered by Blackslee in 1904, when he found that conjugation between thalli of different mating types precedes the formation of zygospores in mucor.

Some fungi exhibit variability with respect to host requirement to complete their life cycle. Rust pathogens, for example, exhibit such a requirement. Autoecious rusts are those, which are able to complete their entire life cycle on the same host. *Melampsora lini*, a fungus which incites rust on *Linum usitatissimum*, produces all its spore stages on linseed. However, heteroecious parasitic fungus, like *Puccinia graminis tritici*, causing stem rust of wheat, requires barberry to produce pycniospores and aeciospores and produces urediospores and teliospores on wheat.

According to the mode of parasitization, the pathogenic fungi have been broadly categorized into two groups, viz., obligate and facultative. Obligate (biotrophs) parasites parasitize only living host cells. On the other hand, facultative parasites (necrotrophs) kill the host cells before progressing through the plant tissues. These can feed saprophytically and can, therefore, survive in the absence of living host cells. Facultative parasites kill the host cells by secreting substances, which either dissolve the cell walls or destroy the cytoplasm. Toxins of the pathogens (pathotoxins) have been studied intensively. *Helminthosporium victoriae*, for example, produces a toxin, victorin (Wood, 1967), which acts on the plasmamembrane of the cells of susceptible host plants but not the resistant ones (Samaddar and Scheffer, 1968).

The ability of a pathogen to infect the host species is known as pathogenicity. It is an inherited characteristic of a pathogen. Virulence is the ability of a pathogen to overcome the resistance of host. Alternatively virulence is also defined as the degree of pathogenicity.

Generally virulence confers upon a pathogen the ability to attack some genotypes of host but not others. However, pathogenic strains of pathogens differ in aggressiveness, i.e. they differ in the amount of disease on a given host genotype, even when carrying the same genes for virulence. Aggressiveness is also called horizontal pathogenicity while the virulence is known as vertical pathogenicity. Aggressiveness is regarded as polygenic in inheritance.

Although morphologically similar, the microorganisms differ in their pathogenic reaction on different host plants. Eriksson in 1894, in Sweden, showed that black stem rust (*Puccinia graminis*) taken from wheat did not infect oat, rye and some other grass species. He recognized six sub-species or formae-specialis, *viz.*, tritici, becalis, avanae, phleipratensis, agrostis, and poae, within the botanical species *Puccinia gramminis* on the basis of their pathogenesis to different grass species. In 1911, Barrus described physiological races (or strains as he called them) on the basis of differing pathogenicity on different varieties of same host species. On the basis of differential reaction on bean varieties, he distinguished two strains of bean anthracnose fungus. Stakman (1914) showed that Eriksson's formae-specialis were not pathogenically homogeneous, but like the bean anthracnose organism included physiological races, varying in their pathogenicity on different varieties of same cereal host species. Physiological races are pathogens of same species with similar or identical morphology but differing in their pathogenic capabilities.

Parasites, whether they are fungi, virus, bacteria, insects, etc. show inherited variability. This should be taken into consideration while breeding for resistance against a given pathogen, because new forms may evolve or increase and attack the previously resistant variety. Many terms have been used to describe different phenotypes and genotypes in the parasitic populations, *viz.*, physiological races, strains, biotypes, pathotypes and variants. Physiological race has been widely used. According to Robinson (1969) pathotype is a more correct term, but is less widely accepted, to describe individuals which have particular pathogenicity in common.

Physiological races are identified by their reaction on a group of host varieties, known as differential hosts. At first the differential hosts were chosen without precise knowledge about the genes for resistance possessed by them. Because the disease reaction observed on a host plant results from interaction between genes for resistance in host plant with genes for virulence/avirulence in the pathogen, a plant breeder must work with specific race. For example, variety *x* may be resistant to race 1 of a pathogen and variety *y* to race 2. Let us assume that the genetic composition of two varieties is *AAbb* and *aaBB*, respectively, where genes *A* and *B* confer resistance to races 1 and 2, respectively. If the varieties *x* and *y* are inoculated with mixture of races 1 and 2, both will be susceptible. Moreover, the results of segregating material shall mislead the breeder. Because in F_2, plants possessing genes *AABB*, *AAB–*, *A – BB*, and *A – B–* shall be resistant to the mixed inoculation, conclusion would be made that crossing of two susceptible varieties yield resistant progeny.

Briggs (1933) showed that with two levels of reactions, viz., resistant and susceptible, one variety can differentiate two physiological races of a pathogen, two varieties can differentiate four and *n* varieties can differentiate 2^n physiological races. This, nevertheless, assumes that each differential host possesses different genes for resistance, specific for a given race.

Like the varieties of crop plants, the pathogenic fungi (and also bacteria, viruses) manifest variability with regard to disease reaction. A great variety of life cycles, patterns of mating, mutations, heterocaryosis, and parasexual cycle provide mechanism to generate variability in the pathogenic fungi. New physiological races arise from these mechanisms and attack the previously resistant varieties.

Transformation, transduction, conjugation, in addition to mutation are the means responsible for genetic variability observed in bacteria. Mechanisms also exists in viruses, nematodes for producing genetic variants.

13.3 GENETICS OF HOST PATHOGEN INTERACTION

The manifestation of disease symptoms on host plants is a result of interaction of genes for resistance/susceptibility possessed by the plants with genes for virulence/avirulence carried by the pathogen. Nevertheless, manifestation of disease is also influenced by the environmental conditions. A susceptible host, a virulent pathogen and environment favourable for development of pathogen lead to disease symptoms on plant parts.

The genetics of disease resistance in the plants was reported for the first time by Biffen (1905). Inheritance of pathogenicity was also investigated. But it was Flor's (1956) studies, dealing with flax (*Linum usitatissimum*) and rust pathogen (*Melampsora lini*), which suggested a correspondence between the genes for resistance in the host plant and genes for virulence in the pathogen. The basic pattern of gene-for-gene relationship is referred to as 'quadratic relationship' (Figure 13.2).

The genes of pathogen are symbolized by *A* or *a*, with *a* for virulence. A resistant or incompatible reaction occurs when the host possesses gene for resistance and the pathogen carries gene *A*. Susceptibility results when either the host plant does not possess genes for resistance or the pathogen carries genes for virulence. Flor's (1955, 1956) analyses of genetics of flax rust showed that for each major gene conditioning resistance in the host there is a specific gene conditioning virulence in the pathogen. In this kind of gene-for-gene relationship, there is no major gene for resistance that cannot be overcome by a corresponding gene for virulence in the pathogen. In flax, according to Flor (1956) at least 26 different major genes for resistances at 5 independent loci have been identified, and for each gene there is a corresponding gene for virulence in flax rust fungus (Day, 1974).

		Genotype of the host plant	
Genotype of pathogen		*rr*	*R-*
	A	susceptible	resistant
	a	susceptible	susceptible

Fig. 13.2: Disease reaction in a common gene-for-gene correpondence. Virulence is a recessive character usually, while resistance is dominant in inheritance.

Since the time of Flor's research, gene-for-gene relationships similar to that in flax rust have been suggested for many combinations of hosts and parasites, including fungi, nematodes, bacteria, viruses and insects (Day, 1974; Sidhu, 1975). Such fungi include rusts of wheat, oat, maize, sunflower and coffee, smuts of wheat, oat and barley. Fungal pathogen causing late blight of potato, downy mildew of lettuce, apple scab, gray mold of tomato, rice blast, northern leaf blight of maize also show gene-for-gene relationship. Bacteria, *Xanthomonas campestris* pv. *malvacerum* on cotton appears to have a gene-for-gene correspondence.

The gene-for-gene correspondence shown above represents the most commonly observed pattern but not the only pattern of interaction between gene(s) of host plants and genes of pathogen. In the common pattern shown in Figure 13.2, susceptible reaction results from either the absence of a gene for resistance in the host plant or from the presence of specific gene for virulence in the parasite. A different pattern is observed, when the pathogen produces a host specific toxin (Leonard, 1984).

Three out of four combinations between the complementary genes shall exhibit resistant reaction (Figure 13.3), unlike the results shown in Figure 13.2. Diseases produced by *Bipolaris victoriae* (Victoria blight of oat), *B. maydis* (southern leaf blight of maize) show similar patterns of gene-for-gene relationship.

		Genotype of the host plant	
		ss	*S-*
Genotype of pathogen	*t*	resistant	resistant
	T	susceptible	resistant

Fig. 13.3: Disease expression in plant host when pathogen produces a toxin. *T* is the gene in pathogen, which enables toxin production

A gene-for-gene correspondence is observed for most of the pathogens, especially where resistance is monogenically or oligogenically inherited. Genes responsible for horizontal resistance probably do not operate on a gene-for-gene basis (Van der plank, 1968, 1978).

However, Perlevliet and Zadoks (1977) and Nelson (1975) proposed that all resistance genes in the host and all virulence genes in the pathogen populations form an integrated system. Major and minor genes for resistance interact on a gene-for-gene basis with major and minor genes for virulence. When only a few genes with minor effects operate, differential interaction can be detected easily and this is called vertical resistance. When a large number of genes with small and cumulative effects determine resistance/pathogenicity pattern collectively, it is difficult to separate the small differential interactions from experimental error. The resistance behaves as horizontal resistance and appears to be race-non-specific. Nelson (1975) assumed that vertical and horizontal resistance genes are one and same, *i.e.* there are no separate genes determining vertical and horizontal resistance. Vertical and horizontal type of resistant reactions are not the actions of different genes, rather they are the expressions of genes with different genetic background. According to Nelson (1975) there are no major and minor genes. The resistance genes function vertically when separate and horizontally when together.

The genetics of resistance to some diseases is influenced by the environmental conditions. Temperature is one factor which determines the dominance nature of resistance. Certain genes are dominant at lower temperatures but recessive at higher temperatures. When two or more pathogens simultaneously infect a host, the genetics of resistance becomes confusing. Tomato plants are monogenically resistant to wilt caused by *Fusarium* provided there is no infection by nematodes. However, when nematodes also infected the tomato plants the resistance appeared polygenic in inheritance (Sidhu and Webster, 1983). Further the tomato plants susceptible to *Verticillium* but resistant to *Fusarium*, were resistant to both the pathogenic fungi when soil harboured both *Fusarium* and *Verticillium* (Sidhu and Webster, 1979).

13.4 RESISTANCE TO VIRUS

Diseases caused by viruses involve the role of vectors, which transmit the viruses to the host plants. For this reason breeding for resistance to diseases caused by viruses requires separate consideration.

Viruses cause very devastating diseases in vegetable crops, pulses and oilseed crops and others. Some examples are leaf curl, cucumber mosaic of tomato, yellow vein mosaic of okra, leaf curl and mosaic of chilli, stunt of gram, sterility mosaic of pigeonpea, yellow mosaic and leaf curl of mung and urd bean, yellow mosaic of soybean. Virus is transmitted to plants via vectors. Aphids constitute the largest group among the vectors of plant viruses. Most plant viruses are transmitted during the insect's active flight within and between the crops or between secondary hosts and main crops. Generally nymphs are more efficient in transmitting both non-persistent and persistent viruses (Thottappilly *et al.*, 1990). Stunt of gram is transmitted/ vectored by aphis *Craccivora* (Nene and Reddy, 1976). *Bemisia tabaci* is the most important species of white flies reported as vector of plant diseases. It transmits several viruses. It has wide host range. Adults and larvae feed by sucking from phloem of the leaves. Thrips also vector the transmission of viruses. Larvae can transmit the virus before they pupate. Adults more commonly transmit the virus. They remain infective throughout their life span, but their transmission ability is variable. Eriophyid mites are extremely small arthropods, which play considerable role in virus transmission. Eriophyids may be disseminated by wind or may be transported by other insects.

Host resistance can be against the virus and/or against its vector.

Combined resistance to both the pathogen and vector(s) is likely to impart durable resistance as compared to resistance to either vector(s) or the virus (Nene, 1995). According to Fraser (1987) resistance to virus means any inhibition of virus multiplication or of expression of its pathogenic effects. Plant resistance to insects could be due to antixenosis, antibiosis or tolerance (see Chapter 14). Antixenosis denotes the presence of morphological or chemical plant factors, which adversely alter the insect behaviour, which leads to selection of an alternate host. It may also be due to possession of unique plant chemicals, which repel or deter the insects from feeding it or selecting the plant for oviposition (non-preference for food, shelter and oviposition). In antibiosis the biology of insect is adversely affected while tolerance refers to ability of host plant to withstand or recover from insect damage.

Vetten and Allen (1983) found antixenosis to vector *Bemisia tabaci* in cowpea and lima bean. Evans (1954) reported that groundnut cultivar Mwitunde exhibited antibiosis to vector *A. craccivora*, which resulted in field resistance of this cultivar to rosette disease.

References

Adams, R.; T. Geissman and J. Edward (1960). Gossypol a pigment of cotton seed. *Chem.Rev.* 60: 555-574.

Allard, R.W. (1960). *Principles of plant breeding*. John Wiley and Sons, Inc., New York, pp. 323-369.

Attavar, M. (2002). Tool for productivity gains. *The Hindu Survey of Indian Agriculture*. M/s Kasturi and Sons, Chennai, 600-002, pp. 145-149.

Barrus, M.F. (1911). Variation in varieties of beans in their susceptibility to anthracnose. *Phytopathology*, 1: 190-195.

Biffen, R.H. (1905). Mendel's law of inheritance and wheat breeding. *Jour.Agric.Sci.*, 1: 4-48.

Briggs, F.N. (1933). A third genetic factor for resistance to bunt, *Tilletia tritici*, in wheat. *J.Genetics*, 27: 435-441.

Briggs, F.N. and P.F. Knowles (1967). *Introduction to plant breeding*. Reinhold Publishing Corporation, NewYork, pp. 341-373.

Day, P.R. (1974). *Genetics of host-parasite interaction. Freeman*, SanFrancisco.

Davis, D. (1964). Host fungitoxicants in selective pathogencity of *Fusarium oxysporum*. *Phytopathology*, 54: 290-293.

Doggett, H.; K.J. Starks and S.A. Eberhart (1970). Breeding for resistance to shoot fly, *Atherigona varia*, Rond. *Crop Sci.*, 10: 528-531.

Ellingboe, A.H. (1981). Changing concepts in host-pathogen genetics. *Annual Review of Phytopathology*, 19: 125-143.

Eriksson, J. (1894). Uber die specialisirung des parasitismus beiden Getreider ostpilzen. *Ber-Beut.Bot.Gessel.*, 12: 292-331.

Evans, A.C. (1954). Rosette disease of groundnuts. *Nature* (London), 173: 1242-1243.

Flor, H.H. (1955). Host-parasite interaction in flax-rust: its genetics and other implications. *Phytopathology*, 45: 680-685.

Flor, H.H. (1956). The complementary systems in flax and flax rust. *Adv. Genet.*, 8: 29-54.

Fraser, R.S.S. (1987). The biochemistry of resistance to viruses. In *Biochemistry of virus infected plants*. Ed. Fraser, R.S.S. Research Studies Press Ltd., England, pp. 103-170.

Friend, J. (1973). Resistance of potato to *Phytophthora*. In *Fungal pathogenicity and the plant's response*. Ed. R.J.W. Byrde and C.V. Cutting. Academic Press, London, p. 383.

Futrell, M.C. and O.J. Webster (1966). Ergot infection and sterility in grain sorghum. *American Soils*, XI (1 & 2): 163-168.

Goulden, C.H., K.W. Neatby and J.N. Welsh (1927). The inheritance of resistance to *Puccinia graminis tritici* in a cross between two varieties of *Triticum vulgare*. *Anatomical Record*, 37: 182.

Hardwick, N.V.; A.D. Greenwood and R.K.S. Wood (1971). The fine structure of haustorium of *Uromyces appendicultus* in *Phaseolus vulgaris*. *Canadian Journal of Botany*, 49: 383.

Hartley, R.; E.C. Jones and T.M. Wood (1976). Carbohydrates and carbohydrate esters of ferulic acid released from cell walls of *Lolium multiflorum* by treatment with cellulolytic enzymes. *Phytochemistry*, 15: 305.

Heath, M.C. (1971). Haustorial sheath formation in cowpea leaves immune to rust infection. *Phytopathology*, 61: 383.

Hijwegen, T. (1963). Lignification, a possible mechanism of active resistance against pathogens. *Netherlands Journal of Plant Pathology*, 69: 314.

Jayaraj, S. (2002). Prudent suppression of pests. In *The Hindu Survey of Indian Agriculture*. M/s Kasturi and Sons, Chennai 600 002, pp. 187-190.

Johnson, R. and C.N. Law (1975). Genetic control of durable resistance to yellow rust — *Puccinia striiformis* — in the wheat cultivar Hybride de Bersee. *Annals of Applied Biology*, 81: 385.

Khush, G.S. (1995). Approaches and strategies in breeding for pest resistance. In *Genetic Research and Education*: *Current Trends and the`Next Fifty Years*. Indian Society of Genetics and Plant Breeding, New Delhi. Ed. B. Sharma *et al.*, pp. 163-179.

Knott, D.R. (1995). Adult plant resistance to *Puccinia* in *Triticum* spp. In *Genetic Research and Education*: *Current Trends and the Next Fifty Years*. Indian Society of Genetics and Plant Breeding, New Delhi. Ed. B. Sharma *et al.*, pp. 287-296.

Kumar, J. and P.S. Phul (1977). How plants fight disease. *Science Today*, March, pp. 16-17.

Leonard, K.J. (1984). Population genetics of gene for gene interaction between plant host resistance and pathogen virulence. In *Genetics — New Frontiers*, Proceedings of the XV International Congress of Genetics. Ed. V.L. Chopra *et al*. Publ.Oxford & IBH Publ.Co., New Delhi, pp. 131-148.

McIntosh, R.A. (1995). Strategies for breeding wheats with adult plant resistance to rusts. In *Genetic Research and Education: Current Trends and the next fifty years*. Indian Society of Genetics and Plant Breeding, New Delhi, Ed. B. Sharma *et al*., pp. 297-305.

Miller, R.J. and D.E. Koeppe (1971). Southern corn leaf blight: Susceptible and resistant mitochondria. *Science*, 173: 67.

Muller, K. and H.Borger (1940). Experimentalle unter chungen uber die Phytophthora-resistenz der kartoffel. Arb. Biol.Reichsanstat Lund-4-*Forstwirtsch*.Berlin, 23: 189-231.

Narayanan, S.S.; V.V. Singh and R. Kothandarman (1990). Cotton genetic resources in India. In *Cotton Scenario in India — A Souvenir*. Eds. A.K. Basu, N.D. Mannikar and S.S. Narayanan, ICAR, New Delhi, 10-28.

Nelson, R.R. (1978). Genetics of horizontal resistance to plant disease. *Annual Review of Phytopatholgoy*, 16: 359-378.

Nene, Y.L. and M.V. Reddy (1976). Preliminary information on chickpea stunt. *Trop.Grain Legume Bull*., 5: 31-32.

Nene, Y.L. (1995). Virus and vector resistance in grain legumes. In *Genetic Research and Education: Current trends and next fifty years*, Ed. B. Sharma *et al*., Indian Society of Genetics and Plant Breeding, New Delhi, pp. 257-270.

Parlevliet, J.E. and J.C. Zadoks (1977). The integrated concept of disease resistance: a new view including horizontal and vertical resistance in plants. *Euphytica*, 26: 5-21.

Popp, W. (1933). A preliminary study of the reactions to stem rust of certain wheat varieties in different stages of development. Proc. World Grain Exhibition and Conf., Canada, 2: 238-243.

Rajaram, S.;R.P. Singh and M.Van Ginkel (1995). CIMMYT's approach to breeding for durable disease resistance in bread wheat: A Global perspective. In *Genetic Research and Education: Current Trends and the Next Fifty Years*. Indian Society of Genetics and Plant Breeding, New Delhi. Ed. B. Sharma *et al*., pp. 325-338.

Ramkrishnan, J.S. (1948). Ergot sclelrotia in *Sorghum vulgare*, Pers. *Curr.Sci*., 17: 218.

Raper, J.R. (1954). Life cycles, sexuality and sexual mechanisms in fungi. In "Sex in Microorganisms", *Amer.Assoc.Advancement Sci*., Wash. D.C.

Robinson, R.A. (1969). Disease resistance terminology. *Review of Applied Mycology*, 48: 593.

Robinson, R.A. (1976). The pathosystem concept. In "Plant Pathosystems." *Springer-verlag*, Berlin: 184.

Russell, G.E. (1978). *Plant breeding for pest and disease resistance*. *Butterworth*, London, 485.

Samadder, R.R. and R.P. Scheffer (1968). Effect of the specific toxin in *Helminthosporium victoriae* on host cell membranes. *Plant Phvsiology*, 43: 21.

Sharp, E.L.; A. Bronnimann and F.H. McNeal (1972). Reaction of selected spring wheat varieties to infection by *Septorium nodorum. Plant Disease Reporter*, 56: 761.

Sidhu, G.S. (1975). Gene-for-gene relationships in plant parasitic systems. *Sci.Prog.*, 62: 467-485.

Sidhu, G.S. and J.M. Webster (1983). Horizontal resistance in tomato against *Meloidogyne-Fusarium* complex: an artifact of parasite epistasis. *Crop Prot.*, 2: 205-210.

Simmonds, N.W. (1983). Strategy of disease resistance breeding. *FAO Plant Prot.Bull.*, 31: 2-10.

Stakman, E.C. (1914). A study in cereal rusts. Physiological races. *Minnesota Agric.Exp.Sta.Bull.* No. 138.

Stakman, E.C. (1954). Recent studies of wheat stem rust in relation to breeding resistant varieties. *Phytopathology*, 44: 346.

Stakman, E.C. and F.J. Piemeisel (1917). A new strain of *Puccinia graminis*: *Phytopathology*, 7: 73.

Starks, K.J.; S.A. Eberhart and H. Doggett (1970). Recovery from shoot fly attack in sorghum diallel. *Crop Sci.*, 10: 519-521.

Thottappilly, G.; H.W. Rossel; D.V.R. Reddy; F.S. Morales; S.K. Green and K.M. Makkouk (1990). Vectors of virus and mycoplasma diseases, an overview. In *Insect pests of tropical food legumes*. Ed. S.R. Singh. John Wiley & Sons Ltd., pp. 323-342.

Van der plank, J.E. (1963). *Plant diseases*: *Epidemics and control*. Academic Press, New York.

Van der plank, J.E. (1968). *Disease resistance in plants*. Academic Press, New York.

Van der plank, J.E. (1978). *Genetics and molecular basis of plant pathogenesis. Springer-Verlag*, Berlin.

Vetten, H.J. and D.J. Allen (1983). Effects of environment and host on vector biology and incidence of two whitefly spread diseases of legumes in Nigeria. *Ann.Appl.Biol.*, 102: 219-227.

Ward, H.M. (1902). On the relations between host and parasite in bromes and their brown rust. *P.dispersa* (Erikss). *Ann.Bot.*, 16: 233-315.

Wood, R.K.S. (1967). *Physiological plant pathology*. Blackwell Oxford.

14

Resistance of Plants to Insect Pests

In India alone annual crop losses due to various pests and diseases, etc. are estimated to be about Rs.6 to 7 thousand crores. Of this, losses from diseases are 26 per cent, from insects and rodents 26 per cent, from weeds 33 per cent and birds, nematodes etc. account for the rest (Nagarajan, 1990). All sorts of crop plants, viz., cereals, pulses, oilseeds, fibre crops, vegetables, sugarcane, etc. are damaged by the insects. In oilseeds alone, for example, losses due to insect attack amount to 20 per cent (Rangarao *et al.*, 1995). Under favourable conditions some insects, such as shoot fly of sorghum (*Atherigona soccata*), can be responsible for as much as 100 per cent destruction of the crop. Insect damage is usually more extensive in the tropical and sub-tropical regions than the temperate because high temperatures tend to increase the insect activity and reproductive potential (Russell, 1978).

Insect pests of various crops could be controlled by various methods, viz., avoidance, direct control measures, biological control and resistant varieties. Avoidance comprises of various precautionary measures such as quarantine, good crop husbandry, crop rotation, use of clean seed, hygiene or sanitation, etc. Identification, multiplication and release of natural enemies of pest insects at appropriate time involve biological control. Practically every crop pest has its natural enemies in the form of parasites, predators and insect pathogens. Nuclear Polyhedrosis Virus (NPV) of *Heliothis* has been used to control pest insects of pulse crops very successfully. There are nearly 300 types of NPVs that are specific to each insect and primarily pathogenic on Lepidopterans. *Trichogramma chilonis* has been used to control stem borer in rice. Use of predators such as *Chrysoperla* is on increase. *Bacillus thuringiensis* is not only used to control the insect pests through sprays in the field, but the cry/bt gene has been integrated into the genetic material of some crop plants, through recombinant DNA technique, to develop transgenic varieties, e.g. MECH 12bt, 162bt and 184bt of cotton, recently available to the farmers of India. These varieties are said to be resistant to bollworm of cotton.

Subsequent to the discovery in 1939, by a Swedish scientist Paul Muller, that DDT has insecticidal property use of chemicals began to control the pest inscets. After the green revolution, use of insecticides to control pest insects in India increased enormously. Some farmers are using it indiscriminately. Starting with the manufacture of BHC in 1953 in India, presently over 39 technical grades of pesticides, including insecticides, fungicides, herbicides, etc., are being produced in the country. Demand for insecticides is on increase. It is estimated that for the year 2000-2001 over 86,000 metric tones of pesticides would be required (Jayaraj, 2002). Major pesticide consuming states in the country are Andhra Pradesh, TamilNadu, Punjab, Gujarat, Karnataka, West Bengal and Uttar Pradesh. Among the insecticide consumer crops cotton ranks first (52 per cent) followed by rice (13 per cent), vegetables (8 per cent) and 27 per cent in other crops (Jeyarajan, 1997). At the global level it is estimated that about 2.2 million metric tonnes of active ingredients of pesticides are used each year (Sarukhan, 2000).

Indiscriminate use of insecticides to control pest insects leads to several harmful effects. Some of them may be:

(i) It causes environmental pollution and hence is not eco-friendly.

(ii) In addition to specific pest-insect being controlled, beneficial insects are also killed. Even other insects are destroyed leading to reduction in biodiversity.

(iii) Insects become resistant to insecticides.

(iv) Minor pests become serious pest in the course of time.

(v) There is a possibility of resurgence of pest-insect.

(vi) Use of insecticide often leads to residual toxicity in the agricultural produce. This is particularly so in fruits and vegetables. For example fruits of tomatoes, okra, brinjal etc., are picked at intervals but farmers do not observe correct spray schedule.

(vii) Cost of production increases.

Use of resistant varieties is ecofriendly, not injurious to human/ animal health, cost effective and relatively stable means to reduce damage by pest insects. Use of resistant varieties in combination with other methods of suppression of insects (avoidance, biological control) may be very effective in reducing the population of pest insect to a low level. The objective of integrated control measures is to maintain the population of pest-insect at levels below those causing economic injury (Stern *et al.*, 1959). Resistant varieties, therefore, form an important component of Integrated Pest Management (IPM). Once the pest has been greatly suppressed or eradicated from an area, the resistant varieties may serve over years as a major deterrent factor for reestablishment of the species (Panda, 1979).

14.1 TYPE OF RESISTANCE OF PLANTS

During the course of evolution of plant species, plants acquired several mechanisms to defend against pest-insects. Currently considerable genetic variability exists in the germplasm of various crop species for resistance to pest insects.

Unintentional selection for resistant plants against a specific insect probably occurred in the very early days of agriculture. The process of natural selection for resistance continued until man exercised his selection pressure (artificial selection) favouring certain plants with desirable attributes. As a result many land races possess built-in genetic resistance to the insect pests. For example, the post-rainy sorghum variety Maldandi, and two other types from Maharashtra, viz., Allu and Dagadi, possess resistance against shoot fly, *Atherigona soccata*.

Painter (1951) defined resistance of plants to insect attack as "the relative amount of heritable qualities possessed by a plant which influence the ultimate degree of damage by the insect". However, Beck (1965) defined resistance as "the collective heritable characteristics by which plant species, race, clone or individual may reduce the probability of successful utilization of that plant as a host by the insect". Obviously the definition given by Beck (1965) may not include tolerance as a mechanism of resistance. The degree of resistance may range from low to high. Immunity, meaning the complete absence of damage by the insect, is rare to come across. The reason being, under heavy infestation levels even the highly resistant variety may experience some damage. Resistance is a relative term and not absolute. Panda (1979) classified resistance against insect into two categories, viz., ecological resistance and genetic resistance. Ecological resistance stems from environmental factors whereas genetical resistance owes to genetic factors present in the plant. Painter (1951) categorized ecological resistance as pseudoresistance because it arises from some temporary shifts in the environmental conditions favourable to otherwise susceptible variety. Panda (1979) further classified ecological resistance into host evasion and induced resistance.

On the basis of host habitation, insects are grouped into three categories, monophagous, oligophagous and polyphagous. Insects, which may live on one taxonomic unit only are designated as oligophagous. On the other hand, insects which avoid all the hosts, except, a particular species or a variety are called monophagous. Polyphagous insects are those, which avoid only a few plants and prefer a wide host range.

For an oligophagous insect finding a right host plant is not enough but contacting it at appropriate stage of development is equally important. At the time of incidence of specific insect the phenologies of both, the plant and the insect should synchronize. Host evasion takes place when growth pattern of a plant is modified, which induces asynchrony between the phenologies of the plant and the insect. Resistance manifested by a plant may change with age of plant. Plant sucking insects, such as aphids, are more sensitive to nutritional changes in their specific host plant. The aphids, which are adapted to tap rich metabolite substances in the phloem of meristematic tissues, can not multiply successfully on senescent tissues.

Asynchronies in phenologies could be induced by proper scheduling of planting dates. Population dynamics of *Atherigona soccata* was studied from 1977 to 1983 at the International Crops Research Institute for Semi Arid Tropics (ICRISAT), Hyderabad. There were two peaks, viz., August and October-November (Taneja and Leuschner, 1985). At the beginning of monsoon season the shoot fly is almost absent and hence early planting is completely free from shoot fly (*Atherigona soccata*). As the plantings get delayed, shoot fly population increases, depending upon the variety, and seedling attack goes upto 100 per cent (Rao, 1972).

Induced resistance represents an alteration of some condition in plant environment, which may lead to a temporary beneficial modification in the host plant. There are many examples in which agricultural chemicals, e.g. growth regulators, insecticides, herbicides etc., can change the physiological processes or nutritional status within the plants (Shaw and Little, 1972; Rodriguez, 1960). The growth regulators promote, inhibit or otherwise modify the physiological processes in the plants, as a result behaviour of plants may change with respect to the insect attack. Susceptibility of plants to insect attack and diseases was found to increase following application of herbicide 2,4-D. Growth regulators such as Maleic hydrazide and chlormequat when applied to plants tends to reduce the fecundity and growth rates of aphids feeding on them (van Emden, 1969). After the seed dressing of sorghum with carbofuran 50 SP, to protect the crop from shoot fly, sorghum plants were observed to grow at a faster rate as compared to the crop grown without seed treatment with carbofuran. Carbofuran has probably hormonic effect on plants (Nayeem and Bapat, 1976). Aphids are sensitive to the levels of nitrogen in the plants but respond negatively to the levels of potassium even in the presence of high N (van Emden, 1966). In certain cases pseudo-resistance is possible through changes in the physiology of host plants by nutritional management or by interference of certain micro-organisms, including systemic resistance (Jayaraj, 2002). Certain fluorescent pseudomonads induced resistance to pests as in case of *Pseudomonas gladioli* in cotton, *P. maltophila* in maize and culture filtrates of actinomycetes possessing acaricidal activity against coconut eriophyid mites. *Azospirillum* biofertilizer induces resistance in sorghum due to production of polyphenols, particularly ortho dihydroxy phenols. Pseudomonads also induced résistance to diseases e.g. in rice to bacterial blight, vegetables to wilt and damping off, pulses to root rot, groundnut to *Cercospora* leaf spot, etc.

Painter (1951) recognized three types of resistance of plants to insect attack, viz., non-preference, antibiosis and tolerance. A fourth type of resistance, pest avoidance, which is expressed as a tendency to escape infestation, should also be considered as a mechanism of resistance. Non-preference refers to the characters of plants, which discourage their use for oviposition, as food or for shelter. It is shown by the plants, which are unattractive and unsuitable for colonization and oviposition. Antibiosis adversely affects the development or reproduction of insects on the host plants. It refers to those attributes of plants, which are preventive, injurious to the normal development of insects. Tolerance enables a host plant to withstand an attack by insects without suffering severe damage. Tolerance can be evaluated by rejuvenation potential, healthy leaf growth, flowering compensation potential, superior plant growth etc., which may be useful in rainfed condition (Narayanan, 1995). Tolerance specifies ability of a host plant to sustain a pest population, damaging to a more susceptible host variety, by rapidly repairing or overcoming the injury and maintaining its growth and yield.

Antibiosis and non-preference disrupt the normal life of insect pest and may sometimes exert a lethal effect. On the contrary, tolerance is a defensive characteristic of plant, which prevents the economic injury without inhospitality to the insect population.

14.1.1 Non-Preference

Any inherited characteristic of a plant which discourages its feeding, its colonization and its use for oviposition by the insect makes it 'non-preferred' by the insects. To describe this type of resistance mechanism, van Marrewijk and De Ponti (1975) proposed a term 'non-acceptance'. Non-acceptance is appropriate because in most of the known examples of this type of resistance, insects do not 'accept' a resistant host plant even if there is no alternative source of food. Resistant plants are thus less acceptable than the susceptible ones. Kogan and Ortman (1978) coined term 'antixenosis' to this type of resistance mechanism. However, non-preference is still popular and in vogue.

Morphological, physiological or biochemical attributes of plants could be responsible for the non-preference by the insects. For example, presence of bloom/waxy coating on leaves and stem is positively related with resistance in castor against jassids (Jayaraj, 1967), thick and corduroy leaves against jassids in groundnut, glandular hairs on floral parts against *Spodoptera* sp. in sunflower. Shoot fly (*Atherigona soccata*) is a serious pest-insect of sorghum. It attacks sorghum from one to four weeks after seedling emergence. White cigar shaped eggs are laid singly on the underside of leaf parallel to the mid-rib. After hatching of the eggs maggots move downwards between 5th and 6th leaf of seedlings till they reach the growing point. The maggots cut around the growing points causing drying of central leaves and typical 'dead hearts'. A number of workers have reported non-preference for oviposition as a mechanism of resistance of sorghum (Blum, 1967; Jotwani *et al.*, 1971; Narayana, 1975). It was observed at the International Crops Research Institute for Semi Arid Tropics (ICRISAT), Hyderabad, that susceptible cultivars are preferred for egg lying in terms of higher number of eggs per plant and higher number of plants with eggs. Significantly higher number of egg laying was observed on susceptible cultivar CSH-1 as compared to resistant cultivars under field conditions (Table 14.1). Under no choice conditions relatively more eggs were laid on resistant cultivars than under multiple choice conditions. The glossy (pale green, shiny leaves) trait was reported in sorghum to be associated with resistance to shoot fly (Bapat *et al.*, 1975, Maiti and Bidinger, 1979). Further investigations indicated that shoot fly incidence was higher on nonglossy lines than glossy ones in post-rainy (rabi) season but glossiness contributed less to resistance during rainy season (Taneja and Leuschner, 1985).

Table 14.1: Incidence of shoot fly on sorghum lines at ICRISAT, 1982*

Pedigree	*Choice*		*No choice*	
	Egg laying %	*Dead hearts* %	*Egg laying* %	*Dead hearts* %
IS 1082	53.1	29.2	85.3	72.7
IS 2122	55.4	40.7	91.3	82.1
IS 2195	63.3	50.5	76.3	73.9
IS 5484	48.1	41.8	72.1	58.7
IS 18551	57.2	42.7	51.6	44.0
CSH-1 (susceptible control)	93.1	92.3	100.0	95.3
S.Em.±	9.9	8.2	7.7	6.8

**Source*: Taneja and Leuschner (1985)

Non-preference of some varieties of raspberry to raspberry aphid (*Amphorophora idaei*) is so strong that the aphids quickly walk off the resistant plants. The size, shape and the colour of plants can contribute to the non-preference by the insects. Red cabbage and red leaved Brussels sprouts are less favoured by butterfly, *Pieris rapae*, and certain other Lepidoptera for oviposition than the green varieties (Dickson and Eckenrode, 1975). Insects are particularly sensitive to the green and yellow parts of the spectrum of visible light and butterflies are presumably attracted more to green and yellow plants than the red plants.

Presence of hairs (pubescence) on leaves is associated with resistance to many insects. Pubescent (hairy) wheat varieties are resistant to cereal leaf beetle (*Oulema melanopus*). Oviposition behaviour of the adult female, the viability of eggs and growth of surviving larvae are all adversely affected by pubescence (Schillinger and Gallun, 1968). Females appeared to be reluctant to lay eggs on hairy leaves and, therefore, had fewer eggs on such leaves. Lukefahr *et al.* (1971) reported that strains of cotton with glabrous leaves were less satisfactory for oviposition of *Heliothis* spp. than those with

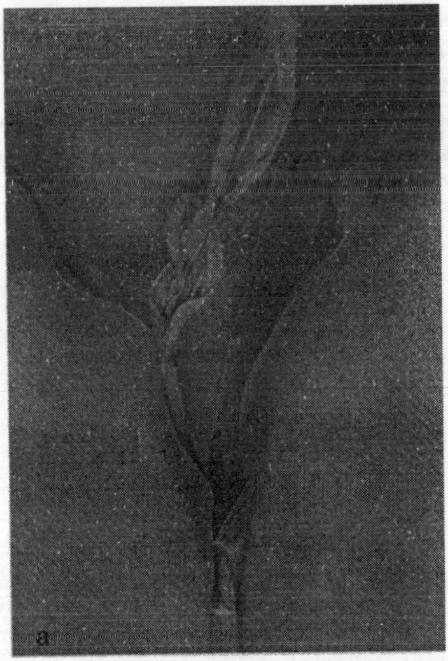

(a)

(e)

Fig. 14.1: Dead heart of sorghum plant caused by (*a*) shoot fly and (*e*) adult of shoot fly (Photographs *courtesy* Dr. H.C. Sharma of ICRISAT, Hyderabad)

hairy leaves. Eggs laid on glabrous leaves were less as compared to those on hairy leaves. *Heliothis zea* and *Pectinophora gossypiella* laid more eggs on pubescent Acala cottons with nectaries than on smooth leaved nectariless strain (Davis and Ellington, 1974). These observations differ from the general observation that pubescence is non-preferred trait. For example, in cotton itself, pubescence is associated with resistance to jassids (*Empoasca* spp.). In turnips, resistance to turnip aphid is associated with pubescence (Barnes and Cuthbert, 1975). Several wild species of *Helianthus*, possessing glandular hairs, sesgui terpenoid lactones and diterpens are reported to be highly resistant to sunflower moth and *Spodoptera* species (Gershenzon *et al.*, 1985). Oviposition of *Chilo suppressalis*, the striped moth borer, is less on the pubescent than on the glabrous rice varieties (Pathak, 1977).

In cotton, several characters, viz., smooth leaf, hairiness, frego bracts, okra leaf, nectariless, and naked seed, contribute to non-preference (Narayanan, 1995). Oviposition on *Gossypium barbadense* (Egyptian cotton), by boll weevil (*Anthonomus grandis*) was lower than on *G.hirsutum* (Maxwell *et al.*, 1969). Non-preference for oviposition by *A.grandis* is apparently associated with frego bract character. About one third as many eggs are laid on frego bracts as on the normal bracted varieties of cotton (Lincoln *et al.*, 1970). Soybean genotypes with high trichome density (i.e. greater than 550 per cm) on leaves, smaller leaf area, high leaf moisture, shorter internodes, and anthozynadin pigment on stem are less preferred by *Melanagromyza sojae* (Talekar and Chen, 1986).

Mechanical obstruction to oviposition or feeding due to toughness or thickness of plant tissues may be responsible for reduced insect infestation.

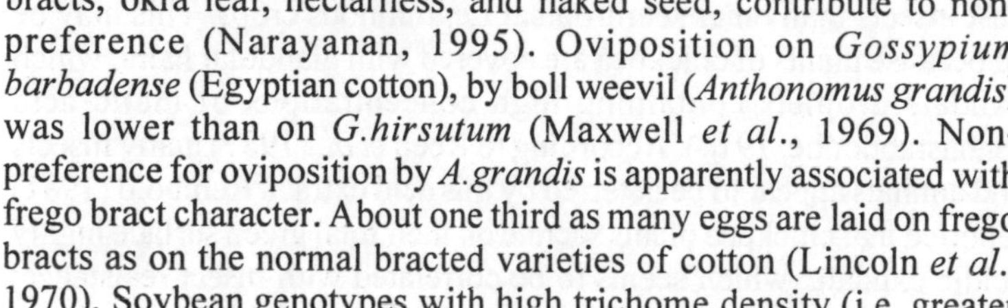

(a) (b)

Fig. 14.2: Trichomes on underside of leaf of sorghum (Photographs *courtesy* Dr. H.C. Sharma, ICRISAT, Hyderabad)

For example, thickness of leaf and toughness of vascular tissues can affect the level of resistance to jassids in cotton (Batra and Gupta, 1970). Wheat varieties with solid stems are resistant to wheat stem saw fly. Ponnaiya (1951) reported presence of irregular shaped silica bodies in the plant tissue of sorghum. These bodies were found from fourth leaf onwards in resistant varieties and from 6th leaf onwards in varieties susceptible to shoot fly. Blum (1968) observed significant differences in lignification and silica deposition in sorghum. Resistance in some rice varieties to *Chilo suppressalis* is associated with thick layers of sclerenchymatous and lignified tissues in stem (Patanakamjorn and Pathak, 1967) or concentrations of silica (Sasamoto, 1961; Djamin and Pathak, 1967). A high silica content in the host plant may interfere with feeding of stem borers by wearing their month parts.

Biochemical plant compounds, such as some essential oils may act as repellent to the pest insect. Chickpea suffers less damage from pest insects than other semitropical Leguminous crops. This may be so because plants of chickpea are covered with glandular hairs, which exudate droplets containing high concentrations of malic acid (Sahasrabuddhe, 1914). According to Reed *et al.* (1987) many insects and animals appear to be deterred by this acid extract. Rembold (1981) reported that chickpea plants secrete on their total green surface highly acidic exudate, which seems to be correlated with insect resistance, mainly through its malic acid contents. Cultivars of chickpea resistant to leaf minor (*Liriomyza cicerina*) had high amounts of malic acid and the susceptible ones had lowest levels (Rembold *et al.*, 1990).

Insects respond to various feeding stimuli while selecting their host plant (Kennedy, 1965). Absence of such stimuli may contribute to non-preference type of resistance. Sugars (glucose, fructose, sucrose) act as feeding stimulant. The corn borer larvae feed more intensively on plants containing highest concentrations of sugars (Beck, 1956). Amino acids also influence the infestation by insects acting as a stimulant (Beck and Hanec, 1958). There were quantitative differences of serine and phenylalanine between the borer resistant and susceptible brinjal (*Solanum melongena*) varieties (Panda and Das, 1975). These amino acids were higher in susceptible than resistant variety. According to Malik (1995) amino acid and sugar are phagostimulatory to aphid attack in case of plants of Crucifereae. The susceptible species had higher amounts of all the constituents as compared to the resistant and tolerant ones. Higher amino acid and sugar attracted more number of aphids (*Lipaphis erysimi*) and accelerated their rate of multiplication.

Host selection by cabbage aphid, *Brevicoryne brassicae*, is stimulated by the presence of mustard oil glucoside, signigrin, in leaves. A low concentration of signigrin is apparently associated with resistance to aphids in brassicas. Mustad oil glucoside, sinigrin, and many essential oils from Umbellifereae act as attractants and feeding incitants for larvae of *Listroders costrirostris obliques* (Sugiyama and Matsumoto, 1959; Dethier, 1941). Plants of Solanum species that are resistant to Colorado potato beetle (*Leptinotarsa decemlineata*) owe their resistance probably to the alkaloids produced. Tomatine appears to act as repellent. Solanine, chaconines and leptines also have adverse influence on larval feeding (Buhr *et al.*, 1958).

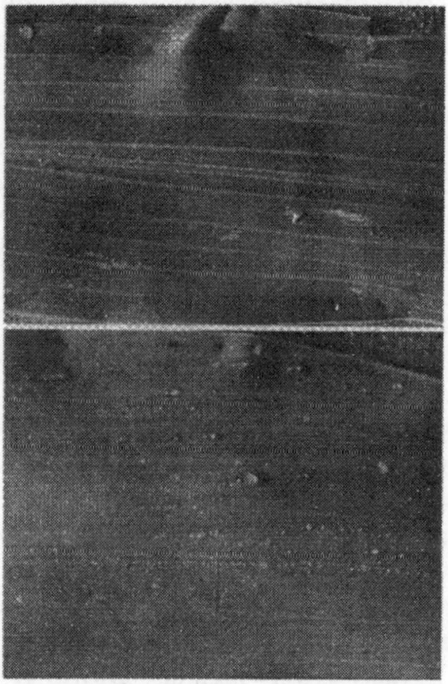

Fig. 14.3: Leaf moisture on sorghum. Susceptible cultivars have more moisture in leaf whorls. (Photographs *courtesy* Dr. H.C. Sharma, ICRISAT, Hyderabad)

14.1.2 Antibiosis

A resistant component is called antibiosis when an insect feeds on resistant plant and suffers from several adverse physiological disorders of temporary or permanent nature (Panda, 1979). Antibiosis that is frequently exploited causes a reduction in pest population, decrease in rate of development of insects, lowered reproductive potential, constitutive or wound induced antibiotics and phytoallexins (El-zik and Thaxton, 1989).

Responsive resistance mechanisms which are usually triggered by contact of a parasite with the host plant is rare in case of plant and the insect. Most of the resistance factors already exist in the host plant before the onset of insect attack.

When an insect feeds on a resistant cultivar, the rate of its reproduction is reduced (Dahms, 1972). The rate of development of nymphs or larvae is often reduced when insects feed on resistant varieties. When the first instar nymphs of *Nephotettix virescens* were caged on resistant rice plants about 90 per cent of them died within a few days compared to about 10 per cent on susceptible plants. The surviving nymphs grew very slowly on resistant plants. Adults of *N. virescens* also had a shorter life span on many resistant varieties.

The resistance of most varieties of rice to *Chilo suppressalis* appears to be due to reluctance of female moths to oviposit on them. Larvae often grew more slowly and suffered a high mortality on resistant plants as compared to susceptible ones suggesting that there is antibiosis (Pathak *et al.*, 1971).

Although non-preference for oviposition seems to be the primary mechanism for shoot fly in sorghum, evidence of some degree of antibiosis is also available (Jotwani and Srivastava, 1970). Survival and development of shoot fly was adversely affected when the insect was reared on resistant varieties (Narayana, 1978). Larval and pupal periods were extended by 8 to 15 days on resistant varieties of sorghum (Singh and Jotwani, 1980). Not only growth and development retarded on resistant plants but survival and fecundity of shoot fly was also adversely influenced. Raina *et al.* (1981) observed that mortality among the first instar larvae was very high and growth of surviving larvae was significantly lower on resistant cultivars.

Antibiosis can stem from toxins present in the resistant plants. The green peach aphid, *Myzus persicae*, does not survive on *Nicotiana gossei* because toxin exudes from leaf hairs and kills the aphids (Thurston and Webster, 1962). Exudates from glandular leaf hairs of many plants from family *Solanaceae*, including several species of *Solanum* and certain tobacco varieties, are toxic to a wide range of insects (Gibson, 1976). Resistance to insect pests in cotton has been attributed to the presence of sub epidermal glands, which contain high concentration of polyphenol pigment gossypol (Lukefahr and Martin, 1966; Lukefahr and Houghtaling, 1969). Larvae of *Heliothis virescens* showed marked distaste for cotton plants with many pigment glands or which had high gossypol content in seeds (Wilson, 1971). Glandless variety Acala is more susceptible to insect pests than those with glands (Benedict *et al.*, 1977). Glandless varieties are more susceptible to *Erias biplaga* than fregobract or nectariless varieties of cotton (Reed, 1974).

Resistance to *Myzus persicae* and *M. euphorbie* shown by *Solanum polyadenium*, *S.tarijense* and *S.berthaultii* is probably associated with presence of glandular hairs on the leaves and stems (Gibson, 1971). When the aphids rupture the cell walls of these glandular hairs during movement over the leaf surface, a sticky liquid exudes from the hair cells and contaminates the limbs of aphids. This effectively immobilizes many of the aphids to the extent that they starve to death.

Leaves of *Lycopersicon hirsutum* f. *glabratum* contain highly active ethanol soluble antibiotic compound that is lethal to insect pests such as tomato fruit worm, tobacco flea beetle (Fery and Cuthbert, 1975). The larvae of cutworm (*Agrotis orthogonia*) were found to be sensitive to amino acid balance but not to differences in carbohydrates in wheat cultivars. Grison (1952, 1958) observed that sugar and lecithin content of potato foliage had marked effect on fecundity of Colorado potato beetles. Old foliage contained more sugar but less lecithin concentration than the younger leaves. When the beetles were fed on older leaves, they laid fewer eggs per day than when fed on younger leaves. According to Grison carbon nitrogen ratio was less important in reproduction than the sugar-lecithin ratio. The European corn borer requires a low sugar-high protein diet early in the larval period but needs a high sugar-low protein diet for later instars (Beck, 1956). These nutritional requirements are met by the insects by moving away from low sugar-high protein leaves to plant tissues rich in sugar during second or third instar. Resistance to Lepidopterous defoliators is known to be associated with high carbohydrates (more than 33 per cent), sterols (more than 20 to 50 per cent) and low nitrogen. Resistant genotypes of soybeans also contain flavonoides, tannins, terpenoides, cyclopropenoid acid, cyclitols (Tester, 1977).

Probably both non-preference and antibiosis were involved in the inheritance of resistance of barley to green bugs (Painter, 1967). More larvae were observed on susceptible cultivars than the resistant ones.

Resistant plants often contain higher concentrations of benzyl alcohol, which is toxic to green bugs (*Schizaphis graminum*). Green bugs are more restless on resistant wheat, barley and oats than on susceptible plants of these crops. This reduces feeding by the pest, decreases fecundity and hence less damage to the crop (Starks and Burton, 1977).

14.1.3 Tolerance

Tolerance is the plant's relative ability to suffer a smaller loss of yield by the dent of its innate ability to grow and reproduce itself or to repair the injury to a marked degree following infestation by the pest insect. Tolerance does not, in any way, restrict or hinder the colonization of the host plant by the insect nor does it adversely influence the development or reproduction of the insect. As a matter of fact a tolerant variety may grow normally and produce higher yield than the susceptible variety, even though both are equally highly infested with the pest insect.

In tolerance type of mechanism, recovery resistance may be a component imparting ability on the host plant to withstand the damage and yield better than the susceptible cultivar. Doggett (1972) reported about a variety, Namatare, indigenous to Uganda, Africa, which displays recovery resistance. Many cultivars of sorghum produce tillers following death of main shoot due to the damage by shoot fly. Usually the shoot fly destroys tillers also. There is evidence that the tillers of resistant or less susceptible varieties are less preferred by the shoot fly for egg laying (Deeming, 1972). Doggett and Majisu (1966) found a variety in Uganda (Africa) called Namatare, which is tolerant to shoot fly as a result of its high tillering ability and ability to recover from early seedling attack (Young, 1972). Although several varieties were evaluated for their reaction to shoot fly attack, many were devastated, but Namatare grew showing little damage. It was later observed that the main stem of Namatare was frequently damaged but the tillers grew up and gave a good yield, within a fortnight of cropping date of undamaged shoots. Length of time for plants to recover is very important. The local people believe that Namatare would not yield properly unless damaged by the shoot fly. Namatare is a beer variety of sorghum. Serena is another sorghum cultivar, which shows recovery resistance.

Dunn and Kempton (1971) identified non-preference, antibiosis and tolerance in some European Brussels sprout varieties. Several varieties were tolerant and suffered little damage due to the infestation by *Brevicoryne brassicae*, although they were heavily infested. On the other hand, Irish Elegance suffered great damage as a result of *B.brassicae* infestation, although the aphid population on this variety was not exceptionally high. Some maize hybrids are tolerant to *Heliothis zea* although they can get heavily infested with earworm larvae in the field; they suffer little damage as a result (Wiseman, McMillian and Widstrom, 1972). Tolerance can be evaluated by rejuvenation potential, healthy leaf growth, flowering compensation potential, superior plant vigour, etc.

Ability of a plant to tolerate injury caused by the insect attack depends upon the pattern of crop growth. When this ability reduces the effect of injury on yield it is known as compensation or compensatory growth (Panda, 1979). Compensation observed following insect injury may result from various causes. In wheat, when the injury occurs early in the life of crop, the surviving plants, especially the uninjured ones, grow larger and give more than normal yield (Bardner, 1968). He observed that surviving plants produced heavier ears than the normal. Similar observations were reported for maize damaged by Lepidopteran stem borer by Harris (1962). Tolerance may result when the harvested organs of plants are attacked by the pest insect but the organs are produced in excess. This mechanism may be seen in crops with indefinite growth, such as many varieties of cotton (Panda, 1979). In such cases, the insect attack merely decreases the amount of physiological shedding and delays harvest.

Plant tolerance is best expressed in relation to insects with sucking mouth parts such as aphids, leaf hoppers etc. The damage, which these insects cause includes loss of fluid, loss of food such as carbohydrates, proteins etc., clogging of the conducting tissues by stylet sheath material and introduction of toxic material (Panda, 1979).

Non-preference, antibiosis and tolerance to the pest insects manifested by resistant cultivars are sometimes expressed simultaneously in a single host. For example, both non-preference for oviposition and antibiosis are said to operate against shoot fly in sorghum (Blum, 1972) and chickpea (Reed *et al.*, 1987).

14.2 GENETICS OF RESISTANCE

Information about the mode of inheritance of resistance against a specific pest-insect in the resistant source of cultivars is a prerequisite before embarking on an actual breeding programme. The choice of an appropriate breeding procedure or the technique of genetic engineering to be employed for transferring the desired gene(s) depends upon the genetic architecture for resistance in the source material.

As stated in the chapter on disease resistance (chapter 13), inheritance of resistance to pest-insect in various sources is largely governed oligogenically or polygenically. Nevertheless, in a few cases resistance is monogenic in inheritance. The genes concerned may be dominant or recessive.

Although the mechanisms for resistance are described as non-preference, antibiosis and tolerance (and even escape), one should bear in mind that these are the mechanisms observed in resistant plants by various workers. However, the underlying cause is genetic in nature. Whether it is pubescence or presence of silica bodies or sclerenchymatous cells or phytoalexins, everything is inherited. The chemicals produced in plants, said to be responsible for resistance, may be the end product or an intermediary product in the chain of metabolic pathway. When one gene is involved, it may imply that from the precursor one step is required to produce a given metabolite. When two or more genes are involved it suggests that several steps are involved before the desired end product is produced. Gossypol is a prominent pigment found in the glands distributed throughout the leaves, stem, root cortices and floral parts of most cotton varieties (Adams *et al*., 1960). Gossypol is identified as phytoalexin, toxic to 14 forms of *Fusarium oxysporum* (Davis, 1964) and reported to be injurious to cotton boll worm, tobacco bud worm and rodents (Lukefahr and Martin, 1969). In cotton, genes G12 and G13 with modifiers confer glanded nature and ability to produce higher terpenoids (Narayanan, 1995). Nectariless attribute of cotton is governed by two recessive genes ne1 and ne2. Hairiness is controlled by gene H_1 for hirsuit and H_2 for pilose besides H_3, H_4 and H_5. Individual gene does not produce the desired attribute from scratch but is responsible for the end result. So many other genes may be involved in the background. Whether it is physiological responses, biochemical expressions, morphological appearance, everything is genetically determined. Genes determine potentialities but the realization of their potentialities depends also upon the environment in which the genes perform their function. For more insite into this aspect a student may refer literature on biochemical genetics.

14.2.1 Monogenic Resistance

In many, if not all, the self-pollinated crops resistance to insects is often controlled by a single gene, e.g., Hessianfly (*Mayetiola destructor*) in wheat, green bug (*Schizaphis graminum*) in wheat and sorghum. Resistance to aphids in raspberry is also monogenic in inheritance (Russell, 1978). Resistance to fruit fly is monogenic in watermelon and pumpkins.

Actually it is very simple to transfer a single gene for resistance, especially if resistance is governed by dominant gene, to a desirable line. In the F_2 generation distinction between resistant and susceptible plants can be easily made. But it is disadvantageous because when mechanism for resistance is governed by a single gene, whether dominant or recessive, resistance breaking biotypes can also easily develop. Nevertheless, some monogenically controlled resistances to insects have been very durable and the insects have not been able to overcome the resistance by resistance breaking biotypes (A biotype is a population of an insect capable of damaging and surviving on plants previously known to be resistant to other populations of same insect), even when resistant varieties were exposed to pest attack for many years. For example, jassid resistant cotton and leaf beetle resistant cereals are being widely grown for many years and hence the resistance should be classified as durable, although resistance is monogenic (Russell, 1978).

14.2.2 Oligogenic Resistance

When more than one gene pair, effect of each of which is large and measurable and not much influenced by the environmental agencies, governs the resistance to insects, is known as oligogenically inherited. Several mechanisms could be responsible for resistance when inheritance is oligogenic. Each gene may determine a separate mechanism. Else the same mechanism could be shared by a number of genes. Resistance in rice cultivars to brown plant hopper was reported to be governed by at least four genes, *Bph1*, *bph2* (Athwal *et al.*, 1971), *Bph3* and *bph4* (Lakshminarayana and Khush, 1977). For green leaf hopper (*Nephotettix impicticeps*) five genes have been identified in rice cultivars, viz., *Glh1*, *Glh2*, *Glh3* (Athwal *et al.*, 1971) and *glh4* and *Glh5* (Siwi and Khush, 1977). Two non-allelic genes with dominant effect, *GM1* and *GM2* control the resistance in rice to gall midge (Chaudhary *et al.*, 1986). The oligogenically governed resistance is relatively more stable and durable because chances of development of race breaking biotypes are less as compared to monogenic resistance. But incorporation of several genes responsible for resistance is some what difficult as compared to monogenic resistance.

14.2.3 Polygenic Resistance

When resistance is inherited polygenically, variation for resistance is observed in degrees. It may range from high to low. Gene models, mating designs are employed to study the inheritance of polygenically determined resistance. Because the phenotypic effect of each gene is mimicked by the nonheritable agencies, it is usually a practice to estimate the heritability of resistance, i.e. how much variation observed for resistance is heritable in nature and how much is governed by environmental agencies (see Chapter 6). Moreover, gene models provide cumulative effect of gene action, and not the effect of individual gene. Genes at other loci also tend to influence each other's effects. Hence the inheritance is complex in nature.

The mating designs employed to estimate the magnitude of components of variances responsible for the inheritance of resistance are based on two models. In one model, the geneticist assumes that the lines crossed to produce experimental population are samples from an hypothetical large population (random effects) and hence information pertains to the population. In fixed effects model, the lines crossed are deliberately chosen and information gained from the mating design refers to the lines and not to the population. Hence it is not correct to generalize the results obtained by geneticists regarding inheritance pattern when fixed effects model is used.

Because polygenically governed resistance implies that the resistance is controlled by a number of genes and the effect of each is very small, transfer of such resistance to a desirable cultivar is relatively difficult. However, because several genes control the mechanism of resistance there is less possibility of emergence of race breaking biotypes. Hence polygenic resistance which may be against several races/biotypes, is relatively stable and durable. Monogenic resistance is usually race specific/biotype specific.

Actually relatively less research work has been done in the past to study the inheritance of resistance in plants to pest-insects as compared to fungal diseases. The genetics of virulence in insects is also less studied. Consequently, the interaction of genes for virulence in insects with genes for resistance/susceptibility in plants is not yet well known. Moreover, whether the genes for resistance, to a specific insect, observed in different cultivars are common or different is also not well understood. Because the biotypes are less frequent in insects, information about vertical and horizontal resistance is scanty. However, Pathak (1995) described vertical and horizontal resistance against insects. Further he mentions that a close relationship exists between insects and host plants analogous to the gene-for-gene relationship proposed by Flor (1956, 1971) in flax. In other words, for every major gene for resistance in host plants there is a corresponding matching gene for attacking a plant species in the insects. A very good example of gene-for-gene relationship between the genes for resistance in host plant and genes for ability to attack a plant in insect is that of resistance of wheat to Hessian fly (*Mayetiola destructor*). Wheat varieties carrying genes H_1 and H_2 are resistant to Hessian fly in California and less resistant in Kansas. These cultivars are susceptible in other regions of United States of America. This could be explained on the grounds that the genotypes of Hessian fly are different in Kansas and other regions of USA and hence the biotypes could overcome the resistant genes H_1 and H_2. Eight biotypes of Hessian fly have been identified in USA (Hatchett and Gallun, 1968). Great plains, one of the biotypes, can not attack and survive on varieties of wheat carrying any of the eight genes for resistance. Biotypes E and F are able to attack wheat cultivars carrying resistant genes H_3, H_4 and H_5, H_6, respectively. No biotype has been identified, as yet, which can overcome the resistance controlled by all the H genes for resistance. This relationship may be true for polygenic resistance in plants.

14.3 FACTORS AFFECTING EXPRESSION OF RESISTANCE

Phenotype is known to be determined by the genotype of an organism and environment around it. If two or more cultivars are raised in two or more environments, genotype × environment interaction also determines the phenotypic expression.

Microenvironment around the plants influences the reaction of host plant to the insects. Close row-to-row spacing of rice was more prone to attack by borer and gall midge than rice transplanted in wider rows (Prakasarao, 1972). On account of close row to row spacing very high relative humidity was recorded at the base of plants, a factor conducive for proliferation of gall midge. On the other hand, relative humidity at the base of plants with widely spaced rows was much lower. It is observed that in general rate of evaporation of plants at the top is twice as much as at half its height and only negligible near the base, where soil is damp.

Nutrient status of the plants also influences the degree of resistance/susceptibility. The use of fertilizers has been suggested to decrease resistance of plants to insects attack. However, in maize, reverse holds true. For example, highly fertilized and vigorously growing maize is least susceptible to the European corn borer. Potassium deficient plants are prone to be more susceptible to aphids. van Emden (1971) observed that modification in the nutritional environment of a plant can alter their susceptibility to pest attack.

Temperature influences various physiological processes within the plants. This may influence the reaction of plants to insect attack. Pea aphid (*Acyrthosiphon pisum*) exhibited higher rate of reproduction on pea plants at lower temperatures (Dahms and Painter, 1940). Resistance of wheat cultivars to Hessian fly was less pronounced at high temperatures (Cart- Wright *et al.*, 1946). In general, low temperatures exert a negative influence on resistance of plants to insect attack. However, some varieties can maintain considerable resistance at lower temperatures (Panda, 1979).

Presence of a disease in plants can influence the incidence of insect attack. Colorado potato beetle preferred to lay more eggs on potato leaves, which were decaying and infected with virus. Smith (1939) reported a high correlation between per cent stem rust incidence on leaves and damage by *Melanoplus mexicanus* in wheat.

Varieties resistant in one region can be susceptible to same insect in other region because of the biotypes.

14.4 GENETIC VARIABILITY IN INSECTS

Genetic variability is present among insects for their virulence. For example, several biotypes have been reported in some insects, e.g. Hessian fly (8), raspberry aphid (3), green bug (3), pea aphid (3), brown plant hopper (4). Figures in parentheses denote the number of known biotypes of the insect. Appearance of insecticide resistant forms of many insects is an evidence favouring emergence of biotypes. Insects have great potential for genetic variation, which can lead to development of biotypes. Nevertheless, appearance of resistance breaking biotypes is generally less among the insects as compared to the fungi and bacteria. Although the reason for this is not clear, it may be possible that insects have not been exposed to the same intensive selection pressure as the fungi, to develop resistance breaking biotypes. According to Russell (1978) the reason for relatively few cases of true resistance breaking variants of the insects could be that insects usually have few propagules than the fungi or bacteria, in which genetic variation may be manifested. Millions of pathogenic bacterial cells or sufficient mycelial growth of fungi is usually supported by the host plant. Hundreds of thousands of fungal spores can be present on host plant infected with fungal pathogen. Hence, there is immense opportunity for development of biotypes in bacterial or fungal populations. Moreover, in addition to mutation and sexual reproduction there are several exotic means in microorganisms to generate variability e.g. transduction, transformation in bacteria and parasexual cycle, heterocaryosis in fungi. Insects are short of such mechanisms. Moreover, resistance of plants to insects is usually not responsive but owes to resistance mechanism already existing in the plants, before the infestation by insects.

Biotypes arise most frequently among the aphids, probably because they reproduce parthenogenically and sustain relatively larger population on host plants. They also have a relatively shorter life cycle. In addition to the true race breaking biotypes, Russell (1978) distinguished three other forms of biotypes. The true biotypes can attack the previously resistant cultivars. They are adapted to attack specific host plants and manifest interaction between genotypes of plant and insects. Biotypes seldom develop when resistance is due to tolerance or non-preference (Gallun, 1972).

References

Adams, R.; T. Geissman and J. Edward (1960). Gossypol a pigment of cotton seed. *Chem.Rev.* 60: 555-574.

Athwal, D.S.; M.D. Pathak; E.M. Bacalangoo and C.D. Pura (1971). Genetics of resistance to brown plant hoppers in *Oryza sativa. Crop Sci.*, 11: 745-750.

Bapat, D.R.; K.A. Nayeem and M.H. Markande (1975). Correlation studies of HCN content and shoot fly attack in sorghum. *Sorghum Newsletter*, 18: 50.

Bardner, R. (1968). Wheat bulb fly, *Leptohylemyia coarcita*, Fall, and its effect on the growth and yield of wheat. *Ann.appl.Biol.*, 61: 1-11.

Barnes, W.C. and F.B. Cuthbert (1975). Breeding turnips for resistance to the turnip aphid. *Hort. Science*, 10. 59.

Batra, G.R. and D.S. Gupta (1970). Screening of varieties of cotton for resistance to jassids. *Cotton Growing Review*, 47: 285.

Beck, S.D. (1956). The European corn borer, *Pyrausta nubilalis* (Hubn) and its principal host plant. II The influence of nutritional factors on larval establishment and development. *Ann.Entomol. Soc. Am.*, 49: 582-588.

Beck, S.D. (1965). Resistance of plants to insects. *Annual Review of Entomology*, 10: 207.

Beck, S.D. and W.Hanec (1958). Effects of amino acids on feeding behaviour of the European corn borer, *Pyrausta nubilalis* (Hubn.). *J.Insect Physiol.*, 2: 85-96.

Benedict, J.H.; T.F. Leigh; W. Tingey and A.H. Hyer (1977). Glandless Acala cotton: More susceptible to insects. *California Agriculture*, 31:14.

Blum, A. (1967). Varietal resistance of sorghum to the sorghum shoot fly (*Atherigona varia soccata*). *Crop Sci.*, 7: 461-462.

Blum, A. (1968). Anatomical phenomena in seedlings of sorghum varieties resistant to the sorghum shoot fly (*Atherigona varia soccata*) *Crop Sci.*, 8: 388-391.

Buhr, H.; R. Toball and R. Schreiber (1958). Die Wirkung von einigen pflanzlichen sounderstoffen insbesondere von Alkaloiden, aufdie Entwicklung der Larven des kartoffelkafers. (*Leptinotarsa decemlineata*, Say). *Entomol.exp.appl.*, 1: 209-224.

Cartwright, W.B.; R.M. Caldwell and L.E. Compton (1946). Relation of temperature to the expression of resistance in wheats to Hessianfly. *American Soc.Agron.Jour.*, 38: 259-263.

Chaudhary, R.C.; G.S. Khush and E.A. Henrich (1984). Varietal resistance to rice stem borer in Asia. *Insect Sci.Applic.*, 5: 447-463.

Dahms, R.G. (1972). The role of host plant resistance in integrated insect control. In *Control of sorghum shoot fly*. Proceedings of an International Symposium, Nov. 1-3, 1971. Eds. Jotwani, M.G. and W.R. Young. Oxford and IBH, New Delhi, pp. 152-167.

Dahms, R.G. and R.H. Painter (1940). Rate of reproduction of the pea aphid on different alfalfa plants. *J.Econ.Entomol.*, 33: 482-485.

Davis, D. (1964). Host fungitoxicants in selective pathogenicity of *Fusarium oxysporum*. *Phytopathology*, 54: 290-293.

Davis, D. and J. Ellington (1974). Breeding Acala cottons for insect resistance. *Agricultural Experiment Station, New Mexico State University*, Research Report No. 286.

Deeming, J.C. (1972). A review of taxonomy of African shoot flies of sorghum. In *Control of sorghum shoot fly*. Proceedings of International Symposium, Nov.1-3, 1971. Ed.M.G. Jotwani and W.R. Young. Oxford and IBH, New Delhi.

Dickson, M.K. andC.J. Eckenrode (1975). Variation in *Brassica oleracea* resistance to cabbage looper and imported cabbage worm in the green house and field. *Journal of Economic Entomology*, 68:757.

Dethier, V.G. (1941). Chemical factors determining the choice of food plants by *Papilio* larvae. *Am.Naturalist*, 75: 61-73.

Djamin, A. and M.D. Pathak (1967). Role of silica in resistance to Asiatic borer, *Chilo suppressalis* (Walker) in rice varieties. *Journal of Economic Entomology*, 60: 347.

Doggett, H. (1972). Breeding for resistance to sorghum shoot fly in Uganda. In *Control of sorghum shoot fly*. Proceedings of International Symposium, Nov.1-3, 1971. Ed.M.G. Jotwani and W.R. Young. Oxford and IBH, New Delhi, pp. 192-201.

Doggett, H. and B.N. Majisu (1966). Record of research. *E. African Agr.Forestry Res.Organ.* Ann.Rept., 1965, p. 86.

Dunn, J.A. and D.P.H. Kempton (1971). Differences in susceptibility to attack by *Brevicoryne brassicae* (L.) on Brussels sprouts. *Annals of Applied Biology*, 68: 121.

El-zik, K.M. and P.M. Thaxton (1989). Genetic improvement for resistance to pests and stresses in cotton. In *IPM systems and cotton production.* Eds. Frisbie, R.E., K.M.El-zik and L.T. Wilson. John Wiley and Sons, N.Y.USA, 191-224.

Flor, H.H. (1956). The complementary genetic systems in flax and flax rust. *Adv.Genet.*, 8: 29-54.

Flor, H.H. (1971). Current status of the gene-for-gene concept. *Annual Review of Phytopathology*, 9: 275-296.

Fery, R.L. and F.P. Cuthbert (1975). Antibiosis in *Lycopersicon* to the tomato fruit worm (*Heliothis zea*). *Journal of the American Society for Horticultural Science*, 100: 276.

Gallun, R.L. (1972). Genetic interrelationships between host plants and insects. *J.Environ.Qual.*, 1: 259-265.

Gershenzon, J.; M. Rossiter; T.J. Mabry; C.E. Rogers; M.H. Blust and T.L. Hopkins (1985). Insect antifeedant terpenoides in wild sunflower: a possible source of resistance to sunflower moth. In *Bio-regulators for pest control*. Ed. P.A. Hedin, Wahsington, D.C., 433-446.

Gibson, R.W. (1971). Glandular hairs providing resistance to aphids in certain wild potato species. *Annals of Applied Biology*, 68: 113.

Gibson, R.W. (1976). Glandular hairs on *Solanum polyadenium* lessen damage by the Colorado beetle. *Annals of Applied Biology*, 82: 147.

Grison, P.A. (1952). Relation entre l' etat physiologique de la plant hote, *Solanum tuberosum, et al.*, fecondite du doryphore, *Leptinotarsa decemlineata* Say. *Trans. Intern.Congr.Entomol.* 9th Amsterdam.

Grison, P.A. (1958). Action des lecithines sur la fecondite du Doryphore. *Academic des Science Compt. Rend.*, 227: 1172-1174.

Harris, K.M. (1962). Lepidopterous stem borers of cereals in Nigeria. Bull.ent.Res., 53: 139-171.

Hatchet, J.H. and R.L. Gallun (1968). Frequency of Hessian fly, *Mayetiola destructor*, races in field populations, *Annals of Entomological Society of America*, 61: 1446.

Jayaraj, S. (1967). Studies on the resistance of castor plants (*Ricinus communis* L.) to the leaf hopper, *Empoasca flavescens* (F.) (*Homoptera*: *Jassidae*). *Z.Angew.Ento*., 59: 117-126.

Jayaraj, S. (2002). Prudent suppression of pests. In *The Hindu Survey of Indian Agriculture*, Publ. M/s Kasturi and Sons, Chennai 600 002. Ed. N.Ravi, pp. 187-190.

Jeyarajan, R. (1997). Low cost component of IPM. In *The Hindu Survey of Indian Agriculture*, Publ. M/s Kasturi and Sons, Chennai 600 002. Ed. N.Ravi, pp. 165-167.

Jotwani, M.G.; G.C. Sharma; B.G. Srivastava and K.K. Marwaha (1971). Oviposition response of shoot fly, *Atherigona varia soccata* (Rondani) on some promising resistant lines of sorghum. In *Investigations on insect pests of sorghum and millets (1965-1970)*. Final Technical Report, Div. of Entomology, IARI, New Delhi, pp. 119-122.

Jotwani, M.G. and K.P. Srivastava (1970). Studies on sorghum lines resistant against shoot fly (*Atherigona varia soccata*, Rondani). *Indian Journal of Entomology*, 32: 1-3.

Kennedy, J.S. (1965). Mechanism of host plant selection. *Annals of applied biology*, 56: 317.

Kogan, M. and E.F. Ortman (1978). Antixenosis — a new term proposed to define Painter's non-preference modality of resistance. *Bullent Soc.Am.*, 24: 175-176.

Lakshminarayana, A.and G.S. Khush (1977). New genes for resistance to the brown plant hopper in rice. *Crop Sci*., 171: 96-100.

Lincoln, C.; G.Dean; N.A.Wadde; Y.C. Yearian; J.R. Phillips; N.P. Tugwell and L. Roberts (1970). Resistance of Frego-type cotton to boll weevil and boll worm. Paper presented at Annual meeting of the Entomological Society of America, S.E. Branch, January, 1970.

Lukefahr, M.J. and D.F. Martin (1966). Cotton plant pigments as a source of resistance to the boll worm and tobacco bud worm. *Journal of Economic Entomology*, 59: 176.

Lukefahr, M.J. and J.E. Houghtaling (1969). Resistance of cotton strains with high gossypol content to *Heliothis* spp. *Journal of Economic Entomology*, 62: 588.

Lukefahr, M.J.; J.E. Houghtaling and H.M. Graham (1971). Suppression of *Heliothis* populations with glabrous cotton strains. *Journal of Economic Entomology*, 64: 486.

Maiti, R.K. and F.R. Bidinger (1979). A simple approach to identification of shoot fly tolerance in sorghum. *Indian Journal of Plant Protection*, 7: 135-140.

Malik, R.S. (1995). Amino acid and sugar content of the plant in relation to mustard aphid (*Lipaphis erysimi*, Kalt.): Comparative resistance in Cruciferous species. In *Genetic Research and Education: Current trends and the next fifty years*. Eds. B. Sharma *et al.*, Indian Society of Genetics and Plant Breeding, New Delhi, pp. 238-242.

Maxwell, F.G.; J.N. Jenkins; W.L. Parrot; W.T. Buford (1969). Factors contributing to resistance and susceptibility of cotton and other hosts to the boll weevil, *Anthonomus grandis*. *Entomologia Experimentalis et Applicata*, 12: 801.

Nagarajan, S. (1990). Integrated approach vital. In *The Hindu Survey of Indian Agriculture*, Publ. M/s Kasturi and Sons, Chennai 600 002, pp. 155-161.

Narayana, D. (1975). Characters contributing to sorghum shoot fly resistance. *Sorghum Newsletter*, 18: 21-22.

Narayanan, S.S. (1995). Resistance to *Heliothis* and other serious insect pests in *Gossypium* spp. In *Genetic Research and Education: Current trends and the next fifty years*. Eds. B. Sharma *et al.* Indian Society of Genetics and Plant Breeding, New Delhi, pp. 195-210.

Nayeem, K.A. and D.R. Bapat (1976). Hormonic effect of carbofuran on growth and development of sorghum. *Sorghum Newsletter*, 19: 59-60.

Painter, R.H. (1951). *Insect resistance in crop plants*. Publ.Mac Millan, New York.

Painter, R.H. (1967). Plant resistance to insects as applied to breeding vegetable crops. In *Proceedings of the 17th Horticultural Congress*, 3: 259.

Panda, N. (1979). *Principles of host plant resistance to insect pests*. Hindustan Publishing Co., New Delhi.

Panda, N. and R.C. Das (1975). Antibiosis factor of resistance in brinjal varieties of shoot and fruit borer, *Leucinodes orbonalis*. *South Indian Horticulture*, 23(1&2): 43-48.

Pathak, M.D. (1977). Defense of the rice crop against insect pests. *Annals of the NewYork Academy of Sciences*, 287: 287.

Pathak, M.D.; F. Andres; N. Galacgac and R. Raros (1971). Resistance of rice varieties to striped rice borers. *Technical Bulletin*, 11: IRRI. Los Banos, Philippines.

Pathak, R.S. (1995). Gene systems of plant resistance to insects. In *Genetic Research and Education: Current trends and the next fifty years*. Eds. B. Sharma *et al.* Indian Society of Genetics and Plant Breeding, New Delhi, pp. 211-217.

Patanakamjorn, S. and M.D. Pathak (1967). Varietal resistance of rice to the Asiatic rice borer, *Chilo suppressalis* (Lepidoptera: *Crambidae*), and its association with various plant characters. *Annals of the Entomological Society of America*, 60: 287.

Ponnaiya, B.W. X. (1951). Studies in the genus sorghum I. Field observations on sorghum resistance to insect pest *Atherigona indica*, M. *Madras University Journal* B., 21: 96-118.

Prakasarao, P.S. (1972). Ecology and control of *Tryporyza incertulas*, Walker and *Pachydiplosis oryzae* (Wood Mason) Mani in rice. Ph.D. Thesis Utkal University, Bhubaneswar, India.

Raina, A.R.; H.Z. Thindwa; S.M. Othino and R.T. Corkhill (1981). Resistance in sorghum to the sorghum shoot fly: Larval development and adult longevity and fecundity on selected cultivars. *Insect Science and its application*, 2: 99-103.

Rangarao, V.; D.R.C. Bakhetia; M. Lakshminarayana and H. Basappa (1995). Breeding for insect pest resistance in oilseed crops. In *Genetic Research and Education: Current Trends and the next fifty years*. Indian Society of Genetics and Plant Breeding, New Delhi, Ed. B. Sharma *et al.*, pp. 180-194.

Rao, N.G.P. (1972). Discussion, control of sorghum shoot fly. In *Sorghum in seventies*, Ed. Rao, N.G.P. and L.R.House, pp.222-223. Oxford & IBH, New Delhi.

Reed, W. (1974). Selection of cotton varieties for resistance to insect pests in Uganda. *Cotton Growing Review*, 51: 106.

Reed, W.; C. Cardona; S. Sithanatham and S.S. Lateef (1987). Chickpea insect pests and their control. In *The chickpea*, Eds. Saxena M., C. and K.B. Singh. Publ.CAB International, Wallingford, Oxon, U.K., pp. 283-318.

Rembold, H. (1981). Malic acid in chickpea exudate, a marker for *Heliothis* resistance. *International chickpea Newsletter*, 4: 18-19.

Rembold, H.; P. Wallner; A. Kohne; S.S. Lateef; M.Grune and Ch.Weigner (1990). Mechanism of host-plant resistance with special emphasis on biochemical factors. In *Chickpea in Nineties*, Proceedings of the second International Workshop on Chickpea Improvement, 4-8 Dec., 1989. ICRISAT, Patancheru 502-324, India. Eds. Walby B.J. and S.D. Hall, pp. 191-194.

Rodriguez, J.G. (1960). Nutrition of host and its reaction to pests. *Amer.Assoc.Advan.Sci.*, 61: 149-167.

Russell, G.E. (1978). *Plant breeding for pest and disease resistance*. Publ. Butterworths, London, p. 485.

Sahasrabuddhe, D.L. (1914). The acid secretion of gram plant (*Cicer arietinum*). Imperial Agricultural Research Institute, Pusa Bulletin, 45: 12.

Sarukhan Jose (2000). Precautionary principle. In *The Hindu Survey of Environment*, Publ. M/s Kasturi and Sons Ltd., Chennai 600 002. Editor Ravi, N., pp. 155-161.

Sasamoto, K. (1961). Resistance of rice plant supplied with silicate and nitrogenous fertilizers to the rice stem borer, *Chilo suppressalis*, Walker. In *Proceedings of the Faculty of Liberal Arts and Education, No. 3*, Yamanashi university, Japan.

Schillinger, J.A. and R.L. Gallun (1968). Leaf pubescence of wheat as a deterrent to the cereal leaf beetle, *Oulema melanopus*. *Annals of Entomological Society of America*, 61: 900.

Shaw, G.G. and C.H.A. Little (1972). Fertilization and spruce bud worm development. In *Insect and Mite nutrition*, Ed. Rodriguez, J.G. Amsterdam. North Holland, pp. 702, 589-597.

Singh, S.P. and M.G. Jotwani (1980). Mechanism of resistance in sorghum to shoot fly. III Biochemical basis of resistance. *Indian Journal of Entomology*, 42: 551-566.

Siwi, B.H. and G.S. Khush (1977). New genes for resistance to the green leaf hopper in rice. *Crop Sci.*, 17: 17-20.

Smith, R.W. (1939). Grass hopper injury in relation to stem rust in spring wheat varieties. *Amer.Soc.Agron.Jour*. 31: 818-821.

Starks, K.J. and R.L. Burton (1977). Green bugs: a comparison of mobility on resistant and susceptible varieties of four small grains. *Environmental Entomology*, 6: 331.

Stern, V.M.; R.M. Smith; R. Van den Bosch and K.S. Hagen (1959). The integrated control concept. *Hilgardia*, 29: 81-101.

Sugiyama, S. and Y. Matsumoto (1959). Olfactory responses of the vegetable weevil larvae to various mustard oils. Studies on the host plant determination of the leaf feeding insects. *Nogaku Kenkyu*, 46: 150-157.

Talekar, N.S. and B.S. Chen (1986). The bean fly pest complex of tropical soybean. In *Soybean in tropical and sub tropical cropping systems*. Proc. Symposium, Tsukuba, Japan, Sept.26 to Oct.1, 1983. Ed. Shanmugsundaram. AVRDC Taiwan, China, pp. 257-271.

Taneja, S.L. and K. Leuschner (1985). Resistance screening and mechanism of resistance in sorghum to shoot fly. In *Proceedings of the International Sorghum Entomology Workshop*. July, 15-21, 1984. Texas A & M University, College Station, Texas, ICRISAT, Patancheru, 502-324, India, pp. 115-129.

Tester, C.E. (1977). Constituents of soybean cultivars differing in insect resistance. *Phytochemistry*, 16: 1899-1901.

Thurston, R. and J.A. Webster (1962). Toxicity of *Nicotiana gossei*, Domin to *Myzus persicae* (Sulzer). *Entomol.exp.appl.*, 5: 233-238.

van Emden, H.F. (1966). Plant insect relationship and pest control. *World Rev.Pest Control*, 5: 115-123.

van, Emden, H.F. (1969). Plant resistance to *Myzus persicae* induced by plant growth regulator and measured by aphid relative growth rate. *Entomol.exp.appl.*, 12: 125-131.

van Emden, H.F. (1971). Plant insect relationships in relation to control. *Pans.*, 17(2): 251-252.

van Marrewijk, G.A. M. and O.M.B. De Ponti (1975). Possibilities and limitations of breeding for pest resistance. *Medede lingen van de Fakulteit Landbouwwettnschappen Rijksuniversiteit, Genet.*, 40: 229.

Wilson, F.D. (1971). Response of tobacco budworm larvae to cotton seedlings carrying various combinations of gland determining alleles. *Crop Sci.*, 13: 107.

Wiseman, B.R., W.W. McMillian and N.W. Widstrom (1972). Tolerance as a mechanism of resistance in corn to corn ear worm. *Journal of Economic Entomology*, 65: 835.

Young, W.R. (1972). Sources of resistance to the sorghum shoot fly, *Atherigona varia soccata*, Rond. In *Control of sorghum shoot fly*. Proceedings of International Symposium. Nov. 1-3, 1971. Eds. M.G. Jotwani and W.R. Young. Oxford & IBH Publications, New Delhi, pp. 168-179.

15

Breeding Resistant Varieties

Breeding methods employed for incorporating genes for resistance to either disease causing organisms (*e.g.* fungi, bacteria, virus) or to insects do not differ in fundamental ways from those adopted to improve agronomic traits. Hence, for self-pollinated crops, where the end product is essentially homozygous plants at most of the loci but the variety could be little heterogeneous, pedigree method, back-cross method or bulk method of handling the segregating populations could be preferred. When there is variability for disease or insect resistance in land races or indigenous varieties pure line selection or mass selection could be employed to isolate agronomically desirable plants with resistance to disease/insect under consideration. By virtue of mating system, individuals in the populations of cross-pollinated crops are heterozygous at most of the loci and may be homogeneous (as in case of hybrids) or heterogeneous (as in case of composites, synthetics). Hence the methods of breeding for cross-pollinated crops differ from those employed for self-pollinated crops. Mass selection, recurrent selection could be applied as breeding methods in case of cross-pollinated crops. Nevertheless, backcross method can also be used. Pedigree method has also been used to improve inbred lines of maize (Allard, 1960).

There is however, a specific requirement, which must be satisfied, while developing disease and/or insect resistant varieties. For distinguishing between the resistant and susceptible plants to a specific race of a pathogen or a specific insect, presence of the pathogen or insect is essential. Heavy pressure of insect or abundant inoculum of the pathogen should exist in the experimental field. In the absence of pathogen or insect, differentiation between resistant and susceptible plants is impossible. A breeder has to create epiphytotic conditions while screening the material. For screening against specific insects either 'hot spots' have to be identified where severe and uniform infestation of a given insect occurs regularly. Alternatively the insect is reared on artificial diet and the plants are infested artificially with larvae of specific instar. Because creation of epiphytotic or artificially rearing the insect and infestation of plants requires specialized knowledge, plant pathologist and/or entomologist should essentially form a part of resistance breeding programme.

Breeding for resistance obviously begins with search for resistance conferring genes, i.e. resistant sources.

15.1 SOURCE OF RESISTANCE

At the outset a breeder should search the locally adapted varieties of the crop concerned for resistant genes. Local varieties are the most desirable source of resistant genes as compared to the exotic varieties from other countries of the world. Even resistant cultivar from other state of same country may be desirable. There are obviously two reasons for this negative option. Firstly, the local varieties are expected to carry a negligible amount of undesirable linkages than the exotic sources and secondly the exotic source of resistance for a specific disease/ insect may be associated/or carry susceptibility to a disease/insect to which the variety to be improved is not susceptible. Thus unintentionally a breeder may induce susceptibility in the variety to be improved for a specific pathogen.

Leppik (1970) reviewed the role of centres of gene diversity as a source of disease and pest resistance. In his opinion systematic exploration of primary and secondary gene centres of a particular vegetable may provide additional gene pools for resistance breeding. However, it is not necessary that resistance is found only in the centres of origin of crop (Harlan, 1972).

When desirable resistance can not be found in locally adapted material a breeder should search exotic material. A large number of strains of linseed, such as R.R.9, 10, 37, 38, etc., were developed as a result of cross between Australian varieties, which were immune to rust, and N.P. varieties developed at Pusa (Richharia, 1957). Exotic germplasm was used frequently in the improvement programmes of cowpea (Asthana and Chandra, 1995). An exotic variety Black eye in combination with Type 2 led to the development of variety Pusa Phalguni. Black eye was also used to develop Gujarat cowpea 3 (G.C 82-7) in combination with variety V 16 (Amba), which is a mutant of Pusa Phalguni. Similarly a dwarf and powdery mildew resistant variety HFP 4 was evolved by crossing Type 16 of field pea with an exotic variety EC 109196. A gram variety C 235, which is tolerant to blight and wilt, was crossed to a Greek variety, E 1007, resistant to *Ascochyta* blight. This cross led to development of gram variety Gaurav.

Three wheat varieties were evolved from a cross between an improved Indian variety, N.P. 165, and a Kenya wheat, which exhibited high degree of resistance to black and brown rusts and loose smut (Pal, 1957). The varieties so developed are N.P.797, 798 and 799

Indigenous varieties, which are kept in seed stock shelf, because of replacement by the new high yielding varieties, could also be the source of resistance. Nevertheless, such lines should be chosen with caution that they were not among the 'break down' varieties to any disease and/or insect at any time in the past.

Germplasm lines, which represent the world collection of varieties, serve the source of resistance when some new disease or race of a disease or insect appears. Satisfactory level of resistance, sometimes, is not available in cultivated species. Related species or genera have, at times, provided genes for resistance. Although fertile progeny from inter-specific or inter-generic crosses could be obtained, use of these sources of resistance is difficult through conventional breeding methods. This is especially so when resistance is polygenic in inheritance. Wild species also possess genes for resistance. Resistance to *Fusarium*, which causes wilt in tomato, was incorporated in *Lycopersicon esculentum* from *L.pimpinellifolium*. Material generated from this cross formed the basis to develop resistant varieties of tomato (Russell, 1978). Jambhale and Nerkar (1986) developed a Yellow Vein Mosaic (YVM) variety of okra, 'Parbhani Kranti', from an inter-specific cross of *Abelmoschus esculentus* cv. 'Pusa Sawani' with *A. manihot*. Punjab Padmini, Arka Anamika and P7 varieties of okra also have *A. manihot* as one parent (Ram, 1998).

Cross-compatibility, hybrid sterility, hybrid inviability, etc. cause obstruction in successful use of wild relatives. However, several devices have been developed to utilize resistant genes from wild relatives. *Gossypium somalense* G (H) possesses resistance against *Heliothis*, *G. somalense, G. raimondii* W & P, *G. harknesii*, B, *G. bickii*, P, *G. armourianum* K and some *G.arboreum*, L. races to different boll worms (Narayanan, 1995), *Helianthus mollis, H. tuberosus* and *H. solicifolius* to carrot beetle in sunflower (Rogers and Thompson, 1978). *Lycopersicon hirsutum*, a wild relative of tomato, carries genes for resistance to fruit borer and leaf curl. *Solanum demissum*, a wild relative of potato is known to bear genes for resistance against Colorado potato beetle and late blight. Resistance from alien species or genera may not be more durable as compared to resistance from close relatives (Kulkarni and Chopra, 1989).

When the scientists fail to find genes for resistance in germplasm, local material, related species, wild species, genes for resistance could be obtained through the use of mutagens and somaclonal variation. Somaclonal variation is the variability among plants regenerated from tissue culture (Larkin and Scowcroft, 1981).

Recombinant DNA technique has been used to covalently integrate alien gene(s) from micro-organisms, animals, unrelated plants into the desired crop plants. Such covalent integration of an alien gene in a plant from these sources through recombinant DNA technique is called transgenesis, and the plant so developed is transgenic plant. A gene called Bt or cry is carried by a bacteria, *Bacillus thuringiensis*, which causes death of Lepidopterous larvae. There are various forms of Bt or cry genes, such as cry 1A, cry 1C, etc. Transgenic cotton varieties carrying the Bt gene are said to be resistant to bollworms, potato varieties resistant to Colorado potato beetle and maize varieties to European corn borer. Tomato varieties carrying cry gene are completely free from caterpillar (Radhakrishnan, 2002). Thus microorganisms could also be considered as a source of resistant genes (see Section 15.4).

15.2 THE BREEDING METHODS

For resistance breeding, methods like pedigree, backcross and bulk could be employed to handle the segregating material from a cross of a desirable variety of self-pollinated crop with a resistant donor. This is particularly so when the donor parent comes from local material, old indigenous variety or exotic variety from same species. Nevertheless, for segregating populations involving related species or genera or wild species some special techniques, *e.g.* embryo rescue, doubling of the chromosome number etc. may have to be adopted.

In those countries or states of the developed countries where land races or the indigenous varieties have not been scientifically exploited, pure line or mass selection of desirable plants showing resistance to a desired disease and/or insect could be employed. This will give quick results. Because the variety is adapted to a given geographical region and popular with the farmers, it will not take much time to spread widely in cultivation. Remember that in case of self-pollinated crops, the end product of any selection/hybridization followed by selection will result in a variety comprising of plants homozygous at practically all the loci.

Mass selection, various methods of recurrent selection, development of composites or synthetics or hybrids are the popular methods used for cross-pollinated crops. By virtue of mating system, inbreeding plants of these crops leads to loss of vigour, reduced fecundity, appearance of various lethal or morbid types. Hence, the end product of any breeding programme should aim at development of a heterozygous population, which may be homogeneous (as in case of hybrids) or heterogeneous (as in case of composites or synthetics and double cross, double top cross hybrids). Backcross method can also be used successfully to transfer a resistant gene from a donor in cross-pollinated crops. However, the sample of gametes taken from the recurrent parent (a desirable variety to be improved) should be large enough so that it represents the gene frequency characteristic of that variety. Otherwise, the resistant variety developed may not carry the genotype of original variety. When backcross programme is correctly executed, the beauty of this breeding method is that it reconstructs the exact genotype of a recurrent parent but in addition carries a gene(s) for resistance to a disease/insect from the donor.

In case of self-pollinated crops, either the pedigree or backcross method is chosen for resistance programme when adequate resistance is not found in the commercially grown varieties, but is available in those types which cannot be used commercially because of their unsuitable agronomic attributes. With either method, one of the parents is chosen on the basis of its superior agronomic or horticultural characteristics while the donor parent is selected on the basis of demonstrated high level of resistance. The reader should note that the intensity of resistance is somewhat lost in the newly evolved variety as compared to the donor parent and this is the reason for selecting that line as donor which exhibits high level of resistance.

When the resistant parent is expected to contribute some desirable agronomic attributes to the segregates, in addition to the desired gene(s) for resistance, pedigree method of handling the segregating material would be appropriate breeding method. Backcross method would be the best choice when the donor parent is undesirable for agronomic traits but carries desirable gene for resistance only.

Bulk method could be an option in those regions where the specific disease or insect under consideration occurs regularly and uniformly. That is, it is a constant feature of the region.

At CIMMYT, Mexico, three-way or limited backcrosses are made between carefully selected high yielding lines with resistant lines from germplasm (Rajaram *et al.*, 1995). The F_1s are evaluated for agronomic attributes and disease reaction. The F_1s with poor agronomic features and susceptible to the disease are discarded. Major selection pressure for resistance is applied from F_3 generation onwards. It is assumed that by F_3 generation the segregating population attains enough homozygosity. Advanced lines are then evaluated at more than one location, not only for disease resistance but also for desirable agronomic characters. Plants with desirable agronomic characters along with high level of resistance under Mexican environments are selected and subsequently distributed throughout the world for international evaluation.

Usually breeders do not prefer to expose the F_1 and F_2 generation plants to 'disease epiphytotics or to heavy insect pressure. This is so because the breeder is not willing to lose agronomically superior plants.

Sufficient plants are grown in the F_2 generation. Agronomically superior plants are selected. In the F_3 generation (seed of plants selected in F_2 generation) seed is divided into two parts, one portion of each F_3 selection is subjected to screening in the epiphytotic condition/heavy insect pressure and the other portion is used to carryout evaluation at several endemic research locations. Selections are made from F_3 to F_6 generation for the plants which are superior for agronomic features and highly resistant to the disease/insect under consideration. Promising lines are then evaluated in multilocational experiments.

The Indian Council of Agricultural Research launched all India Co-ordinated Commodity Improvement Projects after 1957. All India Coordinated Maize Improvement Project was the first to start. Each crop project has several research stations in the country. The team of scientists working at most of the centres comprises of a breeder, entomologist, plant pathologist and agronomist. As a result, the germplasm and the segregating material is screened and evaluated by the entomologists and plant pathologists at each of the locations. Such an evaluation takes care of genetic variants of pathogens/insects and variation in soil types, temperature, relative humidity, light, rainfall pattern, etc.

Although multilocational testing may not be a perfect system for identification of resistant material, evidence from research at the CIMMYT, Mexico for thirty years indicated that multilocational testing helped to confirm genetic diversity in the germplasm (Rajaram *et al.*, 1995).

Studies in epidemiology strongly suggests that the plant breeders must always be ready to face the increase in races/biotypes to which their resistant varieties are susceptible. Races of fungi arise via hybridization, heterocaryosis, parasexuality or by mutation. Efficient plant breeding programmes must, therefore, include plans to deal with the threat of new races before they reach devastating state.

Earlier, the variety which succumbed to a new race, was replaced by a new variety bred for resistance to new race. Such a breeding programme was initiated after the emergence of threat of new race. Khush (1995) proposed four approaches to prolong the useful life of a variety with vertical resistance. The reader will recall that when a variety is more resistant to some races of a pathogen it is called vertical. It means this variety will succumb when new race, for which it does not carry gene for resistance, appears. The approaches are:

(i) Deployment of genes over time,
(ii) Deployment of genes over space,
(iii) Pyramiding of vertical genes, and
(iv) Multiline varieties.

Deployment of genes over time is based on the principle of crop rotation to control certain soil borne pathogens. In the first year a resistant variety is grown which carries one of the several vertical genes for resistance to a given pathogen. Next year another variety with another vertical major gene is grown. This system does not permit the pathogen to build up sufficient inoculum of a particular race in sufficient amount to be responsible for generating an epiphytotic.

Different varieties with diverse major genes for vertical resistance are grown in particular areas, when a crop is grown over a wide region and the disease inoculum can travel a long distance and affect the crop in other area. Races, which survive in one zone, are ineffective in another. This concept is deployment of different genes over the space.

Incorporation of diverse major genes for resistance to a pathogen/insect in one and the same variety is the concept of pyramiding of genes. The concept of pyramiding of genes was given by Watson and Singh (1953), which is based on the hypothesis that it is unlikely that mutation will simultaneously occur at more than one locus at a time in the pathogen so that it overcomes the resistance of variety comprising of pyramided genes. It is believed that greater the number of major genes for resistance in a variety greater would be its longevity. Such varieties may be more successful in areas where endemic terminates than in an area where the varieties are in constant contact with the pathogen, which would eventually force evolution of virulent races. The main limitation of this approach is that it may lead to the evolution of 'super races'. If the super race is produced, which may overcome the combined resistance, then several resistance genes may be simultaneously rendered ineffective and may lead to quick exhaustion of available genes for resistance (Khush, 1995).

Most of the rust resistance programme of wheat is based on development of a true breeding variety carrying a specific major gene for vertical resistance to a specific race of pathogen. This is a short term approach to control the attack of rust because such varieties become susceptible to a new race of rust pathogen even before it is put through the test. Consequently, these varieties become ineffective. Biotypes also appear in case of insects, and hence a previously resistant variety 'breaks down'. In nature several mechanisms operate to keep the disease and its host in balance *viz.*, inter-specific diversity, intra-specific diversity, host resistance and stabilizing tendencies in pathogen populations (Browning and Frey, 1969). Following the demonstration by Biffen (1905, 1912) that resistance to a disease is an inherited character of a plant, plant breeders began to use resistant genotypes of host plants. Thus began the man guided management system for host resistance against a specific disease. As a result both the intra and inter-specific heterogeneity in the host plants was lost as an effective mechanism to control the disease (Frey *et al.*, 1979).

Rosen (1949) proposed that the plant breeders should return to the development of heterogeneous populations of plants to control diseases. In order to provide phenotypic uniformity for agronomic characters, such as plant height, maturity, size of kernel, colour of kernel, quality of produce, etc., and heterogeneity for resistance conditioning genes in a common cultivar Jensen (1953) proposed the development of multiline cultivars. A multiline variety is a population of plants that is agronomically uniform but heterogeneous for genes that condition reaction to a disease organism (Gill *et al.*, 1979). Multiline approach is concerned with management of genes conditioning vertical resistance. Several scientists contributed to the multiline theory in its present form (Jensen, 1952; Borlaug, 1953; Suneson, 1960; Van der Plank, 1963, 1968; Browning and Frey, 1969). No matter what race is present, there would be something resistant to it in the variety. Because the race can attack only one or a few constituent lines of a multiline variety, its increase would be slow and less damaging even to the susceptible lines. Borlaug (1958) proposed similar approach, except that the component lines would be developed by the backcross method, and consequently would be phenotypically similar but would have different genes for resistance. According to his comprehensive programme for developing multiline varieties of wheat and keeping the plants completely resistant to stem rust, an isoline is replaced as soon as a race of pathogen could attack it. Borlaug (1958) proposed to blend 8 to 10 isolines, each with diverse stem rust resistance conditioning genes into a phenotypically uniform multiline. This is known as the 'clean crop approach'. There is a difference in the philosophy on disease control through multiline varieties, which resolves around the question regarding the partial or complete resistance of component isogenic lines to the pathogen. As per the 'clean crop approach' advocated by Borlaug and his co-workers, which is being practiced at CIMMYT to control the yellow, brown and black stem rusts of wheat, all the component lines of multiline variety should be resistant to the prevalent races of the disease pathogen to be controlled. The aim of this approach is to keep the crop, as far as possible, completely free from the disease. In the second approach, the "dirty crop approach", which was advocated by Frey, Browning and Simons (1975) none of the component lines is completely resistant to all the races of the pathogen. Thus the multiline composed from these lines would be rusted to some extent. It is suggested that in order to stabilize the race structure of the pathogen some susceptibility is a must (Suneson, 1960; Leonard, 1969). The loss of the crop is reduced through the development of synthetic tolerance or population resistance, as suggested by Browning and Frey (1969).

For developing a multiline variety a widely adapted, high yielding and agronomically desirable variety is chosen as a base line. Diverse sources of cultivars/genotypes possessing major genes for resistance to various prevalent races of the pathogen are identified and used as donor parents. Each of these genotypes possess different gene for resistance. Each of these lines are crossed separately to the base line to initiate a backcross programme. Assuming that ten resistant sources have been identified, then ten backcross programmes will be run separately ending with ten isogenic lines, but each of the line carrying different resistant gene(s). Because in case of multiline variety a number of vertical genes are deployed to provide resistance, each vertical gene protects the variety against certain races of pathogen. In result, although the genes employed are vertical resistance genes, the overall behaviour is like horizontal resistance.

The mechanism through which the multiline variety exerts its effect on disease intensity at any time is assumed to be through reduction of both the initial inoculum (X_0) and the apparent infection rate (r). According to Van der Plank (1963) the amount of inoculum, X at any given time is given by the expression X_0^{ert}, where X_0 is the initial inoculum, r is the average rate of infection and t is the time during which infection has taken place. Evidence available, in case of multiline varieties, indicated that the initial proportion of affected tissue (X_0) is reduced as compared to a pure susceptible line, as all or part of the components are vertically resistant. This is valid for all the races, except those complex ones, which carry virulence genes corresponding to all resistant genes present in the multiline (Parlevliet, 1979). The rate of increase of the proportion of affected tissue, r, is reduced for same reason. Many spores produced do not reach a susceptible component, but land on the resistant tissue. The protection afforded to the crop by multiline variety could be due to:

(i) The epidemic development is slowed down. There is less disease at any one time and so less yield damage.
(ii) The pattern of virulence genes in the pathogen population is stabilized.

The first multiline cultivar of wheat, which was released for commercial cultivation, was Miramar 63. This was released from Columbian programme (Rajaram *et al.*, 1979). A multiline "Tumult" of winter wheat has been released in the Netherlands (Groenewengen, 1977).

Pathak (1995) suggested three strategies for management of vertical genes resistant to insects. These are (*i*) sequential release of cultivars with single gene for resistance, which is being followed for Hessian fly in wheat and brown plant hopper of rice, (*ii*) pyramiding several genes for resistance in a single cultivar, which is being adopted for brown plant hopper in rice (Khush, 1977) and (*iii*) developing of multilines. Because, in case of insects, evaluation is not possible until new biotype evolves, a broad range of material must be collected and maintained for future use. Insects are constantly changing and hence it is very much important to develop new strategies to stabilize resistance.

Horizontal resistance is conferred by mechanisms, which are beyond the parasite's capacity to change. Horizontal resistance is more likely governed by polygenes. Horizontal resistance provides incomplete but more durable disease/insect control. The level of resistance is generally moderate and some disease symptoms do appear. However, yield losses are not significant.

Insect biotypes have not formed when polygenic resistance has been used. Some scientists believe that it is more desirable to concentrate on the development of biotype non-specific varieties with tolerance, which reduce the effect of infestation to an acceptable level rather than aim to breed immune varieties. Tolerant cultivars provide protection against an insect and allow colonization by a small number of insects that provide livelihood to parasites and predators (Pathak, 1995).

Usually donors for polygenic resistance are agronomically poor. Moreover, it takes long time to accumulate polygenes into an improved agronomic background to develop acceptable level of resistance. Selection schemes proposed by Brim (1966), Brim and Stuber (1973) could be employed to accumulate polygenes for resistance in self-pollinated crops. Various methods of population improvement are available to develop populations with polygenic resistance to a disease/insect in case of cross-pollinated crops. Recurrent selection has been successfully used in maize to incorporate resistance to diseases.

15.3 SCREENING OF SEGREGATING POPULATION

For a breeding programme to be successful, it should be possible to easily and positively distinguish between the plants with desirable character from those which are not desirable. In the absence of pathogen causing a given disease and environment favourable for its activity a breeder may not be able to differentiate resistant from susceptible plants. The same thing holds true for distinguishing susceptible and resistant plants to a given insect pest. For identification of insect resistant genotypes, several approaches are practiced, *viz*., field screening in "hot spots", greenhouse screening, 'no choice test', 'multiple choice test', screening after rearing insect on artificial diet, etc.

According to Russell (1978), whenever possible, screening and selection for resistance to a fungal disease should be carried out in field. He pointed out two advantages of field testing over glasshouse and laboratory tests, *viz*., (*i*) In field testing much larger populations can be screened for their reaction to a pathogen, and (*ii*) Plants are tested for resistance under natural conditions. However, Allard (1960) is of the view that accurate distinction between the resistant and susceptible plants is impossible in field because the level of infection is often light and irregular in the natural epiphytotics. Consequently, repetition is necessary for obtaining authentic results so that valuable time and efforts are not wasted.

Screening of plants in the field, under natural conditions of disease development, may be preferred whenever artificial methods of inoculation are not available or when the disease is of minor importance which does not demand the use of expensive and sophisticated methods (Kulkarni and Chopra, 1989).

Plant breeders have to deal with two main groups of epidemic fungal diseases (Russell, 1978) *viz.*, those fungal diseases which cause damage in most of the years and those which are damaging only sporadically. It may be difficult to carry out effective screening in field condition against sporadic diseases, provided breeder depends only on natural disease epidemies. Whenever, environmental conditions are unfavourable for development of a disease, selection is not possible for disease resistant segregates in the field conditions. This means loss of a year, money and efforts. Hence, for such diseases Russell (1978) advocates development of natural epidemics or to initiate artificial epidemic or identify an area where natural outbreak of disease is a regular feature. Natural epidemies of some diseases could be encouraged by raising the relative humidity artificially by using a sprinkler system or wetting the leaves by means of overhead sprays or some other forms of irrigation. Evaluation of the screening techniques for resistance of sorghum genotypes to grain mold fungi at ICRISAT, Hyderabad indicated that screening is possible without inoculation with mold causing fungi (such as *Fusarium moniliforme*) and bagging of panicles if overhead sprinkler irrigation is used from flowering to harvest (Bandopadhyay and Mughogho, 1988).

Artificial epiphytotic in the field could be created by inoculating all or a proportion of the test plants or by inoculating plots of very susceptible varieties which are sown at regular intervals among the test rows.

It is relatively easier to screen segregating material in the field, against those diseases which occur regularly. Field epidemies of a disease through natural spread of a pathogen can be created in areas where the disease is endemic or in areas where environment is expected to favour spread of disease to experimental plots, following inoculation of susceptible lines. While screening for resistance under natural conditions of disease, it is important that chance escapes are not selected as resistants.

Screening for resistance to diseases in the field under natural conditions has three disadvantages (Kulkarni and Chopra, 1989):

(i) While screening for resistance to a specific disease, other diseases may also interfere. Since environmental conditions fluctuate from season to season, interaction between host and pathogen may influence proper screening.
(ii) Distribution of pathogen is non-random in the field. This is particularly true for soil borne diseases.
(iii) Screening against resistance to a specific race of a pathogen is not possible.

Artificial screening techniques have been, therefore, developed for important diseases of major crops. Glasshouse and laboratory tests play an important role in screening of resistant cultivars particularly because conditions can be standardized and there can be no chance escapes. Artificial screening techniques involve developing pure culture of the fungal pathogen, increase of inoculum on appropriate growth media (or a susceptible host), inoculation of plants or plant parts at appropriate stage of the plant, and after incubation, evaluation. During the increase of inoculum and subsequent incubation of inoculated plants care should be taken to eliminate the possibility of contamination with other strains of the pathogen than the one under consideration. One has to be careful that the culture of pathogen has optimal degree of virulence and aggressiveness because some pathogens are known to lose virulence, in the absence of host plant, when cultured for long periods (Kulkarni and Chopra, 1989).

Screening against some necrotrophic (Facultative) fungi can be done in the absence of pathogen concerned. These fungi and many plant pathogenic bacteria produce toxins, which kill the tissues of host plants (Strobel *et al.*, 1972). Pathotoxins of several pathogenic fungi such as *Helminthosporium sacchari*, causing leaf blight of sugarcane and *Helminthosporium victoriae*, causing leaf blight of oats, have been isolated and used to screen plants for resistance against diseases incited by these fungi.

Testing for resistance against insect pests is also carried out in the field or in the glasshouse, as is done for fungal pathogens. Nevertheless, field testing under natural and uncontrolled environmental conditions may be some times misleading due to the experimental errors which creep in. On the other hand, screening in glasshouse or laboratory can be standardized and controlled. Consequently, the results are reliable. For a good host plant resistance programme it is essential to develop an efficient and reliable screening technique which ensures desirable levels of insect pressure, uniformly at most susceptible stage of the crop. These requirements can be met either by selecting a location where the pest occurs regularly (*i.e.* "hot spots") or by testing the material under artificial infestation with laboratory reared insects.

Stem borer (*Chilo partellus*, Swinhoe) is a serious pest of sorghum. It attacks sorghum plants from two weeks after germination until the crop harvest. It produces 'pin holes' on leaves, 'dead hearts' and tunneling into the stem. It completes its life cycle in about a month under optimal conditions in India. Hissar (Haryana), in North India, was identified as 'hot spot' for screening sorghum for stem borer, where severe infestations of this insect-pest occurs regularly (Taneja and Leuschner, 1985).

Most of the books on plant breeding describe methods employed for screening and breeding for diseases caused by either fungi and/or bacteria and viruses. But a passing reference is made to the techniques used to screen for resistance against insect-pests. In this chapter, an example of the method employed for screening sorghum plants against *Chilo partellus* reared artificially, will be presented. The method of artificially rearing the insect, infestation of sorghum seedlings with first instar larvae and evaluation was developed at the International Crops Research Institute for Semi-Arid Tropics (ICRISAT), Patancheru (Taneja and Leuschner, 1985).

Rearing of the insects on artificial diet in the laboratory has been reported for many Lepidopterous stem borers such as *Sesamia inferens* (Chatterji *et al.*, 1969), *Chilo suppressalis* (Ishii, 1971; Kamano, 1973) and *Chilo partellus* (Chatterji *et al.*, 1968; Dang *et al.*, 1970; Lakshminarayana and Soto, 1971; Seshu Reddy and Davis, 1979).

Various artificial diets for rearing *Chilo partellus* were reported but the diet developed at ICRISAT was found to be the most satisfactory (Taneja and Leuschner, 1985). Average moth emergence from this diet was found to be 70 to 75 per cent. The sex ratio of male and female moths was close to 1:1. About 90 per cent moths emerged 25 to 30 days after the egg inoculation on artificial diet.

Egg masses are collected from the sorghum plants growing in the field. To avoid any external contamination, eggs are sterilized during the blackhead stage (*i.e.* about 1 day before hatching) in 10 per cent formaldehyde for 2 minutes, and then washed thoroughly with distilled water before being placed on diet.

The jars used for rearing of the insects are filled with 300 g artificial diet. Approximately 100 eggs at blackhead stage are placed in each jar. The jars are stored in rearing room in dark for 2 days. The temperature of the rearing room is maintained at 28 ± 1°C, relative humidity at 60 to 70 per cent and the light is provided for 12 hours. The larval and pupal stages of the insect are completed in same jar in which the eggs are placed. As mentioned earlier, the first moths (males) emerge 25 to 26 days after the egg inoculation. The females start emerging 2 to 3 days later than the males.

The moths emerging from the jars are collected with the help of a vacuum cleaner attached to a pipe with various outlets. The males and females are collected separately. The males are smaller in size with dark forewings and small pointed abdomen. The female moths after mating lay eggs on butter paper of the egg laying cage. The egg laying cage is an open cylinder 25 cm high and 25 cm in diameter made of galvanized iron wire mesh. A thin georgette cloth is wrapped around the cylinder and uniform holes are made at regular intervals in it. At the time of release of moths in the cylinder a white butter paper is wrapped around the cage. The moths lay eggs on butter paper. Both the ends of cylinder are covered with mosquito net fitted saucers. Butter paper with egg masses is replaced every day. A female moth lays on an average nine to 10 egg masses in a period of four days.

High humidity, about 80 to 90 per cent, is required for embryonic development and hatching. For normal incubation the eggs are stored at 26 ± 1°C and high humidity.

For field infestation of sorghum plants first instar larvae are used. 'Bazooka applicator' with some modifications is used for field infestation. This application requires a carrier along with larvae. Khaskhas (poppy seed) was found to be a satisfactory carrier. Most critical stage of the sorghum plant for larval infestation is around 14 days after germination. With the help of "Bazooka applicator" larvae of first instar are placed in the leaf whorls of plants. Observations are then recorded on plants which comprise of leaf injury, dead hearts formed, stem and peduncle tunneling, etc.

Whenever, screening is executed with insect reared artificially in laboratory, the recording of observation should begin at early vegetative stage (*V*3 in case of soybean) so that resistant plants are identified before anthesis which will facilitate its hybridization, if found resistant, with agronomically desirable cultivars during the same season (Ruffeuer *et al.*, 1987).

Only one disease or one insect pest does not attack crop varieties, in field conditions. Very often several diseases and insects, and these too at different growth stages attack the crop. Even the minor diseases and insect pests contribute to yield losses to a considerable extent. It is, therefore, desirable to breed varieties with a reasonable level of resistance/tolerance to the prevalent diseases and insect pests. This will lead to stability of yield over locations and years.

Improvement for multiple resistances may be executed by a breeding programme in tandem order or simultaneous selection for resistance to prevailing diseases and insects. For achieving the objective, sources with multiple resistances should be identified and skillfully used by the breeder and his team comprising of an entomologist and plant pathologist. Recurrent selection would be an ideal breeding method.

For screening against prevailing insect pest complex, Odulaja and Nokoe (1993) proposed a "maximin-minimax" method which permits classification of varieties into various groups on the basis of per cent yield losses and actual yield potential, *viz.*, resistant and high yielding (most desirable), susceptible and high yielding (tolerant, ideal for integrated pest management), resistant but low yielding (suitable as a donor parent) and susceptible, low yielding (may be used as susceptible check).

It may be in place to recall, at the end of this chapter, that rice variety IR 36 was evolved at I.R.R.I., through breeding which was resistant to four major diseases and five insect pests. Breeding for multiple resistance is, therefore, not an impossible task.

15.4 BIOTECHNOLOGICAL APPROACHES FOR RESISTANCE INCORPORATION

Genes for resistance in the cultivated germplasm are either limited against majority of diseases and some insects, incomplete against some and practically no resistance against few. For example, grassy stunt virus and sheath blight of rice, stem borers of rice, pink boll worm of cotton, pod borer of pigeonpea, karnal bunt of wheat, *Ascochyta blight* of gram, yellow mosaic of mung bean, *Alternaria blight* of mustard, are some for which resistance is moderate or not available in the cultivated germplasm (Khush, 1995; Siddiq and Sarma, 1995). Nevertheless, wild species, unrelated crop species, microorganisms, may be a potential source of resistance to which a breeder could not look with great expectations a few decades ago. With the advances in the field of cellular and molecular biology, an array of biotechnological devices can now be used to isolate genes for resistance after their identification and incorporated into the desired source.

The plant biotechnology approaches could be grouped into two categories, *viz.*, plant tissue culture and plant genetic engineering. The plant genetic engineering encompasses techniques and approaches to introduce foreign genes into the plants. This requires a suitable procedure to regenerate plant species of interest *in vitro*, which tissue culture enables. Tissue culture is the growth of an isolated cell, tissue segment or an organ in an artificial environment, out side parent organism. Tissue culture is possible because of the property of 'totipotency' of cell. Totipotency is the capacity of any cell to regenerate and develop into a whole plant, given proper environment. These techniques could be employed for incorporation of resistance to diseases and insects in the plants.

Embryo rescue, somatic cell hybridization, somaclonal variation are various devices which are available in tissue culture which help to modify the genetic constitution of a plant with respect to resistance against diseases or insects.

(a) Embryo Rescue

Several times desirable genes from inter-specific crosses can not be incorporated in a desirable variety because of crossability barriers, e.g. inability of pollen to germinate, or failure of pollen tubes to grow or perhaps more commonly due to degeneration of endosperm. Major incompatibility may result from abortion of hybrid embryos due to the degeneration of endosperm. Following the pioneering work of Laibach (1925), embryo rescue technique has become popular for transfer of useful genes from related species or genera to a wide variety of crop varieties. Embryo rescue is a technique in which immature developing embryos are dissected out from the developing fruits and grown in a suitable growth medium in which the embryos either differentiate into the shoots and roots or directly develop into plantlets. Incorporation of genes for resistance to leaf and stem rust from *Triticum* species and related genera, such as *Aegilops, Agropyron, Hynaldia* and *Secale* in wheat; resistance to various foliar diseases from *Tripsacum* to maize; resistance to tobacco mosaic virus from *Nicotiana glutinosa*; resistance to *Fusarium* and *Verticillium* from *Lycopersicon pimpinellifolium*; resistance to *Septoria* leaf spot from *L. hirsutum* to tomato and resistance to late blight from *Solanum demissum* to potato are some of the successful attempts (Knott and Dvorak, 1976) to utilize alien genes for resistance through distant hybridization.

(b) Somatic Cell Hybridization

Somatic cell hybridization involves physical union or fusion of protoplasts from two parents. A wall composed predominantly of cellulose encloses every cell in a plant. Individual cells are joined together by pectins. Edward Cocking, a British scientist, in 1960 introduced a method of enzymatic isolation of protoplasts. He used a concentrated solution of enzyme cellulase extracted from the fungus, *Myrothecium verucaria* to degrade the cell walls of tomato root tips. Introduction of commercial enzymes, "onozuka" cellulase and macerozyme in 1963 prompted further research in the field. Pectinase separates the cellular masses into individual cells and cellulase digests the wall leaving behind a living naked cell the "protoplast". The enzymatic isolation, in general, involves incubation in a pectin enzyme (for example macerozyme) to separate the cells, and then cellulase to degrade the cell wall, used sequentially or as a mixture along with a suitable plasmolyticum, *e.g.*, sucrose, sorbital, or mannitol (Chakrabarti, 1976). Sodium nitrate ($NaNO_3$) promotes fusion of protoplasts, without affecting the viability of protoplasts. Protoplasts isolated from two diverse species could thus be fused. The isolated protoplasts can regenerate cell wall in culture and possess capacity for nuclear and cellular division. Because of the inherent totipotency of the isolated cell, it is possible to regenerate a whole plant out of the fused protoplasts. Somatic hybrids have been produced between several species such as *Arabidopsis thaliana* and *Brassica campestris*, potato and tomato, *Nicotiana* and *Lycopersicon,* etc. (Khush and Brar, 1988).

It is now possible to culture and regenerate plants from protoplasts of several recalcitrant species, such as wheat, rice and maize and apply somatic hybridization technique to expand gene pools (Khush, 1995).

(c) Somaclonal Variation

Plant breeding depends for its success on the availability of desirable genetic variability, either natural or experimentally induced. It is now established that frequent genetic changes occur during the process of tissue culture and/or cell culture and many of these changes are manifested as heritable variations among the regenerants. Genetic variability among the plants regenerated from tissue culture is known as "somaclonal" variation (Larkin and Scowcroft, 1981). It is an important source of variability in well adapted genotypes without the involvement of hybridization. Plants regenerated from cell/tissue culture differ in a number of morphological characters such as maturity, yield, reaction to pathogens and insect pests. Although the early success was confined to asexually propagated plants, subsequent investigations showed that somaclonal variation is a rule rather than an exception and is observed in many plant species such as tobacco, maize, rice, oat, barley, sorghum, wheat, brassica, alfalfa, tomato, sweet potato, etc. (Scowcroft and Larkin, 1982). Somaclonal variation was observed in plants regenerated from various culture sources such as callus, protoplast, embryo and meristem.

Actual cause of somaclonal variation is not yet known exactly, nevertheless, it has been established that such variations affect both oligogenic and polygenic characters, and are heritable (Siddiq and Sarma, 1995).

The total magnitude of somaclonal variability is larger than variability induced by mutagens. Moreover, the frequencies of variants for traits encoded by dominant and recessive genes are comparable. Somaclonal variability observed for nuclear borne genes and heritable factors in cytoplasm is also comparable. Hence, somaclonal variability appears to be more advantageous than variation induced by mutagens. In tomato, promising cultivars, DNAP 9 and DNAP 17 have been developed with high total solid contents and resistance to *Fusarium*. In brassica a variety "Pusa Jaikisan" has been developed with improved yield of oil. A somaclonal variant of banana resistant to *Fusarium oxysporum cubens* has been developed (Koundal, 2001). Some useful pest resistant somaclonal variants have been isolated in sugarcane, such as those resistant to Fiji virus and downy mildew (Heinz *et al.*, 1977), maize resistant to *Helminthosporium* (Bretell and Ingram, 1979) and potato resistant to early and late blight (Shephard *et al.*, 1980). Exposure of cell cultures to toxic concentrations of various selective agents, such as culture filtrate and pathotoxins, can increase the selection efficiency for useful somaclones. Selection of cultured cells tolerant to culture filtrate or toxin has been successfully employed to produce pest resistant germplasm. For example, Sacristan (1982) developed *Brassica* resistant to *Phoma lingam*, Thanutong *et al.* (1983) isolated tobacco resistant to *Pseudomonas syringae*, cv. *tabaci*.

(d) Meristem Culture and Virus Elimination

Seeds of virus infected plants rarely contain virus. Consequently, the progeny of sexually reproducing plants is generally virus free, except the new infection. On the other hand, in the case of vegetatively propagated and virus infected crop plants the progeny contains virus particles. The distribution of virus in the plant tissues is uneven, and hence in the virus infected plants there are some regions, which are free from virus. Limmaset and Cornuet (1972) observed that in virus infected plants, the virus concentration decreased towards the growing point. This explains the reason for success of More and Martin (1952) who very successfully regenerated healthy plants from infected dahlias through meristem culture. Cultivation of apical meristems, particularly the shoot apical meristem, is known as meristem culture. Meristematic tissues are composed of cells that are in a state of division or retain power of dividing. A median longitudinal section through the apex of stem when examined under the microscope shows that the apical meristem or the growing region of stem is composed of a mass of small, usually round or polygonal cells which have abundant protoplasm and are in a state of division. The culture of smaller shoot tips (meristem culture), comprising of the apical dome and one or two leaf primordia yields virus free progeny. The technique briefly consists of dissecting meristems, about 0.2 to 0.5 mm thick, and growing them in suitable nutrient medium. Minimum surrounding tissue has to be taken if the object is to develop virus free plants. Meristem culture has been used in a number of crop species for production of virus free plants, *e.g.* banana free from cucumber mosaic virus, cauliflower free from cauliflower mosaic virus, pea from seed borne mosaic virus, potatoes free from paracrinckle virus *X*, *S*, *Y*, leaf roll and spindle tuber virus and sugarcane from mosaic (Siddiq and Sarma, 1995). This technique is also used to develop virus free plants of dahlia, carnation, strawberry, chrysanthemum, orchids, etc. (Khush, 1995).

Remarkable progress has been made to identify and tag the resistant genes in wide variety of organisms. Isozyme markers and Restriction Fragment Length Polymorphism (RFLP) markers have been used to tag the resistant genes. Genes for pest resistance have been tagged with isozyme markers and used as selection criteria in plant breeding. Rick and Fobes (1974) reported tight linkage of gene Mi for nematode resistance in tomato with an isozyme locus Aps1 (acid phosphatase) on chromosome 6. During the last few years comprehensive RFLP maps have been prepared in several plant species and genes for disease and insect resistance tagged with RFLP markers. Various techniques of genetic engineering, which can yield resistant plants, are given below:

(a) Transgenic Plants

The new technology, called genetic engineering involves the direct introduction of desirable synthetic DNA into the desired plant. The genetically modified (*GM*) crops so developed have come to prominence during past few years. Genetic modification is an accurate and effective way of achieving more desirable attributes in plants without the trial and error of traditional plant breeding methods. The genetically modified (*GM*) plants or popularly known as 'transgenic plants' are those that carry gene(s) introduced into them by genetic engineering and express the desired trait(s). Transgenic plant contains gene, which has been artificially inserted, instead of the plant acquiring it through pollination. The inserted alien gene is known as the transgene. It may come from any of the biological organisms, such as unrelated plants or completely different plant species, microorganisms, animals. Prior to genetic engineering, exchange of DNA material was possible only between the individual organisms of same species. A model genetic engineering of plants comprises of four general steps (Radhakrishnan, 2002):

- (i) Selection of a gene(s) whose introduction in plants would be of positive agricultural value,
- (ii) Identification and isolation of such gene(s),
- (iii) Transferring the isolated gene(s) to the plant cell of desired variety of a crop, and

(iv) Regeneration of complete plants from transformed cells or tissues.

Modification of DNA of desirable plants can be accomplished by adding the gene(s) from an alien source through recombinant DNA technique. Reconstruction of the DNA molecule by joining sequences from one source to the sequence of another source of quite different origin produces what is known as recombinant DNA and the technique as genetic engineering (Lewin, 1986). Recombinant DNA molecule is usually obtained with the help of two enzymes, viz., restriction endonuclase and ligase. The former enzyme induces a cut in the double stranded DNA molecule at precise nucleotide sequence and produces well defined DNA fragments. Restriction endonucleases are highly specific enzymes, which cleave DNA strands at particular sequence of bases. The strands of DNA molecule are chopped in a manner which leaves sticky ends at either side of the original DNA duplex molecule. The free sticky ends serve as convenient points to align and join the new segment of DNA with the help of ligases or joining enzymes. What a genetic engineer does is that he introduces cuts in the recipient (such as Ti plasmid of *Agrobacterium*) as well as the foreign DNA (for example DNA of *Bacillus thuringiensis*) with the help of same restriction endonuclease. If the two such cleaved DNA fragments are brought together and treated with ligase, they join despite their different origins, resulting in chimera.

Based on three major gene transfer techniques developed to transform the plants with alien genes, transgenic plants are produced. These techniques are (*i*) Microprojectile bombardment involving shooting the DNA coated tungsten or gold particles into the plant cells, (*ii*) electroporation, which employes a short burst of electricity to get the DNA into the cell, and (*iii*) the *Agrobacterium tumefaciens* mediated transformation (Old and Primrose, 1994). The *Agrobacterium* mediated procedure is commonly used.

The *Agrobacterium tumefaciens* is a soil bacterium and forms tumors, in many higher plants, especially the dicotyledonous plants at the crown separating the stem from root, after infection. This is called as a crown gall disease. Although these bacteria are essential for tumor induction they are not necessary for tumor maintenance and growth. Tumor cells produce various compounds called opines which are metabolized by bacteria. Tumor formation is accompanied by transfer and integration of specific part of tumor-inducing (*Ti*) plasmid into the nuclei of host cells. Plasmid is a replicon stably inherited in an extra chromosomal state, while the replicon is unit of genetic material, which is capable of autonomous replication. The *Ti* plasmid, which is a double stranded circular DNA, is not essential for survival of bacteria. It is present in some bacteria in addition to their own DNA molecule, the genetic material. Strains of *Agrobacterium*, which do not possess this plasmid, are not tumerogenic. The specific part of *Ti* plasmid responsible for tumor induction is the *T*-region. In the plant cells this *T* region of plasmid is called *T* DNA. The tumors are produced by plants upon infection by *Agrobacterium tumefaciens* because of transfer of this oncogenic DNA into the plant cells at wounded site. The *T* DNA carries many genes encoding the biosynthesis of enzymes of hormones and nitrogenous metabolites called opines. The *T* DNA is integrated into the plant chromosome by illegitimate recombination. This property has been utilized by molecular biologists to manipulate plant genomes.

Bacillus thuringiensis a soil bacterium produces spores, which contain a crystalline (cry) protein. The first generation of insect resistant transgenic plants are based on insecticidal crystalline proteins. In the insect gut, the protein breaks down to release a toxin, known as delta-endotoxin. This toxin binds to the intestinal lining, resulting in ion imbalance, paralysis of digestive system and death of insect after a few days (Shetty, 2002). Different versions of the cry genes, also called as Bt, have been identified. They are effective against different orders of insects. A modified version of cry gene has been incorporated into the plants own DNA so that plants own machinery produces the toxin. *Agrobacterium* mediated transformation has led to development of Bt varieties of several crops, like maize, cotton, potato, tobacco, rice, etc. The *Agrobacterium* mediated transformation of plants has three phases (Radhakrishnan, 2002).

(i) Creating a strain of *Agrobacterium* which carries a *Ti* plasmid with a gene that the scientist desires to transfer to the plants, in the present case, say Cry 1A.

(ii) Culture the *Agrobacterium* so created with pieces of plant tissues/cells, so that DNA transfer can occur, for example cotton. And lastly,

(iii) Regenerate a plant from transformed cells. Along with cry gene, a gene responsible for resistance to some antibiotic, such as Penicillin G., Kanamycin, Geomycin, etc. is also placed very close to cry. Cells, which are transformed, are also resistant to antibiotic added to the growth media in which *Agrobacterium* and plant cells are cultured. The antibiotic kills rests of the cells.

It is reported that in 1996, transgenic crops were grown in the world on about 1.7 million hectares, however, in 2000 this area increased to 44.2 million hectares (Radhakrishnan, 2002).

(b) Coat-Protein Mediated Protection from Virus

Baulcombe (1989) discussed strategies for genetic engineering for virus resistance. Two approaches have been used viz., coat protein genes and antisense RNA.

Powell-Abel *et al.* (1986) showed that transgenic tobacco expressing coat protein gene of tobacco mosaic virus (TMV) was resistant to TMV and that the resistance was due to the expressed coat protein. The CP gene not only protects the plants against the virus from which the CP gene has been taken but also other closely related viruses. For example, CP gene of TMV confers resistance to tomato mosaic virus, pepper mild mottle virus, tobacco mild green mosaic virus, etc. (Siddiq and Sarma, 1995). The coat protein mediated virus resistance is achieved by constitutive expression of the CP gene, *i.e.* gene expresses even in the absence of challenge from virus. The already available coat protein in transgenic plants interferes in some manner with multiplication cycle of the virus. However, the precise mechanism of interference at the molecular level is not yet known. Register and Beachy (1988) and Wu *et al.* (1990) believe that the nucleic acid of the virus is not allowed to express by keeping it in a wrapped condition and hence no initiation of infection itself. Essentially the CP mediated resistance is similar to cross-protection.

Other approaches are antisense RNA mediated protein and satellite RNA mediated defense. Reader may refer for more details to Siddiq and Sarma (1995).

(c) Proteinase Inhibitors

Many plants have developed a natural mechanism of defense against herbivorous insects by accumulating enzyme inhibitors and lectins at concentrations that will cause metabolic inhibition upon ingestion by insects. Proteinase inhibitors are proteins, which inhibit the activity of proteinase enzymes. As compared to the Bt toxins, protein inhibitors have a broader spectrum of metabolic inhibition and hence can be used to control many types of insects (Boulter, 1993). Important plant protease inhibitors are cowpea trypsin inhibitor from *Vigna unguiculata*, potato proteinase inhibitor, soybean trypsin inhibitor etc. Lectins are garden pea seed lectin, and phytohaemagglutinin. Transgenic tobacco plants expressing high level of trypsin inhibitor protein encoded by a gene of an African variety of cowpea showed decreased insect damage by *Heliothis virescence*. It was shown that cowpea trypsin inhibitor killed a variety of *Lepidopteran* and *Coleopteran* insects. Lectins also play an important role in resistance against aphids.

(d) Control of Diseases

Chitinase and *â* 1-3 glucanase belong to the so called pathogenesis proteins (PR) because their synthesis increases markedly after attack by the phytopathogens. They have antifungal activity. Broglie *et al.* (1991) demonstrated that tobacco and canola plants having chitinase gene were highly resistant to *Rhizoctonia solani*.

These evidences strongly suggest that a plant breeder is presently equipped with several devices to combat phytopathogens and insect pests. If properly utilized high yielding, disease and/or insect resistant varieties can be evolved using any of these techniques and appropriate breeding procedure.

References

Allard, R.W. (1960). *Principles of plant breeding*. John Wiley and Sons, New York, pp. 321-369.

Asthana, A.N. and S. Chandra (1995). Varietal improvement in pulse crops — current trends and future prospects. In *Genetic Research and Education: Current trends and the next fifty years*. Eds. B. Sharma *et al.* Indian Society of Genetics and Plant Breeding, New Delhi, pp. 925-959.

Bandopadhyay, R. and L.K. Mughogho (1988). Evaluation of field screening techniques for resistance to sorghum grain molds. *Plant Disease*, 72: 500-503.

Balucombe, D. (1989). Strategies for virus resistance in plants. TIG, 5: 50-60.

Biffen, R.H. (1905). Mendel's laws of inheritance and wheat breeding. *J.Agric.Sci.*, 5: 4-48.

Biffen, R.H. (1912). Studies on inheritance in disease resistance. II. *J.Agric.Sci.*, 4: 421-429.

Borlaug, N.E. (1953). New approach to breeding of wheat varieties resistant to *Puccinia gramminis tritici. Phytopath.*, 43: 467.

Borlaug, N.E. (1958). The use of multilineal or composite varieties to control air borne epidemic diseases of self-pollinated crop plants. *Proc.First Intern.Wheat Genet.Symp.* The Public Press Ltd., Winipeg, Canada, pp. 15-27.

Bretell, R.S.S. and D.S. Ingram (1979). Tissue culture in the production of novel disease — resistant crop plants. *Biol.Rev.*, 54: 329-345.

Brim, C.A. (1966). A modified pedigree method of selection in soybeans. *Crop Sci.*, 6: 220.

Brim, C.A. and C.S. Stuber (1973). Application of genetic male-sterility facilitated recurrent selection schemes in soybeans. *Crop Sci.*, 13: 528-530.

Broglie, K.; I. Chet; M. Molliday; R. Cressman; P. Biddle; S. Knowlton; C.J. Mauvais and R. Broglie (1991). Transgenic plants with enhanced resistance to the fungal pathogen, *Rhizoctonia solani, Science*, 254: 1194-1197.

Browning, J.A. and K.J. Frey (1969). Multiline cultivars as a means of disease control. *Ann.Rev.Phytopath.*, 7: 355-382.

Boulter, D. (1993). Insect pest control by copying nature using genetically engineered crops. *Phytochemistry*, 34: 1453-1466.

Chakrabarti, S.N. (1976). Hybrid plants from protoplast fusion. *Science Today*, June, 1976, pp. 19-21.

Chatterji, S.M.; K.H. Siddiqui; V.P.S. Panwar; G.C. Sharma and W.R. Young (1968). Rearing of the maize stem borer, *Chilo zonellus*, Swinhoe, on artificial diet. *Indian Journal of Entomology*, 30: 8-12.

Dang, K.; Anand Mohini and M.G. Jotwani (1970). A simple improved diet for mass rearing of sorghum stem borer, *Chilo zonellus* (Swinhoe). *Indian Journal of Entomolgoy*, 32: 130-133.

Frey, K.J.; J.A. Browning and M.D. Simons (1975). Multiline cultivars of autogamous crop plants. *SABRAO, J.*, 7: 113-123.

Frey, K.J.; J.A. Browning and M.D. Siomons (1979). Management systems for the host genes to control disease loss. *Indian J.Genet.*, 39(1): 10-21.

Gill, K.S.; G.S. Nanda; Gurdev Singh and S.S. Aulja (1979). Multilines in wheat. *Indian J.Genet.*, 39(4): 30-37.

Groenewegen, L.J.M. (1977). Multilines as tool in breeding for reliable yields. *Cereal Res.Commu.*, 5: 125 —132.

Harlan, J.R. (1977). Sources of genetic defense. *Ann.N.Y.Acad.Sci.*, 287: 345-356.

Heinz, D.J.; M. Krishnamurthi; L.G. Nickell and A. Marketzki (1977). Cell, tissue and organ culture in sugarcane improvement. In *Applied and Fundamental Aspects of Plant Cell Tissue and Organ Culture* (eds. J. Reinert and Y.P.S. Bajaj). *Springer-Verlag*, Berlin, 3-17

Ishii, S. (1971). Nutritional studies of the rice stem borer, *Chilo suppressalis*, Walker, and its mass rearing. *Entomophaga*, 16: 165-173.

Jambhale, N.D. and Y.S. Nerkar (1986). Parbhani Kranti a yellow vein mosaic resistant okra. *Hort Sci.*, 21: 1470-1471.

Jensen, N.F. (1952). Intravarietal diversification in oat breeding. *Agron.J.*, 44: 30-34.

Khush, G.S. (1977). Breeding for resistance in rice. *Ann.N.Y.Acad.Sci.*, 28: 296-308.

Khush, G.S. (1995). Approaches and strategies in breeding for pest resistance. In *Genetic Research and Education: Current Trends and the Next Fifty Years*. Eds. B. Sharma *et al.* Indian Society of Genetics and Plant Breeding, New Delhi, pp. 163-179.

Khush, G.S. and D.S. Brar (1988). Wide hybridization in plant breeding. In *Plant breeding and genetic engineering* (ed. A.H.Zakri) SABRAO, Malaysia, 141-188.

Koundal, K.R. (2001). Prospects of biotechnology in increasing agricultural production. *Fertilizer News*, 46(12): 73-78.

Knott, D.R. and J. Dvorak (1976). Alien germplasm as a source of resistance to diseases. *Ann.Rev.Phytopath.*, 14: 211-235.

Kulkarni, R.N. and V.L. Chopra (1989). Breeding for disease resistance. In *Plant breeding — theory and practice*. Ed. V.L. Chopra. Oxford & IBH Publishing Co., New Delhi, pp. 285-307.

Laibach, F. (1925). Das Taubwerden Von Bastardsanen und die kunstliche Aufzucht fruh abster bender Bastardeembryonen. *Z.Bot.*, 17: 417-419.

Lakshminarayana, K. and P.E. Soto (1971). A technique for mass rearing of sorghum stem borer, *Chilo zonellus*. *Sorghum Newsletter*, 14: 41-42.

Larkin, P.J. and W.R. Scowcroft (1981). Somoclonal variation — a novel source of variability from cell cultures for plant improvement. *Theor.Appl.Genet.*, 60: 197-214.

Leonard, K.J. (1969). Genetic equilibria in host pathogen system. *Phytopathology*, 59: 1858-1863.

Leppik, E.E. (1970). Relative resistance of *Cucumis* spp. to diseases and insects. *Plant Introduction Investigations*, 43-54.

Lewin, B. (1983). Genes. Wiley Eastern.

Morel, G. and C. Martin (1952). Guerison de dahlias atteints dune maladie a virus. Compt. *Rend.*, 235: 1324-1325.

Narayanan, S.S. (1995). Resistance to *Heliothis* and other serious insect pests in *Gossypium* spp. In *Genetic research and education: Current trends and the next fifty years*, Eds. B. Sharma *et al.* Indian Society of Genetics and Plant Breeding, New Delhi, pp. 195-210.

Odulaja, A. and S. Nokoe (1993). A maximum-minimum approach for classifying crop varieties into resistance groups based on yield potential and loss. *Intl.J.Pest Mgmt.*, 39: 64-66.

Old, R.W. and S.B. Primrose (1994). *Principles of gene manipulation. An Introduction to Genetic Engineering. Blackwell Sci.Publ.*, London.

Pal, B.P. (1957). Breeding improved varieties of wheat and barley. *Indian Journal of Genetics and Plant Breeding*, 17(2): 148-155.

Parlevliet, J.E. (1979). The multiline approach in cereals to rusts: aspects, problems and possibilities. *Indian J.Genet.*, 39(1): 22-29.

Pathak, R.S. (1995). Genetic systems of plant resistance to insects. In *Genetic research and education: Current trends and the next fifty years*, Eds. B. Sharma *et al.* Indian Society of Genetics and Plant Breeding, New Delhi, pp. 211-217.

Powell Abel, P.A.; R.S. Nelsom; B. De N. Hoffman and S.G. Rogers (1986). Delay in disease development in transgenic plants that express the tobacco mosaic coat protein genes. *Science*, 232: 738-743.

Radhakrishnan, G. (2002). Genetically modified plants — from understanding to applications. In *XI National seed seminar on quality seed to enhance agricultural productivity*. Indian Society of Seed Technology, New Delhi and University of Agricultural Sciences, Dharwar, Eds. Shekhargouda, M. and L. Krishna Naik, pp. 34-36.

Rajaram, S.; B. Skovmand; H.J. Dubin; E. Torres; R.G. Anderson; A.P. Roelfs; D.J. Samborski and I.A. Watson (1979). Diversity of rust resistance of CIMMYT multiline composite, its yield potential and utilization. *Indian J.Genet.*, 39(1): 60-71.

Rajaram, S.; R.P. Singh and M. van Ginkel (1995). CIMMYT's approach to breeding for durable resistance in bread wheat: A Global Perspective. In *Genetic research and education: Current trends and the next fifty years*, Eds. B. Sharma *et al.* Indian Society of Genetics and Plant Breeding, New Delhi, pp. 325-338.

Ram, H.H. (1998). *Vegetable breeding, principles and practices*. Publ.Kalyani Publishers, Ludhiana.

Register, J.C.III and R.N. Beachy (1988). Resistance to TMV in transgenic plants results from interference with an early event in infection. *Virology*, 166: 524-532.

Richharia, R.H. (1957). Plant breeding with reference to oilseed crops. *Indian Journal of Genetics and Plant Breeding*, 17(2): 218-247.

Rick, C.M. and J.F. Fobes (1974). Association of an enzyme with nematode resistance. *Rep. Tomato Gent.Coop.*, 24: 25.

Rogers, C.E. and T.E. Thompson (1978). *Helianthus* resistance to carrot beetle. *J.Econ., Ent.*, 71: 760-771.

Rosen, H. R. (1949). Oat parentage and procedures for combining resistance to crown rust including race 45 and *Helminthosporium* blight. *Phytopathology*, 39: 20.

Ruffeur, T.K.III; R.B. Hammond; R.L. Cooper and St. S.K. Martin (1987). Larval antibiosis screening technique for Mexican bean beetle resistance in soybean. *Crop Sci.*, 27: 598-600.

Russell, G.E. (1978). *Plant breeding for pest and disease resistance*. Publ. Butterworths, London.

Sacristan, M.D. (1982). Resistance to *Phoma lingam* of plants regenerated from selected cell and embryonic cultures of haploid *Brassica napus*. *Theor.appl.Genet.*, 61: 193-200.

Scowcroft, W.R. and P.J. Larkin (1982). Somaclonal variation — a new option for plant improvement. In *Plant improvement and somatic cell genetics* (Eds. I.K.Vasil; W.R. Scowcroft and K.J. Frey). Academic Press, New York, pp. 159-178.

Seshu Reddy, K.V. and J.C. Davis (1979). A new medium for mass rearing of sorghum stem borer, *Chilo partellus*, Swinhoe (*Lepidoptera: Pyralidae*) and its use in resistance screening. *Indian Journal of Plant Protection*, 6: 48-55.

Shephard, J.; F.D. Bidney and E. Shahin (1980). Potato protoplasts in crop improvement. *Science*, 28: 17-24.

Shetty, Shekhar, H. (2002). Genetic modification for crop improvement and the present scenario. In *XI National Seed Seminar on Quality Seed to Enhance Agricultural Profitability*. Eds. M. Shekhargouda and L. Krishna Naik. Indian Society of Seed Technology, New Delhi and University of Agricultural Sciences, Dharwad, Jan., 18-20, 2002, pp. 25-33.

Siddiq, E.A. and N.P. Sarma (1995). Biotechnological approaches for disease resistance in crop plants. In *Genetic Research and education: current trends and the next fifty years*, Eds. B. Sharma et al. Indian Society of Genetics and Plant Breeding, New Delhi, pp. 271-286.

Strobel, G.A.; W.M. Hess and G.W. Steiner (1972). Ultra structure of cells in toxin treated and *Helminthosporium sacchari* infected sugarcane leaves. *Phytopathology*, 62: 339.

Suneson, C.A. (1960). Genetic diversity — A protection against plant diseases and insects. *Agron.J.*, 52: 319-321.

Taneja, S.L. and K. Leuschner (1985). Methods of rearing, infestation and evaluation for *Chilo partellus* resistance in sorghum. In *Proceedings of the International Sorghum Entomology Workshop*, 15-21 July, 1984. Texas A and M University, College Station, Texas USA; ICRISAT, Patancheru 502-324. pp. 175-188.

Thanutong, P.I.; I. Furusawa and M. Yamamoto (1983). Resistant tobacco plants from protoplast- derived calluses selected for their resistance to *Pseudomonas* and *Alternaria* toxins. *Theor.Appl. Genet.*, 6: 209-215.

Van der Plank, J.E. (1963). *Plant diseases, Epidemics and control*. Academic Press, New York.

Van der Plank, J.E. (1968). *Disease resistance in plants*. Academic Press, New York.

Watson, I.A. and D. Singh (1953). The future for rust resistant wheats in Australia. *J.Aust.Inst.Agric.Sci.*, 18: 190-197.

Wu, X.; R.Beachy; T. Wildon and J. Shaw (1990). Inhibition of uncoating of tobacco mosaic virus particles in protoplasts from tobacco plants that express the viral coal-protein gene. *Virology*, 179: 893-895.

16

Polyploidy in Plant Breeding

Besides Mendelian variation, comprising of mutation coupled with or without hybridization between mutant forms, polyploidy also contributes to the evolution of crop species. While describing the reproductive process among plants a reference was made to the somatic and the gametic chromosome number. It was always assumed, in previous chapters, that the somatic cells of plants are diploid, i.e. they possess two complete sets of chromosomes, one received from seed parent and another from pollen parent. As a result of meiotic cell division of certain cells in either anthers or the ovules (Chapter 2) male or female gametes are formed containing half the number of chromosomes in somatic cells. Or put alternatively pollen grains (or egg cell) contain chromosome number equal to one set of chromosomes. Individual chromosomes could be entirely from either seed parent or pollen parent or mixture of both. But the number of chromosomes, under normal situations, does not exceed or is lesser than half the number of chromosomes in somatic cells. The gametic chromosome number is said to be haploid and denoted by *n*. Haploid is defined as the cell with gametic chromosome number. While the somatic chromosome number is expressed as 2*n*.

Thus we assume that somatic nuclei contain two chromosomes of each kind, characteristic of the species while the gametes carry only one chromosome of each kind, characteristic of the species. But deviations from this diploid characteristic number are not uncommon. Polyploid is an organism with other than two basic sets of chromosomes. Basic number is the number of chromosomes in the ancestral diploid ancestors of polyploid. To explain this aspect, let us take an example of wheat (*Triticum* sp.). However, a point to remember is that polyploid could be broadly speaking of two classes viz., autopolyploid and allopolyploid. This will be described later. Common bread wheat is an example of allohexaploid.

The wheat has three groups, viz., einkorn, emmer and aestivum with haploid chromosome numbers 7, 14 and 21, respectively. The diploid einkorn wheat, with 2 sets of chromosomes, each made up of 7 chromosomes, was grown in the Near East as early as 10,000 to 15,000 B.C. These could be *Triticum boeoticum*/*T. aegilopides*. The genomes of einkorn are designated as *AA*. *A* genome or chromosome complement is a complete set (*n*) of chromosomes inherited as unit from one parent. When einkorn forms gametes the number of chromosomes in gametes is 7. This is the haploid (*n*) number. Haploid, the reader will recall, is a cell having gametic chromosome number. On the other hand, an organism with basic chromosome number is denoted by *x* (Figure 16.1). Thus monoploid of wheat shall have one genome of 7 chromosomes.

The diploid and tetraploid (with four sets of chromosomes or 4 genomes) wheats were domesticated almost simultaneously about 8000 B.C. *Triticum monococcum* (diploid) was developed from *T. boeoticum* by selection of types better suited for cultivation. *Triticum dicoccum* was developed from *T.dicoccoides* (Jain, 1989). These are tetraploid wheat species with genomic constitution designated as *AABB*. They possess 28 chromosomes in their somatic cells or 4 genomes of 7 chromosomes each. Their gametic or haploid chromosome number is (*n*) 14 or two genomes viz., *AB*. But monoploid (*x*) number is 7 chromosomes. Mc Fadden and Sears (1947) suggested, on cytological grounds, that the *B* genome (with 7 chromosomes) of emmer group of wheat was derived from *Agropyron triticeum*. However, Riley *et al.* (1958), and Sarkar and Stebbins (1956) postulated that the *B* genome of emmer wheat was derived from *Aegilops speltoides*. Such conclusions are derived from cytological investigations during meiosis of cells from F_1 of emmer wheat and *Aegilops* (or *Agropyron*). Seven chromosomes from einkorn and emmer will form bivalents while other seven chromosomes from *Aegilops* (or *Agropyron*) will be univalents, if *B* genome originated from *Aegilops* (or *Agropyron*).

Riley (1960) suggested possibility that the wild allotetraploid wheat, viz., *T.dicoccoides,* was formed earlier than domestication from a cross of *T. aegilopoides* with *Aegilops speltoides. T. dicoccum* probably arose as a Mendelian variation from *T. dicoccoides*.

Because the tetraploid or emmer wheats have 4 sets of chromosomes in their somatic cells, the gametes bear 14 chromosomes, i.e. half the number in somatic cells. This is the result of meiotic cell division of sporocytes. In other words, the *n* is equal to 14. Emmer is thus a polyploid with 4 sets of chromosomes against only 2 genomes in somatic cells of einkorn. The chromosomes of emmers could, therefore, be depicted in terms of x as $2x_1 + 2x_2$. Assuming, just for better understanding, had the chromosomes of *T. monococcum* ($2n = 2x = 14$) been doubled, an autotetraploid would have resulted with 28 ($4x$) chromosomes in somatic cells. That is $2n$ would have been equal to $4x$ equal to 28 chromosomes.

Genomes per cell		Gametes	
		Genomes	Chromosomes
Diploid $2n = 2x$	AA	A	7
Autotetraploid $2n = 4x$	AAAA	AA	14
Allotetraploid $2n = 2x_1 + 2x_2$	AABB	AB	14
Autotriploid $3x$	AAA		
Autohexaploid $2n = 6x$	AAAAAA	AAA	21
Allohexaploid $2n = 2x_1 + 2x_2 + 2x_3$	AABBCC	ABC	21
Monoploid n	A		

Fig. 16.1: Various ploidy levels

Although the haploid number (n) of this hypothetical autotetraploid would be fourteen chromosomes, the monoploid (x) number would be the basic number of chromosomes or the genomic number, i.e. seven only. Thus x is used to denote the monoploid number. Monoploid is a cell or an organism with basic chromosome number.

In actuality the haploid chromosome number(n) for *Triticum dicoccum* or *dicoccoids*, could be written as $x_1 + x_2$, representing number of chromosomes basic for *Triticum* and *Aegilops*, respectively.

Kihara (1966) proposed that common bread wheat, *Triticum aestivum*, is hexaploid, i.e. somatic cells have 6 genomes or chromosome complements of 7 chromosomes each. According to him common bread wheat, *T. aestivum*, arose as a result of hybridization between cultivated emmer wheat and *Aegilops squarrosa*, followed by doubling the number of chromosomes in F_1. The tetraploid wheat is thus expected to contribute genomes *AB* to the hexaploid common bread wheat and *Aegilops squarrosa* one genome, viz., *D*. After doubling the chromosome number of F_1, the common bread wheat shall have six genomes, viz., *AABBDD*, or 42 chromosomes. The *D* genome from *Aegilops squarrosa* is also composed of 7 chromosomes. The somatic chromosome number of *T. aestivum* is therefore, 42 and haploid or gametic number equal to 21. Once again assume, hypothetically, that this hexaploid arose as a result of multiplication of *AA* genome three times, i.e. somatic chromosome is 42, then although the gametic or haploid chromosome number is 21, the monoploid number would be $x = 7$. And because 42 is six times the monoploid number, polyploid is autohexaploid. Returning to actuality, *T. aestivum* has two genomes, *AB* from emmer (with genomic/monoploid number being described as $x_1 + x_2$) and *D* genome of 7 chromosomes from *Aegilops squarrosa*, the haploid constitution would be described as $x_1 + x_2 + x_3$, and the somatic composition as $2x_1 + 2x_2 + 2x_3$. For an autohexaploid (which is a polyploid having multiple and identical or nearly identical sets of chromosomes of monoploid number) the somatic number would be represented as $6x$, i.e. 6 genomes, with identical chromosomes or nearly identical chromosomes in each genome. As has been pointed out earlier in this section, genomes are usually distinguished by pairing relationships of chromosomes from each genome. Where the genomes differ slightly, they may be distinguished by the same letter with an added subscript, such as A and A_1. With this background about polyploids, haploid and monoploid number of chromosomes, we are now ready to understand two broad classes of polyploids, viz., autopolyploids and allopolyploids. There is another category called aneuploids. Aneuploid is an individual with other than an exact multiple of genomes or chromosome complements which are present in its normal somatic cells.

16.1 AUTOPOLYPLOIDS

A polyploid is an organism with other than two sets of chromosomes per somatic cell. When the cell or an organism carries unusual number of repetition of one or a few particular chromosomes in the genome, the cell or an organism is known as an aneuploid. On the other hand, a somatic cell or an organism containing unusual number of repetitions of chromosome sets (or genomes) is called as euploid. On the basis of differentiation of chromosomes, which make up the genomes, euploids are classified into two categories, viz., autoploid or autopolyploid and alloploid or allopolyploid. Autopolyploid is an organism the somatic cells of which contain more than two sets of chromosomes or genomes.

Each chromosome set is identical or closely similar to each other set. An allopolyploid is an organism whose somatic cells have more than two genomes but the chromosomes of multiple sets are differentiated from each other. Thus species with multiples of different genomes are alloploid or allopolyploid, whereas species with multiples of same genome are autoploid or autopolyploid. The distinction between auto and allopolyploid is not always sharp because hybridization and subsequent doubling of the chromosome number can involve species whose chromosome complements are in various stages of differentiation. The difference between genomes is of degree. Some genomes may exhibit perfect homology or they may fail to pair. Stebbins (1950) has preferred to include those cases as allopolyploids where the combined genomes are so different that their chromosomes fail to pair or only a small number of loosely paired bivalents may be formed. For those allopolyploids, which range between autopolyploids and allopolyploids with very different genomes, Stebbins (1950) used the term segmental alloploid, implying that only segments of chromosomes in the constituent genomes are different. He classified the polyploid status of some principle crops into eight categories (Table 16.1). Each of the eight types of polyploid in this table differs from the others in its cytogenetic characteristics. About 30 to 35 per cent of the angiosperm species are polyploid according to him. Among the grasses about 70 per cent are polyploid.

Table 16.1: Some crops and their polyploid category*

SN	*Type of polyploid*	*Name of crop*
1	Inter-varietal autopolyploid	*Dactylis glomerata* (orchard grass)
2	Autopolyploid varying towards segmental allopolyploid	*Medicago sativa (Alfalfa)*, *Solanum tuberosum* (potato)
3	Segmental allopolyploid	*Arachis hypogea* (Groundnut)
4	Auto-allopolyploid,	*Saccharum sp.* (Sugarcane)
5	Allopolyploid, but nullisomics and genomic substitutions tolerated	*Triticum sp.* (wheat)
6	Allopolyploid, whole chromosome deficiencies tolerated	*Nicotiana tabacum* (tobacco)
7	Strict allopolyploid, only segmental deficiencies tolerated	*Gossypium hirsutum*, *Gossypium barbadense* (cotton)
8	Diploid species but belonging to a genus or of possible ancient amphidiploid origin	*Oryza sativa* (Rice), *Zea mays* (maize), *Helianthus annus* (sunflower), *Lycopersicon esculentum* (tomato)

*Stebbins (1957).

A number of important agricultural crops are autopolyploid. Several forage grasses and ornamental plants are autopolyploids. Some of the varieties of apples, e.g. Baldwin, Gravenstein, Stayman are $3x = 51$ instead of $2x = 34$. Banana is an outstanding example of triploid crop. In this fruit crop maximum vigour is associated with triploid state. Diploid bananas have hard seeds and hence not acceptable to consumers. But although triploids are sterile, due to irregularity in meiosis, fruits are seedless and hence acceptable. This is a way of genetic deseeding. Because bananas are propagated vegetatively seedlessness is not an undesirable character.

Family *Musaceae* has two genera, viz., *Ensete* and *Musa*. *Musa* is divided into five sections. These sections are *Eumusa* ($x = 11$; $2n = 22$ or 33), *Rhodochlamys* ($x = 11$, $2n = 22$), *Callimusa* ($x = 10$; $2n = 20$), *Australimusa* ($x = 10$; $2n = 20$) and *Incertae sedis* ($x = 7$; $2n = 14$). *Eumusa* is edible banana. It is derived from *Musa acuminata* and *M.balbisiana*. A number of triploid varieties of banana found in India have been reported by Majumdar (1989). These are:

*Genomes**	*Varieties*
AAA	Amritsagar, Basrai, Giant Cavendish, Robusta, Chakrakel, Agniswar (Lalkel)
AAB	Poovan (Champa), Rasthali, Nendran, Rajapuri, Sirumalai
ABB	Monthan (Kanchkela), Nallabontha, Thenkunan, Peyan

*Each genome comprises of 11 chromosomes.

Peter (1998) quoted Nishiyama (1971) for assigning various genomes to *Ipomoea batatas* (sweet potato). Accordingly, *Ipomoea batatas* are autohexploid with genomic constitution *BBBBBB*, where each *B* genome comprises of 15 chromosomes. *I.batatas f. trifida* are also autohexaploid, while *I.batatas* var. *littoralis* is tetraploid (*BBBB*), *I. batatas* var. *leucantha* is diploid and somatic cells carry two genomes *BB*. Although *I. triloba* and some other species are also diploid, their genomic designation is *AA*, where each *A* genome carries 15 chromosomes. Thus sweet potatoes occupy probably highest order among known autopolyploids, i.e. autohexaploid ($2n = 90$). The monoploid number is, $x = 15$.

Triploid watermelons, sugar beets, and marigolds are also known. Triploid grapes were developed in Japan (Olmo, 1952). More information about production techniques for triploid watermelons and sugar beets shall be provided later on.

16.1.1 Effects of autopolyploidy on plants

Autopolyploidy leads to increased cell size, at least of the meristematic tissues. Stomata and pollen grains of autotetraploid plants produced through colchicine treatment have increased dimensions as compared to their diploid counterparts. Larger size of cells does not necessarily lead to overall larger plants. The flowers, fruits or some other parts of autopolyploid plants are larger than their diploid counterparts. While the 'gigas' properties are relatively common in autopolyploids, species may differ with respect to the vigour exhibited (increase in size of some plant organs is termed 'gigas' characteristic — Briggs and Knowles, 1967, p.268). The leaves of polyploids are generally thicker and larger in size than their diploid forms.

The ploidy level also influences chemical composition of plant parts. Vitamin A activity in tetraploid maize increased as compared to diploid (Randolf and Hand, 1940). Vitamin C in several vegetables and fruits also increased following chromosome doubling. Tetraploid tobacco had more nicotine in leaves.

Growth rate of polyploids is relatively slower than the diploids (Briggs and Knowles, 1967). The ecological requirements of autopolyploids are different as compared to the diploid. For example, triploid watermelons require higher temperatures for germination of seeds than the diploid seeds.

Self-incompatibility systems observed in diploid species are often weakened or may even disappear in polyploid. This is particularly true for gametophytic form of self-incompatibility.

Autopolyploids are generally reduced in fertility. In order to understand the basic reason for sterility observed in autotetraploids, let us recall our knowledge about the steps involved in meiotic cell division. There are two division phases in meiosis, equational division succeeds reductional division. In the first division, the homologous chromosomes associate in pairs, i.e. synapse. By pachytene the association of homologous chromosomes is complete. There are four homologous chromosomes in each cell of the autotetraploid plant. Hence, at zygotene and pachytene, association shall involve four homologous chromosomes, i.e. quadrivalent. Because during synapsis only 2 chromosomes are associated at one place, 4 chromosomes pair with different homologous chromosomes at different places along their length. All four homologous chromosomes are not associated at same place, along their length. By the end of pachytene each chromosome divides longitudinally and hence there are eight chromatids. Crossing over may involve different homologous chromosomes. Disjunction of quadrivalents at first anaphase is usually two by two, but numerical aberrations sometimes occur. Unequal disjunction of chromosomes in autotetraploids is the main reason for sterility.

When the four chromosomes of autotetraploid are not truly homologous, the 4 homologs manifest selective pairing (Allard, 1960). Differential affinity is noticed in such cases. Homology can vary from complete identity of chromosomes to very weak homology. Former could be noticed when homozygous diploid is doubled, while the latter is expected when F_1 hybrid between sharply differentiated species is doubled.

Although the description is out of place here, chromosome associations in allopolyploid may be mentioned for information of readers. Whenever the chromosomes of two genomes are similar, but not exactly homologous they are termed as homoeologous. The 21 chromosomes of hexaploid wheat (haploid number) have been allocated to homoeologous groups. Each homoeologous group contains a partially homologous chromosome from each of the three genomes, viz., *A*, *B* and *D* (Chapman and Riley, 1966). Chromosomes within a homoeologous group will carry many genes in common, even though the chromosomes originated within different genomes (Poehlman and Borthakur, 1969).

Inviability and sterility may make perpetuation of autoploids impossible in nature. Autopolyploidy combined with allopolyploidy produces 'auto-allopolyploid' and is an important process in evolution of plants (Gardner and Snustad, 1984).

Meiosis is not regular in triploid plants also. Synapsis during meiosis involves association of three homologous chromosomes. They are associated in pairs at different places (trivalent). Segregation of chromosomes is influenced adversely and hence possibility of getting gametes with either $2n$ or n chromosomes is remote. Gametes contain chromosome number ranging between n and $2n$. And these are sterile. Imbalance in chromosome number, probably leads to sterility of gametes.

Actually autopolyploids are not very frequent in nature. They probably did not play significant role in the evolution of plant species as compared to allopolyploids

16.1.2 Induction of polyploidy

Polyploids have been induced experimentally by several methods in various plants. Any thing that interferes with spindle formation during mitosis may result in doubled chromosome number in a cell. The interest in induction of polyploidy in the earlier years was because of several reasons. Most of the agriculturally important plant species are polyploids. Hence doubling of the chromosome number was expected to result in further valuable polyploids. Because the cells of polyploids are larger in size than their diploid counterparts, it was expected that large cell size could lead to larger plants and higher yields. Plant breeders were enthusiastic when techniques were known to double the chromosome number. They expected to revolutionize the plant breeding methods. However, induction of polyploidy has not led us far ahead as expected. Nevertheless, induction of polyploidy has helped to overcome the barriers caused by differences in chromosome numbers between two species in a bid to transfer some desirable character(s).

Autopolyploids may be induced spontaneously or could be induced artificially by doubling the chromosome number of plants with lower levels of ploidy. Spontaneously autopolypioids could arise through:

(i) Doubling of the number of chromosomes in zygote when homologous chromosomes fail to pair, and
(ii) Through fusion of the unreduced gametes.

In either case, if the parent is diploid ($2x = AA$), the resulting autotetraploid shall have genomic constitution $4x = AAAA$. One of the earliest methods to induce polyploidy involved decapitation of stem and subsequently generation of callus formation either with the help of Indole Acetic Acid or without it. Indole Acetic Acid is expected to stimulate re-growth of decapitated part. The callus has some polyploid cells.

The method of inducing polyploidy in plants that has become widely acceptable was developed by A.F. Blakeslee, A G Avery and D.R.Nebel (Gardner and Snustad, 1984). These investigators found that an alkaloid, colchicine extracted from the autumn crocus, *Colchicum autumnale*, could disturb the spindle formation during cell division. When root tips or other growing plant parts were treated with appropriate concentrations of colchicine in appropriate media, the chromosomes of treated cells replicated properly, but spindle formation was inhibited and cytoplasmic phase of cell division did not occur. Instead a 'restitution nucleus' was produced. A restitution nucleus is a nucleus with unreduced or doubled chromosome number that results from failure of mitotic or meiotic cell division. When these cells were propagated, tetraploid plants were produced and tetraploid seeds were obtained. Colchicine is applied to the meristematic regions of plant. There are many opportunities for chimeras to occur, where only portions of their growing tissues are doubled. In such chimeras, if the diploid tissue is aggressive during development, the doubled tissues may never develop to be visibly apparent. Polyploidy may be confined to only epidermis or to only internal tissues (Derman, 1952). In some plants growing area, such as the stem tips and lateral buds, have three distinct layers (Figure 16.2). Cells from each layer are much alike in early stages of development, but subsequently they give rise to separate tissues in stems, leaves, and other organs. The outer layer becomes epidermis (Figure 16.2a); the central layer gives rise to reproductive cells (Figure 16.2b); and the inner layer gives rise to internal parts of stems and leaves. Colchicine applied at appropriate concentrations in appropriate media to the meristematic cells results in transformation of $2n$ cells into $4n$ cells in one or more layers. By propagating tetraploid tissue, it is possible to produce tetraploid plants. When the central layer turns out to be autotetraploid ($4n$), the gametes, which will ultimately be formed will be $2n$ in genetic constitution. When these $2n$ gametes fertilize gametes with unreduced chromosome number a tetraploid shall result. On the other hand, when gamete with reduced chromosome number (n) fertilizes gamete with $2n$ chromosomes, triploid is formed. For seed and seedling treatment concentration of colchicine in aqueous solution can range from 0.1 to 0.4 per cent. The period for which treatment may persist ranges from 2 to 3 hours to 24 hours in case of vegetables (Kalloo, 1988). Colchicine is applied by wetting with aqueous solution, by spraying in an emulsion form or through lanolin paste. It is effective when seeds are germinating, or on young seedlings or to roots or to growing points such as shoots or buds. Concentration of solution, duration of treatment and temperature at which treatment is accomplished influence the result. Colcemid, a synthetic equivalent of colchicine, can also be used for inducing polyploidy (Poehlman, 1987).

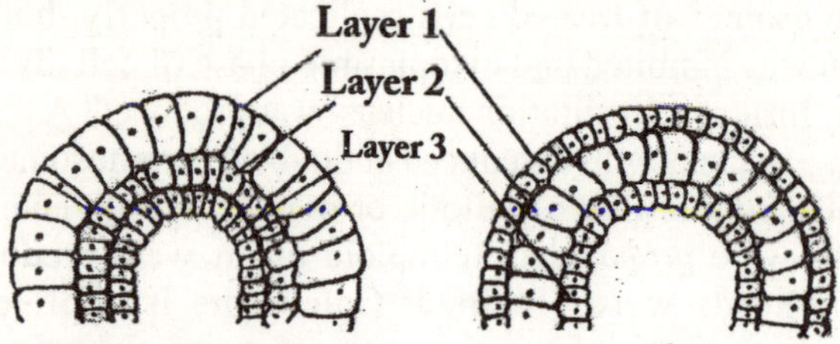

Fig. 16.2: Effect of colchicine on cells of stem meristem

Other chemicals like acenaphthane, chloral hydrate, chloroform, ether etc. have been used to induce polyploidy but these are not as effective as colchicine.

With the discovery that colchicine could be used to artificially produce polyploids, almost at will as it was then thought, production of such plants seemed to extend to the breeder unlimited new vistas of advancement in plant improvement. During those days hopes of immediate success were high because of widespread occurrence of polyploidy as a natural phenomenon. The expectations were reinforced by the observation that many artificially induced polyploids (autopolyploids) have several desirable characters, such as larger seed size and higher contents of various proteins, vitamins and sugars, both in seed and root crops. But undesirable characters were also observed in artificially produced autopolyploids, like fewer flowers and fruits per plant, slow growth, partial sterility.

According to Stebbins (1957) in no case a single autopolyploid variety of a crop out yielded economically valuable new variety, even after selection, and there is little possibility that this will prove true in future, except in case of forage crops.

16.1.3 Use of autopolyploidy in plant breeding

Survey of the literature reveals that autopolyploids are less frequent in nature as compared to allopolyploids. Nevertheless, since 1941 (after Blakeslee and others suggested use of colchicine to induce polyploidy) plant breeders engaged themselves with induction of autopolyploids in a bid to create larger plants with higher yield.

Initially, relatively few induced polyploids were successful as compared to their diploid parents. Following three factors should be taken into consideration as a guide for production and utilization of autopolyploids in plant improvement:

(i) Autopolyploid manifests greater vegetative growth but reduced seed production. This implies that autopolyploid induction would be more useful for vegetative parts of the plants, such as forage, or root, but not the seed.
(ii) Autopolyploids produced from diploids with lower chromosome number have been relatively more successful.
(iii) Performance of a diploid species in a given environment may not be a correct indication regarding the performance of corresponding autotetraploid produced from it. Cross-pollinated species may be more successful as parents for autopolyploid induction than self-pollinated species (Poehlman, 1987). The reason is obvious. Because of cross-pollination greater genetic variability will be generated by cross-pollinated plant species and consequently a balanced polyploid genotype may emerge.

Successful autopolyploid programmes were developed in few crops. A few examples are presented below:

Triploid Sugar Beets

As mentioned earlier polyploidy leads to enhanced vegetative growth. Hence this attribute may be useful for root crops. The roots of triploid sugar beet are larger in size than diploids, but they maintain the sugar content of diploids and hence yield more sugar per unit area. Triploids are highly sterile and do not produce seed. Triploid sugar beets are produced by interplanting diploid and tetraploid. Tetraploid is used as seed parent and triploid seed is therefore harvested from tetraploid. Autotetraploid beets have smaller roots as compared to diploids.

Triploids can be produced when tetraploid (individuals with 4 genomes in somatic cells) is crossed with diploid or when something goes wrong in meiosis and unreduced gametes are produced (with $2n$ chromosomes) which unite with gametes carrying n chromosomes (haploid chromosome number). During the zygotene and pachytene stages of meiotic cell division three homologous chromosomes (trivalent) synapse but in such a way that at a given position only two homologues are together. At anaphase I, two homologues may go to one pole and one to another. But considering all the chromosomes within a cell, the 4 meiotic products may contain from *n* to 2*n* chromosomes, with all integral values between them. Gametes containing either *n* or 2*n* chromosomes may be functional, but the gametes that are deficient for either of the chromosomes or have additional chromosomes are sterile.

Triploid Watermelons

Seedlessness would be advantageous in watermelons. Diploid watermelons (*Citrullus lanatus*) have 22 chromosomes per somatic cell and are fully fertile, and produce large number of seeds per fruit. Natural parthenogenesis is not known for watermelons. Hence Kihara's (1951) technique of producing seedless, triploid watermelons can be utilized. The autotetraploid lines (produced by doubling the chromosome number of a diploid) are planted alternately with diploids in isolation. Tetraploids are employed as seed parent. The triploid seed produced on tetraploids is viable. However, when diploid is used as female, the programme of triploid seed production is unsuccessful. Because of meiotic irregularities, triploids produce no true seeds but rudimentary structures similar to cucumber seeds (small and white) are formed in triploid fruits. For raising a successful commercial crop of seedless triploid watermelon, it is essential to interplant diploid variety because fruit setting on triploids depends upon stimulus provided by the pollen. Main difficulty with triploid watermelon seed production is maintenance of tetraploid seed parent. Moreover, the triploid seedless fruits are irregular in shape and tend to be hollow. Germination of triploid seed is difficult. Production of triploid seed requires hand pollination of tetraploid flowers and hence cost of triploid seed is prohibitive.

Because the gametes produced by triploid plant are largely unbalanced with respect to chromosome number, triploids are inviable. If, however, triploid plants could be vegetatively propagated, viz., by cuttings, they could be advantageous. European pears are propagated via grafting and budding and hence could be maintained in triploid state. Tulips can also be propagated in triploid form through vegetative bulbs.

Autopolyploids of snapdragons (*Antirrhinum*), marigolds (*Tagetes*) and other garden flowers have been successful but in these also polyploidy accompanied inter-varietal hybridization and selection, before it was successful (Stebbins, 1957).

Singh (1993) reported that a triploid (3*x*) clone of tea (*Camelia assamica*) has been recently released for cultivation in India. Cultivar TV 29 has larger shoots and hence produces greater biomass and yields more cured leaves per plant.

Pusa Giant Barseem, an autotetraploid cultivar, was evolved at the Indian Agricultural Research Institute, New Delhi. It yields about 20-30 per cent more fodder than its diploid counterpart.

In addition to induction of autopolyploidy for improvement in yield, the autopolyploids could be hybridized to generate more genetic variability.

16.1.4 Monoploids

A monoploid plant has one genome characteristic of a species. When monoploid plants are viable, they are smaller in size and less vigorous than their diploid counterparts. Because chromosomes do not have homologous partners, the chromosomes go to any pole irregularly at cell division. Consequently gametes with less than *n* chromosomes are usually produced. Such deficient gametes are non-functional.

Monoploids of diploid plant species possess only one genome per somatic cell. But the haploid plant, i.e. plants with gametic chromosome number, of a polyploid shall carry more than one genome per somatic cell against only one found in monoploid/haploid plant of a diploid. Hence haploid plants of polyploids are called polyhaploids.

Naturally occurring haploid plants are reported for many crops such as tobacco, maize, sorghum, cotton etc.

Monoploids are of interest to plant breeders because doubling the chromosome number of monoploid leads to formation of diploid homozygous at all the loci. The method of developing diploid homozygous line was suggested for maize as early as 1943 by Einset. Chase (1952) described the technique used for isolation of monoploids. Because they develop parthenogenetically (egg cell may develop without fertilization to give rise to a haploid plant) i.e. are expected to be maternal in origin, their recognition is facilitated if the parents are genetically distinct and male carries a dominant marker gene having expression in seed or seedling stage. Markers used may be like purple plant colour genes or purple plumule or scutellum marker of maize, etc. Because of disturbed meiosis, a monoploid plant is usually sterile. Provided all the chromosomes in somatic cell of monoploid migrate to one pole at first meiotic division, haploid gametes are produced. Union of such gametes will give rise to a plant homozygous for all its genes. The chromosome number of monoploid could be doubled with the use of colchicine. This method saves time required to produce homozygous lines. Inbred line development is time consuming.

16.2 ALLOPOLYPLOIDS

Allopolyploids have in their somatic cells more than two different genomes. The chromosomes in the different sets are differentiated from each other. In order to explain this aspect let us take an example of allotetraploid cotton species. Among the several old worlds cotton species, i.e., those, which belong to Asia, Africa and Australia, the somatic chromosome number (i.e. $2n = 2x$) is 26. The somatic chromosome number of some of the new world diploid species is also 26. But the chromosomes of old world species are relatively longer in size than those observed in diploid new world species. The three new world cotton species, viz., *Gossypium hirsutum, G. barbadense* and *G. tomentosum* are tetraploid (i.e. $2n = 4x = 52$) and cytological studies reveal that 26 chromosomes are longer in size and rest 26 shorter. Skovsted (1937) proposed that the new world tetraploid cotton species are allotetraploid which probably originated from hybridization of old world diploid cotton species with new world diploid cotton species, followed by doubling the chromosome number of F_1 between these species. Harland (1940) experimentally demonstrated the probable origin of new world allotetraploid cotton by crossing *Gossypium arboreum*, an Asiatic diploid with genomic constitution designated as *AA*, with *G. thurberi* an American wild diploid species with genomic constitution *DD*, followed by doubling the chromosome number of F_1 hybrid with colchicine. When this amphiodiploid (an amphidiploid is an allotetraploid which carries in each of its somatic cells diploid chromosomes from two different species) was crossed with new world allotetraploid, partially fertile hybrid was produced. There is high degree of homology among cotton species from same geographic area and with identical chromosome number. There is a suggestion that allotetraploid new world cotton species evolved from hybridization of *Gossypium herbaceum*, var *Africanum*, another Asiatic cultivated diploid species with genomic constitution *AA*, with *G. raimondii* a wild new world diploid with genomes *DD*, followed by doubling the chromosomes of F_1 and natural selection (Koutu and Mandloi, 2002). *Gossypium arboreum* and *G. raimondii* are the progenitors of some new world tetraploid cottons, according to Stephens (1947).

If the two genomes, contributed by the two progenitor parents to the amphidiploid, are highly differentiated with respect to chromosomes (i.e. genic contents or organization), the chromosomes from two genomes will fail to associate during zygotene. Consequently it will be sterile. However, after doubling of the chromosome number each chromosome will have a pairing mate, and meiosis shall be normal. The cell undergoing meiosis shall show regular bivalent associations between homologous chromosomes (resulting from doubling of chromosome number) and disjoin regularly at first as well second anaphase. Such amphidiploids are fertile. But chromosome homology alone may not be responsible for sterility manifested by some allopolyploids. Some amphidiploids are sterile despite regular cytological behaviour of chromosomes from different genomes while others are fertile with irregular pairing (Allard, 1960). The sterility of polyploids may be physiological in nature and associated with genetically governed imbalance.

Potato is said to be an autotetraploid. Potato is native of South America. Many wild species still occur in South America, specifically in the Andes of Peru and Bolivia. Two contrasting forms were observed, one in Chile and the other in the Andean region of Peru and Bolivia (Peter, 1998). The potato belonging to Chile is botanically *Solanum tuberosum* var. *tuberosum* while that found in Andean region of Peru and Bolivia is *Solanum andigenum* or *Solanum tuberosum* var. *andigenium*. The two species are crossable. *Solanum tuberosum* has in its somatic cells $2n = 24, 36, 48$ chromosomes (Peter, 1998). Basic number (x) is 12. The *S.andigenum* is an autotetraploid ($2n = 4x = 48$). It is believed that *Solanum tuberosum* is a straight autotetraploid of diploid species *Solanum stenotomum* ($2n = 24$) but there are evidences that *S. tuberosum* has an alloletraploid origin. Hybridization between *S.stenotomum* and *S. sparsipilum* followed by doubling of the chromosome number led to origin of *S.tuberosum* (Ram, 1998; Peter, 1998). *Arachis hypogea* is reported to be segmental amphidiploid with $2n = 4x = 40$ chromosomes. (Nigam *et al.*,1990), its basic chromosome number (x) being 10. Groundnut behaves like a diploid during meiosis. Stebbins (1957) assigned groundnut to the segmental allotetraploid category while potato and alfalfa as autopolyploid varying towards segmental allopolyploid.

A Russian cytologist, G.D. Karpechenko (1927) artificially synthesized an amphidiploid from hybridization of two vegetables, belonging to different genera but with same chromosome number ($n = 9$). He crossed radish (*Raphanus sativus*) and cabbage (*Brassica oleracea*), although distantly related. The F_1 hybrid had 9 chromosomes from *Raphanus* and 9 from *Brassica.* Since the chromosomes were not homologous, the F_1 hybrid was sterile. However, some unreduced gametes were formed and in F_2 population Karpechenko was able to recover some tetraploid plants. The fertile plant was called *Raphanobrassica*, which had 18 chromosomes of *Raphanus* and 18 of *Brassica.* This experiment has significance because it showed that a fertile inter-specific hybrid could be produced artificially. Further, it also indicated the possibility of incorporating desirable gene (*s*) from different species. Unfortunately *Raphanobrassica* had the foliage of radish and root of cabbage, and not as expected, i.e. root of radish and foliage of cabbage.

Johnson grass (*Sorghum halepense*) was derived from *Sorghum bicolor* ($n = 10$) and *S. virgatum* ($n = 10$), followed by doubling the chromosome number of F_1 (Bhatti *et al.*, 1960). However, Doggett (1970) is of the opinion that the origin suggested by Bhatti *et al.* (1960) is less probable. Instead Doggett (1970) suggested that *S.halepense* arose from crossing of a sorghum of the *Arundinacea* group and one closely related to *S.propinquum*, followed by chromosome doubling.

Several species of *Brassica* have amphidiploid origin as shown by U (1935). For example, amphidiploid of *Brassica nigra* ($n = 8$, *B* genome) and *B. oleracea* ($n = 9$, *C* genome) led to origin of *B.carinat*a with *BBCC* genomic constitution and $2n = 34$ chromosomes. *B.oleracea* and *B.campestris* ($n = 10$, *A* genome) hybridized to produce an amphidiploid *B.napus* ($n = 19$ and genomes *AACC*). Amphidiploid of *B.nigra* and *B.campestris* resulted in evolution *B.juncea* ($n = 18$, *AABB* genomes).

The examples described above suggest very strongly that allopolyploidy rather than autopolyploidy has played significant role in the origin of many crop species.

16.2.1 Allopolyploids in Plant Improvement

In contrast to the autopolyploids, allopolyploids have not only played significant role in the origin of new crop species but also helped to understand the probable origin of various crop species and to transfer certain characters to cultivated crop species where ploidy level was a barrier.

Experiments by Harland (1940) with cotton, Goodspeed and Clausen (1928) with tobacco, U (1935) with Brassica, McFadden and Sears (1947), Sarkar and Stebbins (1956) and Kihara (1966) with wheat and several others helped to understand role of allopolyploidy in the evolution of new crop species which are presently being grown on large areas. These experiments provided an inspiration to the breeders to use allopolyploidy as bridging species to incorporate some desirable attribute from wild species or other species into cultivated species, which differ in ploidy level.

The transfer of a desirable gene from one species to another across the barrier imposed by the sterility in inter-specific hybrid is more difficult. This is even more difficult when the barrier is due to differences in ploidy level between the two species under consideration. However, induction of polyploidy following hybridization for creation of a bridging species comes to the rescue of breeder. One of the best known examples is transfer of gene for resistance to mosaic disease from *Nicotiana glutinosa* ($2n = 24$) to *Nicotiana tabacum* (Valleau, 1952). *Nicotiana tomentosa* ($2n = 24$) and *N.sylvestris* ($2n = 24$) are thought to be the progenitors of *N.tabacum*. Cultivated tobacco is susceptible to tobacco mosaic virus. *N.glutinosa* carries resistance to mosaic. But because of differences in chromosome number *N.tabacum* and *N.glutinosa* yield sterile F_1. An amphidiploid of *N.tabacum* × *N.glutinosa* was developed known as *N.digluta*. F_1 between *N.tabacum* and *N.digluta* is reasonably fertile because it possesses a complete set of chromosomes from *N.tabacum* and haploid set of chromosomes from *N.glutinosa*. In later generations, the chromosome from *N.glutinosa* carrying resistance to mosaic could be isolated by selection. A single chromosome carrying resistance could be substituted for a chromosome in *N.tabacum*. The process by which a non-homologous chromosome is substituted into the genome of another species is called as '*alien chromosome substitution*'. In this same way resistance to fire disease was transferred to *N.tabacum* from *N.longiflora* (Stebbins, 1957). Such transfers of desirable genes from one ploidy level to another level, involving different species, are known for various crops.

Jambhale and Nerkar (1986) developed a Yellow Vein Mosaic resistant cultivar, Parbhani Kranti, from an inter-specific cross of *A.esculentus* cv. Pusa Sawani with *A.manihot*, an African species. Jambhale and Nerkar (1982) synthesized an amphidiploid from a cross *Abelmoschus esculentus* ($2n = 130$) × *A.manihot* ssp. *manihot* ($2n = 194$), by using colchicine treatment. The amphidiploid had

$2n = 324$ chromosomes and formed 162 bivalents. As against 7 per cent fertility of F_1, amphidiploid showed 82 per cent fertility. Backcrossing to Pusa Sawani and selfing resulted in Parbhani Kranti. Punjab Padmini, another variety of okra, resistant to Yellow Vein Mosaic was bred from cross *A.esculentus* × *A.manihot* spp. *manihot*. Some potato cultivars (*Solanum tuberosum*) are susceptible to late blight (C.O. *Phytophthora infestans*). *Solanum demissum* ($2n = 72$) is resistant to this disease. *S.tuberosum* carries 48 chromosomes in somatic cells. F_1 between these species is sterile. Only univalents are observed during meiosis. However, a bridging species viz., *S.rybinii* was crossed with *S.demissum* and the F_1 of this inter-specific cross was used as seed parent to hybridize it with *S.tuberosum*. Fertile progenies resulted (Peter, 1998). *S.rybinii* Juz. Et.Buk is a wild semi-cultivated species belonging to group *Phureja* and subgroup *Amarilla*. It has $2n = 24$ chromosomes. Similarly when *S.acaule* was crossed with *S.tuberosum*, the F_1 was sterile. But bridging species is developed by crossing *S.acaule* with *S.simplicifolium* and this F_1 is crossed with *S.tuberosum*. Fertile progeny is obtained. *S. acaule* is a tetraploid ($2n = 48$) and *S. simplicifolium* is a diploid ($2n = 24$). If the resistant wild species has a lower chromosome number than the predominant popular cultivated variety, transfer of desirable gene can be accomplished by doubling the chromosome number of wild species. An example of such transfer is the cultivated strawberry and *Fragaria vesca* (Darrow, 1950 and Scott, 1951). Because triploids are sterile, blooming in case of triploid plants continues for longer period. Commercial hybrids of marigold (*Tagetes* sp.), which have in their somatic cells $3x = 36$ chromosomes, are produced by crossing diploid female, *Tagetes erecta* ($2n = 24$) with tetraploid female, *T.patula* ($2n = 48$). The triploid hybrid holds flowers on plants for a relatively longer period (Swarup, 1995).

Man made amphidiploid cereal *Triticale* has been developed. This involved *Triticum* sp. and *Secale cereale*. The somatic chromosome number of *Secale cereale* is 14. Tetraploid and hexaploid wheat species were employed in the development of triticale. The objective in the synthesis of this new cereal was to combine desirable characteristics of two species, viz., high productivity, wide adaptability, good plant type and grain characteristics of wheat and high sink capacity, tolerance to stress conditions, disease resistance and superior nutritional quality of rye (Sandha and Upadhyay, 1986). Systematic breeding work was launched in different countries in sixties. More than 55 triticale cultivars have been released in different countries (CIMMYT, 1984). Primary octaploid and hexaploid triticales were synthesized by crossing agronomically desirable, dwarf and disease resistant varieties of wheat with quality grains (both *aestivum* and *durum*) with desirable rye types. Embryo culture technique was followed in *durum* × rye crosses to obtain F_1 hybrid seedlings. To this cross, *durum* contributed two genomes *AB* with 14 chromosomes and one genome *E*, was contributed by *secale cereale*. Two genes located on chromosomes 5*B* and 5*A*, viz., kr_1 and kr_2, of wheat, facilitate crossing between wheat and rye. Dominant genes reduce crossabilty. Different *aestivums* and *durums* differed with respect to crossabiltiy with rye. C 306 exhibited highest crossabiltiy. Shriveled seed is one of the basic defects in triticale. Triticales evaluated in hills and plains of India showed better performance in hills (Sandha and Upadhyay, 1986).

Raphanobrassica was also attempted but could not succeed in efforts.

The somatic chromosome numbers reported for some of the crop species are presented in table 16.2. This will show how wide spread is polyploidy in these plant species.

16.3 ANEUPLOID

Aneuploidy refers to an individual, which carries chromosomes in its somatic cells, which are not exact multiples of genomes it contains. For example, the somatic chromosome number of tomato (*Lycopersicon esculentum*) is 24. The haploid and also the basic number (x) is 12. We have seen in sections 16.1 and 16.2 that both autopolyploids and allopolyploids possess in their nuclei exact multiples of the genomes. For example, einkorn wheats have 2 genomes per cell, emmers 4 and *aestivums* six genomes per somatic cell. *Ipomoea batatas*, which are autohexaploid have six *A* genomes per somatic cell. However, if an organism has lesser or more chromsome(s) in their somatic cells than the exact multiple of genomic number expected, it is an aneuploid (Figure 16.3).

During meiotic cell division of tomato, for example, the 24 chromosomes form 12 bivalents at zygotene. At metaphase I the 12 bivalents (each chromosome is now longitudinally divided, thus each chromosome has two chromatids attached to same centromere) align themselves on equatorial plate in the cells. Sometimes the two chromosomes of a bivalent fail to separate (non-disjunction) and go to same pole. Suppose it happens in case of only one chromosome bivalent and rest i.e. 11 bivalents, behave normally. As a consequence, two kinds of gametes are formed, viz., half with $n-1$ chromosomes and other half with $n+1$ chromosomes. In other words, about half will be with 11 and other half will carry 13 chromosomes. When $(n-1)$ gametes fertilize gametes with n chromosomes, the genetic constitution of tomato plant will be $2n-1$, i.e. 23 chromosomes. One chromosome less than the usual 2 basic chromosome number. $2n-1$ individuals are called monosomic. Because tomato has 12 different chromosome pairs per cell, there can be 12 monosomics. For instance, in one trisomic, chromosome 1 may be missing in another chromosome 2 may be lacking and so on. Morphologically each monosomic will be distinct not only from its own diploid parent but also amongst different monosomics for different chromosomes.

Table 16.2: Somatic chromosome numbers ($2n$) in some plant species

SN	*Crop species*	*2n*	*SN*	*Crop species*	*2n*
1	Abelmoschus esculentus	64,72, 130	46	Linum usitatissimum	30
2	Abelmoschus manihot	60,68	47	Luffa acutangula	26
3	Abelmoschus moschatus	72	48	Lycopersicon esculentum	24
4	Allium cepa	16	49	Lycopersicon pimpinellifolium	24
5	Allium sativa	16	50	Lycopersicon peruvianum	24
6	Amarathus tricolour	34	51	Lycopersicon hirsutum	24
7	Annona squamosa	56	52	Malus malus ($3x = 51$)	34,51,
8	Arachis hypogaea	40	53	Mangifera indica	40
9	Avena sativa	42	54	Momordica charantia	22
10	Beta vulgaris	18	55	Musa sapientum ($3x = 33$)	22
11	Brassica campestris	20	56	Nicotiana tabacum	48
12	Brassica carinata	34	57	Oryza sativa	24
13	Brassica napus	38	58	Pennisetum americanum	14
14	Brassica nigra	16	59	Pisum sativum	14
15	Brassica juncea	36	60	Prunus americana	16
16	Brassica oleracea var. botrytis	18	61	Prunus avium	16
17	Brassica oleracea var capitata	18	62	Prunus persica	16

(Contd.)

(Contd.)

SN	*Crop species*	*2n*	*SN*	*Crop species*	*2n*
18	Capsicum annum	24	63	Psidium guajava ($3x = 33$)	22
19	Capsicum frutescence	24	64	Pyrus communis ($3x = 33$)	34
20	Cajanus cajan	22	65	Raphanus sativus	18
21	Cicer arietinum	14,16	66	Ricinus communis	20
22	Carthemus tinctorius	24	67	Saccharum officinarum	40
23	Citrus grandis	18	68	Saccharum barberi	41,45, 46,58, 62
24	Citrus limonia	18	69	Saccharum sinense	58,59
25	Citrus sinensis	18,36	70	Secale cereale	14
26	Citrus reticulata	18	71	Sesamum indicum	26
27	Citrullus lanatus	22	72	Seteria italica	18
28	Coccinia cordifolia	24	73	Solanum melongena	24
29	Cucumis sativus	14	74	Solanum tuberosum	48
30	Cucumis melo	24	75	Sorghum versicolor	10
31	Cucurbita moschata	40,48	76	Sorghum bicolor	20
32	Coriendrum sativum	22	77	Sorghum halepense	40
33	Cyamopsis tetragonolaba	14	78	Trichosanthes angina	22
34	Daucus carota	18	79	Trichosanthes dioica	22
35	Frageria grandiflora	56	80	Triticum monococcum	14
36	Glycine max	40	81	Triticum dicoccum	28
37	Glycine soja	40	82	Triticum durum	28
38	Gossypium arboreum	26	83	Triticum aestivum	42
39	Gossypium barbadense	52	84	Triticum spelta	42
40	Gossypium herbaceum	26	85	Triticum sphacrococcum	42
41	Gossyplum hirsutum	52	86	Triticum timopheevi	28
42	Guizotia abyssinica	30	87	Vigna mungo	22,24
43	Hordeum vulgare	24	88	Vigna radiata	22,24
44	Ipomoea batatas	90	89	Vitis sp.	38,40, 76
45	Laginaria siceraria	22	90	Zea mays	20

When the gametes carrying ($n + 1$) chromosomes fertilize gametes with n chromosomes, the tomato plant shall be $2n + 1$, i.e. one chromosome is present thrice rather than twice. Such individuals with 2 complete genomes but 1 chromosome present in excess are called as trisomic. Like the 12 monosomics, there could be 12 trisomics. If non- disjunction takes place for another chromosome also, trisomy will be observed for 2 different chromosomes and such a trisomic will be called double trisomic. Likewise when an individual lacks two different chromosomes as compared to its basic chromosome number or from the two genomes it carries, individual shall be called double monosomic (Figure 16.3). When one pair of chromosome is missing, i.e. $2n - 2$, individual is known as nullisomic. However, if one chromosome is present four times, rather than usual two, organism is called as tetrasomic. Because the chromosome is either missing or is present in excess than the expected genomic number per cell, the phenotype of monosomics, nullisomics are different than their respective diploid counterparts. Aneuploids are generally less vigorous as compared to normal plants. This may be due to physiological disturbances caused by genic unbalance. In monosomic all genes present on one chromosome are absent. Because in nullisomics a chromosome pair is missing, all genes located on the chromosome are absent. It has adverse effect on nullisomics. The meiotic cell division in aneuploids is disturbed and hence they arc partly or highly sterile.

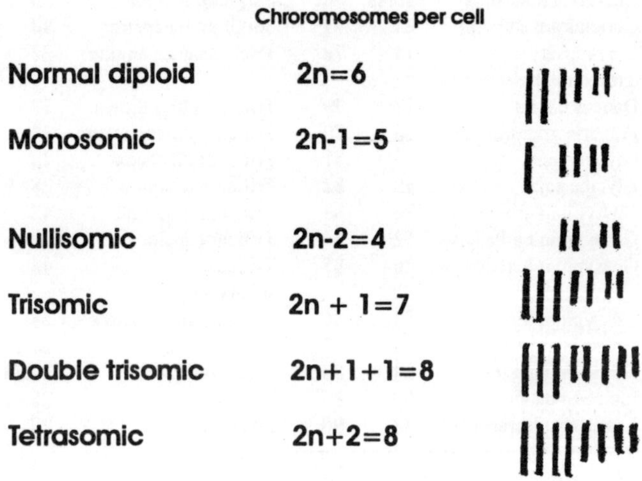

Fig. 16.3: Various aneuploids

Monosomics and nullisomics are functional in some polyploid species that carry several genomes. For example, they are tolerated in hexaploid wheat ($2n = 6x = 42$) bccause homoeologous chromosomes from different genomes may carry genes whose function is similar to genes on those chromosomes involved in monosomics or nullisomics. Hexaploid plants have greater buffering effect against loss of a chromosome pair due to presence of homoeologous chromosomes in other genomes.

In wheat, if nullisomics are selfed, they breed true for the basic chromosome number (Hays *et al.*, 1955). Some nullisomics are male-sterile or female-sterile. Nullisomic crossed to normal plant yields monosomic progeny.

Monosomics produce preponderance of gametes with $(n - 1)$ chromosomes, because univalent chromosome is frequently lost during meiosis. As compared to pollen grains with n chromosomes, pollen grains with $(n-1)$ chromosomes are seldom functional. Theoretically monoploid would produce gametes (n) and $(n-1)$ in equal proportions. Selfing the monosomic plant can be depicted as shown below:

		Male gametes		
Frequency of gametes		0.5 (n)	0.5 $(n - 1)$	Chromosomes
Female gametes	(n) 0.5	$2n$ 0.25	$(2n - 1)$ 0.25	Frequency of zygotes
	$(n - 1)$ 0.5	$(2n - 1)$ 0.25	$(2n - 2)$ 0.25	

Thus about 25 per cent plants would be disomic, 25 per cent nullisomic and 50 per cent monosomic. But in wheat 24 per cent normal, 73 per cent monosomic and only 3 per cent nullisomics are observed following selfing of monosomics (Hays *et al.*, 1955). On the other hand, monosomic wheat crossed to normal one produces about 75 per cent monosomics and 24 per cent normal plants.

Primary trisomics have two complete genomes, normally observed in diploids, plus an additional chromosome. Since it may be any chromosome, there could be n trisomics. For example because a diploid tomato plant carries 12 pairs of chromosomes there could be 12 different trisomics. Trisomics can be tolerated by non-polyploid plants also, where monosomics and nullisomics are not viable. Trisomics have been utilized to associate genes with particular chromosomes in barley, maize, sorghum, tomatoes and some other crops.

Aneuploids of various kinds arise due to several reasons. They arise spontaneously, but at low frequency. Triploid plants are another source of aneuploids. The chromosome numbers in microspores and megaspores will vary from x to $2x$, with most of the gametes with intermediate chromosome number being inviable. Viable microspores and megaspores usually have one or two complete chromosome complements. Some times gametes with $x + 1$ and $2x - 1$ or $2x - 2$ and $x + 2$ chromosomes may be viable. A gene, such as asynaptic gene, '*as*' on first chromosome of maize (Hays *et al.*, 1955), may be responsible for lack of association of homologous chromosomes during first meiotic cell division. Due of failure of synapsis the homologous chromosomes of a diploid may go to any pole and may produce gametes with unreduced chromosome number.

16.3.1 Use of Aneuploids in Plant Improvement

Aneuploids are not directly of any use in agriculture. However, certain kinds of aneuploids can be used for genetic studies and transfer of some desirable character(s) to varieties to be improved.

Nullisomics and monosomics have been used in crops like wheat and tobacco to detect location of genes on chromosomes. Detection of presence of a gene, under consideration, is made possible by the aberrant ratio for segregation in F_2, for the character. Alternatively location of a gene on a chromosome can be inferred from absence of a character associated with missing of a pair of chromosomes. For example, purple colour of aleurone is determined by *Pr* gene on chromosome 5 of maize. If plants nullisomic for chromosome 5 show other phenotype, say red colour, we can conclude that gene essential for purple aleurone is present on chromosome 5.

For studies to associate a gene with certain chromosome, nullisomics and monosomics are used. It is essential that either nullisomics or monosomics have been developed for all the chromosomes of the crop. Twentyone nullisomics have been developed in variety Chinese spring of wheat.

Detection of gene with the use of monosomic will be explained for wheat. Nullisomics are not used because they are difficult to handle due to their reduced vigour and poor fertility. Suppose a recessive gene located on an unknown chromosome of wheat is to be associated with a chromosome, which hosts it. Assume that the 21 monosomic lines carry dominant allele of the gene under consideration. Cross the line under consideration to all the 21 monosomic lines, individually. All the monosomic plants detected in F_1 will manifest the dominant phenotype, except one, which will exhibit recessive phenotype. As shown below about 50% progeny of monosomic × normal is expected

Monosomic × normal plant

		Gametes of normal		
		n	n	Chromosome number
Gametes of monosomic	n	$2n$	$2n$	
	$n-1$	$2n-1$	$2n-1$	

to be hexaploid (normal) and 50 per cent monosomic. But in reality about 24 per cent plants are normal and 73 per cent monosomic (Hays *et al.*, 1955). Only one critical monosomic line will manifest recessive trait, because the dominant allele is absent. The chromosome of the critical monosomic line is the carrier of recessive gene. The explanation is simple. The chromosome bearing recessive gene in the critical monosomic line is contributed by normal line. In normal sibs of this cross recessive characters will not be observed because the chromosome from monosomic carries dominant allele. If the gene to be located from normal line is dominant then recourse is to raise F_2, because all the 21 F_1s shall show the dominant character. The monosomic lines should carry recessive allele of the dominant gene under consideration. Cross the normal line to all the monosomics. The F_1 will be dominant in phenotype. Isolate all the monosomic F_1s and raise F_2 progeny of each line separately. All lines will segregate as per expectation, i.e. 3 dominant to 1 recessive, but the critical line does not segregate, because its F_1 parent does not carry homologue of the chromosome on which dominant gene under study is situated.

Trisomics can also be used to detect linkage of gene with certain chromosome. A line carrying gene under consideration, say dominant gene, is crossed to all the trisomics. F_2 generation of each cross is grown and studied. All F_2s will segregate in Mendelian fashion as expected, but the critical line shall deviate from expected segregation.

Nullisomic and monosomic lines can be employed for substitution of an entire chromosome with a desirable gene into the genome of a variety to be improved.

References

Allard, R.W. (1960). *Principles of plant breeding.* John Wiley and Sons, Inc., New York.

Bhatt, A. G.; J.E. Endrizzi and R.G. Reeves (1960). Origin of Johnson grass. *J.Hered*, 51: 107.

Briggs, F.N. and P.F. Knowles (1967). *Introduction to plant breeding*. Reinhold Publishing Corporation, New York.

Chapman, V. and R. Riley (1966). The allocation of chromosomes of *Triticum aestivum* to the A and B genomes and evidence on genome structure. *Canadian Journal of Genetics and Cytology*, 8: 57-63

Chase, S.S. (1952). Production of homozygous diploids of maize from monoploids. *Agron.J.*, 44: 263-267.

CIMMYT (1984). Annual Report, pp.64.

Darrow, G. (1950). Polyploidy and fruit improvement. *Sci. monthly*, 70: 211-219.

Derman, H. (1952). Polyploidy in apple. *J.Hered.*, 43: 7-8.

Doggett, H. (1970). *Sorghum*, p. 24 Publ.Longmans.

Einset, J. (1943). Chromosome length in relation to transmission frequency of maize trisomes. *Genetics*, 28: 349-364.

Gardner, E.J. and D.P. Snustad (1984). *Principles of genetics*, 2nd Ed. John Wiley and Sons, NewYork, pp. 461-489.

Goodspeed, T.H. and R.E. Clausen (1928). Interspecific hybridization in *Nicotiana* VIII. The *sylvrestris-tomentosa-tabacum* triangle and its bearing on origin of *Tabacum*. *Uni. of Calif.Pub. in Bot.*, 11: 245-256.

Harland, S.C. (1940). New polyploids in cotton by use of colchicine. *Tropical Agriculture*, 17: 53-54.

Hays, H.K.; F.R. Immer and D.C. Smith (1955). *Methods of plant breeding*. Mc Graw Hill Book Company, Inc., NewYork, pp. 39-40.

Jain, K.B.L. (1989). Cereals: wheat and barley. In *Plant breeding — theory and practice*. Ed. V.L. Chopra, Oxford and IBH Publishing Co.Pvt.Ltd., New Delhi, pp. 77-99.

Jambhale, N.D. and Y.S. Nerkar (1982). Induced amphidiploidy in cross *Abelmoschus esculentus* (L.). Moench × *A.manihot* (L.) spp. manihot. *Genetica Agri.*, 36: 19-22.

Jambhale, N.D. and Y.S. Nerkar (1983). Induction of polyploidy in A. spp. hybrids. *Progressive Horticulture*, 15: (1/2): 157-158.

Jambhale, N.D. and Y.S. Nerkar (1983). Interspecific transfer of resistance of yellow vein mosaic disease in okra. *Journal of Maharashtra Agricultural University*, 8: 197.

Jambhale, N.D. and Y.S. Nerkar (1986). Parbhani Kranti, a yellow vein mosaic resistant okra. *Horticultural Science*, 21(6): 1470-1471.

Kalloo, G. (1988). Polyploidy in breeding vegetable crops. *Vegetable Breeding, Vol.I.* C.R.C. Press, Florida, pp. 171-186.

Karpechenko, G.D. (1927). Bull.Appl.Bot., Genet. And Plant Breeding, 17: Quoted by Briggs F.N. and P.F. Knowles (1967).

Kihara, H. (1951). Triploid watermelons. *Proc.Am.Soc.Hort.Sci.*, 58: 217-230.

Kihara, H. (1966). Factors affecting the evolution of common wheat. *Indian J.Genet.*, 26A: 14-28.

Koutu, G.K. and K.C. Mandloi (2002). Cotton. In *Genetic improvement of field crops*, Eds. Singh, C.B. and D. Khare. Scientific Publishers (India), Jodhpur, pp. 247-253.

Mc Fadden, E.S. and E.R. Sears (1947). The genome approach to radical wheat breeding. *Jour.Amer.Soc.Agron.*, 39: 1011-1026.

Majumdar, P.K. (1989). Fruit trees. In *Plant breeding — theory and practice*. Ed. V.L. Chopra, Oxford and IBH Publishing Co.Pvt.Ltd., New Delhi, pp. 335-359.

Nigam, S.N.; M.J. Vasudev Rao and R.W. Gibbon (1990). Artificial hybridization in groundnut. In "ICRISAT" information bulletin, No.29, ICRISAT, Hyderabad, A.P. p. 26.

Nishiyama, I. (1971). Evolution and domestication of the sweet potato. *Botanical Magazine*, Tokyo, 84: 377-387.

Olmo, H.P. (1952). Breeding tetraploid grapes. *Proc.Amer.Soc. Hort. Sci.*, 59: 285-290.

Peter, K.V. (1998). *Genetics and breeding of vegetables*. ICAR, New Delhi.

Poehlman, J.M. and D.N. Borthakur (1969). *Breeding Asian Field Crops*, Oxford and IBH Publishing Co., Calcutta.

Poehlman, J.M. (1986). *Breeding field crops*. 3rd ed. AVI Publishing Company, Inc, Westport, Connecticut, USA.

Ram, H.H. (1998). *Vegetable breeding*. Principles and practices. Kalyani Publishers, Ludhiana.

Riley, R. (1960). The diploidization of polyploid wheat. *Heredity*, 15: 407-429.

Riley, R.; J. Unrau and V. Chapman (1988). Evidence on the origin of the B genome. *J.Hered.*, 49: 91-98.

Randolph, L.F. and D.B. Hand (1940). Relation between carotinoid content and number of genes per cell in diploid and tetraploid corn. *J.Agric.Res.* 60: 51-56.

Sandha, G.S. and M.K. Upadhyay (1986). Progress in Triticale Research. In *Twenty five years of coordinated wheat research, 1961-1986*. Eds. Tandon, J.P. and A.P. Sethi. Wheat Project Directorate, IARI, New Delhi, pp. 235-252.

Sarkar, P. and G.L. Stebbins (1956). Morphological evidence concerning the origin of the B Genome in wheat. *American Journal of Botany*, 43: 297-304.

Scott, D. (1951). Cytological studies on polyploids derived from tetraploid *Fragaria vesca* and cultivated strawberries. *Genetics*, 36: 311-325.

Singh, B.D. (1993). *Plant breeding. Principles and methods*. Kalyani Publishers, Ludhiana.

Skovsted, A. (1937). Cytological studies in cotton. IV chromosome conjugation in interspecific hybrids. *Journal of Genetics*, 34: 97-134.

Stebbins, G.L. (1950). *Variation and evolution in plants*. Columbia University Press, New York.

Stebbins, G.L. (1957). Genetics, evolution and plant breeding. *Indian Journal of Genetics and Plant Breeding*, 17(2): 129-141.

Stephens, S.G. (1947). Cytogenetics of *Gossypium* and problems of the origin of New World cottons. *Adv.Gen.*, 1: 431-442.

Swarup, Vishnu (1995). Heterosis breeding in flowers. In *Hybrid research and development*. Eds. Rai, M. and S. Mauria. *Indian Society of Seed Technology*, New Delhi, pp. 103-187.

U, N. (1935). Genome analysis in Brassica with special reference to the experimental formation of *B.napus* and peculiar mode of fertilization. *Jap.Jour.Bot.*, 7: 389-452.

Valleau, W. (1952). Breeding tobacco for disease resistance. *Econ.Bot.*, 6: 69-102.

17

Mutation in Plant Breeding

Inheritance of various characters of plants is determined by genes, which are located on chromosomes. Genes are faithfully transmitted by the parents to their offspring through gametes. The chromosomes of higher organisms, as we know, are almost always made up of deoxyribonucleic acid (DNA). The DNA is a polymer of four different kinds of nucleotide base pairs. During meiotic cell division, which is involved in the gamete formation, as well as in mitosis, which is responsible for most of the cell division, the DNA is precisely replicated. The semi-conservative replication of DNA is catalyzed by DNA polymerase III in prokaryotes and DNA polymerase α in higher organisms (eukaryotes). The DNA polymerase contains built-in 3' → 5' exonuclease activity to proof read the progeny DNA molecules and correct the mistakes made during the initial polymerization reaction. Thus mechanisms exist to facilitate faithful transmission of genetic information from one generation to the next. Despite this, mistakes in replication do occur. Sudden heritable change in the structure of genetic material is called mutation. A plant which exhibits the novel phenotype resulting from mutation, is called a mutant. In broader sense mutation can be defined as any sudden heritable change in the genotype of an organism, which cannot be explained as due to recombination of previously existing genetic variability. The reader should note that recombination, merely rearranges the genetic variability into new combinations and selection (natural and/or artificial) preserves the combinations, which are best adapted to a given environment. However, mutation produces new alleles. It is because of mutations that alternative alleles of a gene exist. For example, in the absence of mutation all the pea plants would have produced red flowers. It is because of mutation at the red flower locus that we find different flower colours in peas. Without mutation all individuals of a species would have same phenotype. The reader knows very well that genetic variability is essential not only for improvement of plants but also for adaptation in changing environments. Mutation is then an important phenomenon. But if mutations occurred too frequently, they would totally disrupt the transmission of precise genetic information from one generation to the next. Mutations provide new raw material for evolution.

Mutation may involve changes in chromosome number (such as euploidy or aneuplody), chromosome structure (such as translocation, inversion, duplication and deficiency) or changes in nucleotide base sequences that make up a gene. Mutations which involve changes in single base pairs of nucleotides of DNA (or RNA) are called as point mutation. This may include substitution of one base pair for another, or duplication or deletion of single base pairs within a number of base sequences that make up a gene. Mutations resulting from changes in chromosome number or gross chromosomal structure are known as chromosomal mutations.

Mutation played significant role in the evolution of various plant species. Cabbage, cauliflower, broccoli, Brussels sprouts and knolkhol, for example, have descended as mutations from common kale like ancestor, the wild cabbage (*Brassica oleracea* L. var. *sylvestris*) still found in western and southern Europe and North Africa (Allard, 1960; Giles, 1941; Horne, 1954). These vegetable crops would not have been available to the mankind in the absence of mutations.

Data gathered from mutation induction work in *Drosophila melanogaster* and plants during the last several decades clearly indicated that the mutations, which are induced artificially, are similar to those that occur spontaneously (Swaminathan, 1967). Moreover, many genes governing the characteristics in which commercial varieties require improvement occur in one or another known variety of the crop and a number of other desirable traits are known to exist in rapidly expanding world germplasm. These are, probably some of the reasons, in addition to the fact that induced mutations like spontaneous ones are always deleterious, for loss in interest of the breeders to employ mutation breeding for improvement. But Gottschalk and Wolff (1983) reported that more than 500 varieties of various crop species developed by means of induced mutations have been officially released

17.1 KINDS OF MUTATIONS

Mutations could be spontaneous or induced. Spontaneous mutations occur in natural populations and are those which occur without known cause. The frequency with which they occur is relatively low. Measurements of spontaneous forward mutation frequencies for various genes of bacteriophage and bacteria range from about 10^{-8} to 10^{-10} detectable mutations per nucleotide pair per generation. For eukaryotes, on the other hand, estimates of forward mutation rates range from about 10^{-7} to 10^{-9} detectable mutations per nucleotide pair per generation (Gardner and Snustad, 1984). While comparing mutation rates per nucleotide with mutation rates per gene the average gene is usually assumed to be 1000 nucleotide base pairs long. The observed frequency of mutation varies from gene to gene and organism to organism. Some mutator genes, like *Dt* (for dotted on chromosome IX of maize) promote mutation in genes at other loci, irrespective of whether they are on same or other chromosome within same nucleus.

Induced mutations are those, which result from action of mutagenic agent to which the organism is exposed. Operationally it is impossible to distinguish between a particular mutation as to whether it occurred spontaneously or that the mutation was induced by a mutagenic agent. What is regarded as a spontaneous mutation could be actually induced because all plants in nature are subjected to low dosages of solar radiations. Likewise, a mutation isolated following exposure of a plant to a mutagenic agent may happen to be a spontaneous mutation rather than arising from mutagenic treatment.

Solar radiations are one cause of mutation that we can hardly do without. Fortunately ultraviolet rays in sunlight are of very low energy and do not penetrate tissues far enough to expose gonads or germ cells of higher organisms to produce mutagenic effects.

The gene mutations may be recessive in inheritance or dominant. From the results of experiments organized to induce mutations, using X-rays, gamma rays, Neutrons, radioisotopes, etc. Gottschalk and Wolff (1983) inferred that about 99 per cent of the mutations induced in various crops were recessive. Hence mutations are not discernible in M_1 generation. Only about 1 per cent induced mutations are dominant. They have been reported in rice (Okuno and Kawai, 1978), rye (Kuckuck and Peters, 1979) and peas. If a recessive mutation occurs in somatic tissue of a plant treated with mutagen, it will not be immediately observable because it will be sheltered by dominant allele. However, in M_2 generation it will become homozygous, upon self fertilization of M_1 plants, and hence can be observed.

Mutations could be classified according to the magnitude of their effect on phenotype. Gaul (1964) classified the mutations into two categories viz., macro-mutations and micro-mutations. Mutations, which produce drastic effects, which can be recognized with certainty in single plants are macro-mutations. These could be transpecific or intra-specific. Mutations with small effects, which can be seen clearly only by careful observation or statistical means in a group of plants, are micro-mutations. They may be manifest or cryptic. The micro-mutations could be controlled either by single or several genes. This distinction, as macro and micro-mutation, is an arbitrary one, for convenience only because usually there is a continuous gradation from micro-to macro-mutation. Among those listed by Gaul (1964) as macro-mutations are mutations affecting culm length, growth habit, heading date, spike density, anthocyanin development etc. Based largely on screening procedures which are adopted, the extent of pleiotropic or associated effects and phylogenetic value of induced mutants, following classification was suggested by Swaminathan (1967):

Micro-Mutations

All mutations, which can be identified and fixed only through the biometrical procedures, are micro-mutations.

Visible Mutations

All mutations, which can be discerned by 'eye inspection' or by use of appropriate screening procedures, such as through artificial epiphytotics of disease or the adoption of biometrical estimations etc. These could be sub-grouped into lethal and viable.

Macro-Mutations

All mutations in which the change though inherited as a single unit of recombination leads to several phenotypic consequences. Such mutations may be the result of a locus involved being a compound one or due to tight linkage or genuine pleiotropy.

Systematic Mutations

All mutations, which simulate an already existing taxon or necessitate a creation of new systematic unit by virtue of character affected being a key one are called systematic mutations.

Mutations have certain general characteristics, viz., they are recurrent (i.e. the mutations seen today also occurred in the past and will appear in future), mostly deleterious, largely recessive, and random (i.e. any gene can mutate although some genes show more mutation frequency than others).

The mutation of a wild type form of gene to another form that results in a mutant phenotype is known as 'forward mutation'. The designation of wild type and mutant is, however, arbitrary. It should be noted that mutational events are reversible. That is subsequent mutation may occur which restores the original wild type phenotype by mutation of a mutant form. This is referred to as back mutation or reverse mutation. The restoration of wild type form may occur either due to true back mutation at same site in the gene as the original mutation, restoring the wild type nucleotide sequence or by occurrence of a second mutation at different location in the genome. Mutations of latter kind are designated as suppressor mutations. These suppress the effects of original mutation (Figure 17.1). Suppressor mutations may occur at distinct site in the same gene as original mutation or in different genes, even in different chromosomes.

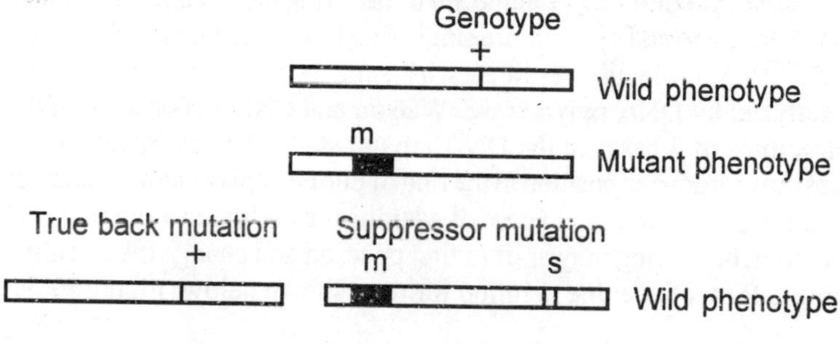

Fig. 17.1: Back mutation

17.2 MOLECULAR BASIS OF MUTATION

Before we learn about the various mutagens and their effects to produce mutations, it would be desirable to understand the molecular basis of how the mutations occur. The experiments of Frederick Griffith; Avery, MacLeod and McCarty and Hershey and Chase suggested that deoxyribonucleic acid (DNA) or ribonucleic acid (RNA) is the genetic material that is responsible for inheritance of organisms. Watson and Crick in 1953 deduced the correct structure of DNA. The DNA has a double helix. Each strand is a polymer of various nucleotides such that the base in one strand pairs with complementary base in other strand. There are four nucleotides, which differ from each other in nitrogen bases, while sugar (deoxyribose) and phosphate are same in all the nucleotides. The stable forms of four bases; two purines (adenine and guanine) and two pyrimidmes (thymine and cytosine) are shown below (Figure 17.2):

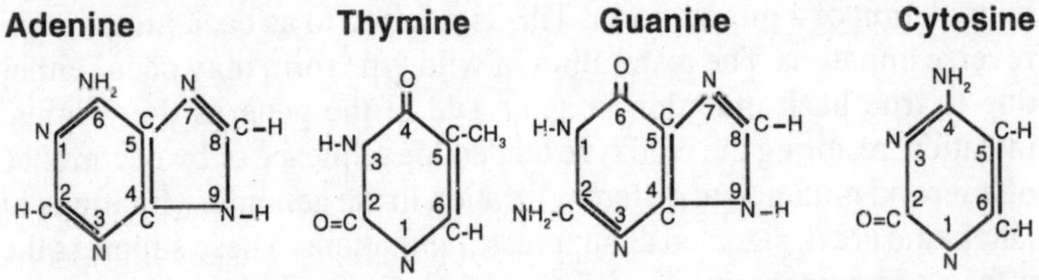

Fig. 17.2: Structural formulas of four bases of nucleic acids

The adenine always base pairs with thymine and guanine with cytosine. Adenine and guanine are double ring bases called as purines while thymine and cytosine are single ring bases known as pyrimidines. The DNA molecule replicates by semi-conservative replication catalyzed by DNA polymerase. Watson and Crick proposed that the structures of 4 bases in the DNA are not static. The hydrogen atoms can move from one position in the ring of purine or pyrimidine to another position. For example, in case of adenine the hydrogen atom in the 6th position of the ring may shift to first position and change the so called amino form of adenine to imino form, as shown below (Figure 17.3):

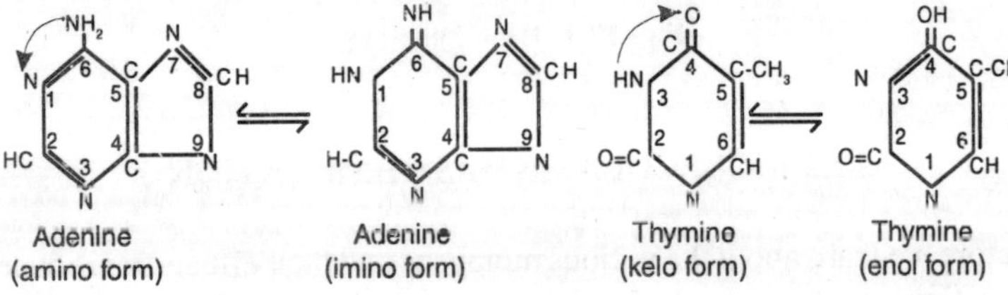

Fig. 17.3: Tautomeric forms of adenine and thymine bases of DNA

In this way keto forms of thymine and guanine may change to enol forms and cytosine to imino form. Such chemical fluctuations were designated by Watson and Crick as tautomeric shifts. Although tautomeric shifts are rare, they are important in DNA metabolism because they change the base pair potential of bases. If the base existed in altered from during semi-conservative replication of DNA or it was being incorporated into nascent DNA chain, mutation may result, because when adenine is present in imino form it would base pair with cytosine rather than thymine, its regular partner. Like wise, guanine in enol form would base pair with thymine instead of cytosine. Due to tautomeric shift adenine changes to imino form and consequently base pairs with cytosine. During semi-conservative replication of such a changed DNA, the base sequence will shift to GC rather than original AT. Similarly when guanine changes to enol form and base pairs with thymine, during replication there will be a change in base pair from GC to AT. This leads to mutation.

Mutations resulting from tautomeric shifts in the bases of DNA involve a replacement of one purine in one of the strands of DNA with other purine and replacement of pyrimidine in the complementary strand with other pyrimidine. Chemical changes in which one purine replaces another purine and pyrimidines are similarly switched have been called as transitions. Mutations resulting from such changes are called transition mutations. When the base pair substitution involves substitution of purine for pyrimidine or *vice versa*, the resulting mutations are known as transversion mutations.

The third type of mutation involves addition or deletion of one or a few base pairs of polynucleotide strands of DNA. Base pair additions or deletions are collectively designated as frame shift mutations. The mutation is called frame shift because codon (a triplet of nucleotides, consecutively located on a polynucleotide strand of DNA, specifying an amino acid) is altered.

All the three types of point mutations mentioned above are present among spontaneous mutations (Gardner and Snustad, 1984).

17.3 MUTAGENIC AGENTS

In 1927, H.J. Muller first demonstrated that X-ray treatment markedly increased the frequency of sex-linked recessive mutations in *Drosophila melanogaster*. Shortly afterwards Stadler announced that X-ray treatment of the maize and barley plants was effective to obtain high mutation rates. Subsequently, Auerbach and Robson (1947) suggested that certain chemical compounds could also be employed to induce mutations. Thus physical mutagens and chemical mutagens have been the principle agents used to increase mutation frequency in plants. Physical mutagens comprise of ionizing and non-ionizing radiations.

17.3.1 Physical mutagens

The physical mutagens include ionizing and non-ionizing radiations. Let us recall the structure of an atom to understand ionization. An atom is made up of two parts, viz., the central nucleus and an extra-nuclear part. The nucleus of an atom comprises of positively charged proton and neutron with mass equal to proton but with no charge. The extra nuclear part has negatively charged electron. Because an atom has equal number of protons and electrons the atom is neutral. An hydrogen atom has positively charged proton in nucleus and one orbital electron. The number of protons and electrons determines the atomic weight whereas atomic number is determined by proton and neutron. Neutron has not so much to do with chemical activity of atom. Chemical activity is due mainly to electron. An isotope of hydrogen, deuterium, has one proton and one neutron in the nucleus and one electron. It is called heavy hydrogen atom because its atomic weight is two against common form, which has atomic weight equal to one. An isotope is a form of same element with similar atomic number but different atomic weight due to different nuclear structure. Ionization is the process, which is responsible for breaking a neutral atom into cation and anion. Ionizing radiations are of two kinds, viz., particulate such as alpha rays, beta rays, fast neutrons and thermal neutrons and non-particulate, e.g. X-rays, gamma rays. Ultra violet is a non-ionizing radiation.

That portion of the electromagnetic spectrum containing wavelengths that are shorter and of higher energy than visible light can be subdivided into ionizing radiation and non-ionizing radiation. X-rays were used to induce mutations most extensively in the earlier experiments because X-ray equipment was widely available and was easy to operate. Gamma rays are commonly used ionizing radiation in plant improvement programmes. Gamma rays are obtained from radioactive isotopes like Cobalt 60 or Caesium-137 (Sharma and Chopra, 1989).

Ionizing radiations are of high energy. When a biological material is irradiated, the high energy rays (photons) hit the orbital electron of atom. The photon energy is transferred to electron in the form of kinetic energy and electron is released (ejected), leaving positively charged free radicles or ions. The ejected electron, also known as photoelectron, has tremendous energy and causes further ionization along the path it collides with other molecules. Consequently, further ejection of electrons results. The net result is that a core of ions is formed along the track of each energy ray as it passes through the matter of living tissues. Biological damage due to radiation could be direct or indirect. Direct effect would be caused when DNA is the direct target and radiation energy is directly absorbed by it. On the other hand, when the initial absorption of energy is by other molecules in the cell, indirect effect would result. Ionizing radiation can cause disruption of continuity of one of the strands of DNA or both the strands. Cross-linking of DNA-DNA may also result from exposure of plant cells to ionizing radiations. A base from the polynucleotide strand may be removed causing change in pairing properties.

The ionizing particle deposits energy in straight path while passing through the matter. The radiation that produces little ionization per micron of its path are called sparsely ionizing radiations, e.g. X-rays and gamma rays. On the other hand, those ionizing particles, which produce several ionizations per micron of its path are known as densely ionizing radiations, e.g. alpha particles and neutrons. The densely ionizing radiations probably act in mechanical way. They cause high proportion of chromosomal aberrations. However, these radiations also produce some point mutations.

The ultraviolet rays are non-ionizing. These do not possess sufficient energy to induce ionizations. They are, nevertheless, readily absorbed by certain substances, such as purines and pyrimidines. Because of their low energy they penetrate tissues only slightly. Ultraviolet rays dissipate their energy to the atoms they encounter while passing through the material and raise the electrons in outer orbital to higher energy levels. This state of atom is called as excitation. Molecules containing atoms either in ionic forms or in excited state are chemically more reactive than those containing atoms in their stable state. Several lines of evidences indicate that thymine dimerization is probably the major mutagenic effect of ultraviolet rays.

17.3.2 Chemical mutagens

Various chemicals could be employed to induce mutations in plants. The chemical mutagens have been classified into two categories depending upon whether they are mutagenic to both replicating and non-replicating DNA (such as alkalyating agents and nitrous acid) and those, which are mutagenic only when DNA is replicating (such as acridine dyes, base analogues). The most important chemical mutagens are ethyl methane sulphonate (EMS), diethyl sulphate (DES), ethyl imine (EI), propane sultone, N-methyl-N-nitroso urethane (MNU) and some related substances (Gottschalk and Wolff, 1983). An unusually powerful mutagens is sodium azide (Nilan *et al.*, 1973). The chemical mutagens are comparable in mutagenic efficiency with physical mutagens.

Alkylating Agents

A series of alkylating agents have been tried for their mutagenicity in barley. According to Natarajan and Ramanna (1966) the alkylating agents fall into two categories, *viz.*, mono-functional and bi-functional. They differ in their reactive alkyl groups, *i.e.* methyl, ethyl, butyl etc. and their non-alkylating prosthetic groups. While the prosthetic groups do not have any significant differential response, different reactive groups give differential mutagenic response (Rao and Natarajan, 1965). Alkylation refers to substitution of an alkyl group (e.g. C_2H_5 of ethyl methane sulphonate) for an hydrogen atom in the nitrogen base of DNA. As a consequence, the base pairing potentials of the bases are altered and transition results. The effect of ethyl methane sulphonate is ethylation at 7th *N* position and at 6th oxygen positions. Such an alkylated guanine pairs with thymine rather than cytosine, which is its regular natural base pairing partner. Some alkylating agents, particularly the bi-functional, i.e. those with two reactive alkyl groups, cross-link the DNA stands and induce chromosome breaks and several kinds of chromosomal aberrations. These mutagens induce all types of mutations, including transitions, transversions, frameshifts and chromosome aberrations, with different frequencies, depending upon alkylating agent.

Nitrous Acid (HNO_2)

Nitrous acid is a very potent mutagen which acts directly either on replicating or non-replicating DNA. It causes oxidative deamination of bases, which contain amino groups, such as adenine, guanine and cytosine. Conversion of amino group into keto group leads to change in hydrogen bonding potential of bases. Adenine and thymine form two hydrogen bonds while guanine and cytosine form three hydrogen bonds during regular base pairing. Adenine is converted to hypoxanthine; cytosine to uracil and guanine is deaminated to form xanthine. Consequently, they do not base pair with their regular partners. Hypoxanthine base pairs with cytosine, as is known uracil base pairs with guanine. The xanthine, however, base pairs with cytosine, like guanine. Deamination of adenine leads to $AT \rightarrow GC$ transition and hence mutation. Similarly deamination of cytosine results in transition mutation. It is worthy of noting that nitrous acid induces transition in both directions, i.e. $AT \rightleftharpoons GC$.

Base Analogues

Analogue is a compound whose molecular structure is similar to some natural compound and which in a biological system is active as an antagonist to it. Thus base analogues that are mutagenic have structures very similar to natural bases so that they are metabolized and incorporated into DNA during replication. But they are different than natural bases in that they increase the frequency of mismatching and consequently mutation. Two commonly employed base analogues are 5-Bromouracil and 2-Aminopurine. 5-Bromouracil (*5BU*) is an analogue of base thymine. This base analogue causes transition fom *AT* to *GC*. The purine base analogue, 2-Aminopurine also induces transition mutation.

Acridine Dyes

Acridine dyes, such as proflavin, acridine orange etc., are powerful mutagens, which induce frame shift mutations.

17.4 MUTATION INDUCTION

Mutation breeding has been suggested by Gottschalk and Wolff (1983) as supplementary to conventional breeding methods. According to them mutation breeding is especially valid for annual diploid and allopolyploid autogamous crops. It is also suitable for vegetatively propagated plant species, including several ornamentals. In allogamous crops it is more time consuming and still remains in infancy. In case of cross-pollinated crops it is necessary, according to Gottschalk and Wolff (1983), to make the initial material homozygous for a large number of gene pairs through inbreeding or sib-mating. Inbred lines should be used for mutagen treatment. In maize, inbred lines are regularly produced. But in other allogamous crops, this may not be an easy task. This is particularly so for self-incompatible crops.

In the sexually propagated crops, seeds are usually considered as suitable material for mutation induction. Treatment of dormant seed is a routine procedure, however, Sharma and Chopra (1989) are of the opinion that seeds soaked in water and at a stage corresponding to the DNA synthesis is an ideal material. Poehlman (1987) also supports choice of dormant seeds for induction of mutations. Radiations may be more suitable for treatment of seeds, because treatment is carried out in dry condition and the seeds can be handled like the untreated seeds. On the other hand, when seeds are subject to chemomutagenesis, soluble solutions of chemical are used and consequently seeds are to be handled very carefully as compared to dry seeds. Another disadvantage, following application of potent chemical mutagens to seeds is their negative effect on viability and fertility of M_1 plants. These negative effects of chemicals are stronger than corresponding effects of mutagenic rays (Gottschalk and Wolff, 1983).

In case of vegetatively propagated crops like sugarcane, potato, sweet potato, the treatment of mutagen could be given to cuttings, bud wood, suckers or tissues normally used for propagation. Because of the problems, like non-flowering or poor flowering and sterility, in some vegetatively propagated crops mutation breeding appears to provide good opportunities for crop improvement. Some ornamental plants and fruit crops can also be included in this category. Radiations are most commonly used mutagens in vegetatively propagated plants than chemical mutagens. Any portion of plant that may give rise to a bud may be exposed to mutagen such as tubers, rhizomes, cuttings, shoot apices. All these plant parts carry multicellular meristematic tissues, which originate from single meristem cell. In vegetatively propagated plants intra-somatic competition, known also as diplontic selection, operates. This may be an hindrance to the success expected in mutation breeding of vegetatively propagated crops. This phenomenon is more intense in operation in vegetatively propagated crops as compared to crops propagated through seeds.

The mutated cell is often at competitive disadvantage as compared to the normal cells. Consequently, the normal cells reproduce more rapidly and crowd out the mutant cell. Some times the mutant cell is excluded from its contribution to gamete formation, being very small. This exclusion of mutant cells by normal cells is known as diplontic selection. However, whenever, competitive disadvantage of mutant cells, as compared to normal cells, is seen to operate at the gametic level, i.e. gametes bearing normal cells are more functional than the gametes carrying mutant cells, haplontic selection is said to be operating. Nayar (1989) proposed to use dormant or semidormant plant propagules to reduce the effect of intra-somatic selection. According to him, under dormant or semidormant condition the number of meristematic cells would be relatively less.

Treatment of seeds and vegetative propagules with mutagens commonly produces chimeras. A chimera is a part of plant, which has different genetic constitution as compared to other parts of same plant. The chimerical structure of plants often has negative effects on the performance, especially for those crops with relatively low propagation coefficient. The number of initial cells present in the embryonic growing points is a characteristic feature of each species. The initials in the embryonic growing points are responsible for producing shoot system of plants. Most species of plants contain several initial cells in their growing points. A distinct mutation will occur in one of them, the others can likewise mutate but not with regard to same gene. If a plant has several initial cells and one of them mutates, a chimerical M_1 plant may arise. This can also occur in those cases where young embryo has several lateral growing points besides apical main growing point. Those plant parts, which develop from mutant initial cell, are heterozygous for mutant gene whereas others are genetically normal.

The problem of chimerism can be avoided by treating pollen grains and by using them for pollinating emasculated flowers (Gottschalk and Wolff, 1983). For potato, a successful method was proposed by Nayar and Dayal (1970) to eliminate chimerical condition. In this method tubers cut longitudinally are exposed to gamma rays. The other half portion of the tuber is left as control. The sprouts emerging from the irradiated tuber are removed periodically and planted individually. Through this method frequency of mutant plants recovered in M_2 generation could be enhanced upto 50 per cent. Another method for overcoming intra-somatic selection is to treat the cultured tissues with mutagen and isolation of individual cells and growing them into plants (Nayar, 1989). In sugarcane the target of irradiation is always the vegetative bud (Sreenivasan, 1989). But *in-vitro* technique could be useful to restrict or avoid formation of chimera.

According to Gottschalk and Wolff (1983) haploid cells and suspension cultures of diploid or allopolyploid crops would be an ideal material for application of mutagens. Each gene is present only once in such single cells and hence heterozygosity for mutant gene does not complicate their identification in M_1 itself. If haploid cells could be regenerated then mutation breeding would be more successful. Mutants derived from single cells have been isolated in *Datura inoxia* (Weber and Lark, 1979), *Nicotiana sylvestris* (Malepszy *et al.*, 1977).

The environment in which mutagen treatment is administered will influence the effect of treatment. Oxygen and temperature change when associated with irradiation may significantly alter the frequency of mutations. Low oxygen tension reduces frequency of mutations. Moisture, temperature and oxygen level may be controlled conveniently and without injury to seeds. Large number of seeds could be treated at time and stored after treatment without injury to seeds.

Lawrence (1964) is of the view that repeated irradiation of high performance crop plants together with moderate to intense selection pressure would be advantageous. The mutants produced by mutagenic treatment of seeds of plants are some times treated again or even several times with same mutagen. Brock (1964) is of the view, however, that it would be better to make full use of the irradiated variation by cycles of recombination with selected lines rather than repeat induction of variation by successive cycles of radiation and selection.

An important feature of mutation induction is that different varieties of same crop can differ markedly from each other with regard to their susceptibility to different mutagens (Gottschalk and Wolff, 1983). For instance, Sharma and Sharma (1979) observed that a specific variety of *Lens culinaris* was relatively resistant to gamma ray and NMU (N-methyl-N-nitrosourethane or urea) treatment, whereas two other cultivars were sensitive. Ahmed and Godward (1981) and Al-Rubei and Godward (1981) reported similar findings in *Cicer arietinum* and *Phaseolus vulgaris*, respectively.

Considerable expenditure in space, time and money is necessary in order to obtain a specific group of mutations or even a single desired mutation (Gottschalk and Wolff, 1983). For example, three blast resistant rice mutants were isolated in Japan out of 51530 plants in M_2 (Yamasaki and Kawai, 1968).

17.5 DOSE OF MUTAGEN

For successful mutation breeding programmes the requirement is effective and efficient dose of mutagen, which has to be employed to induce expected mutations. A mutagen, which induces high degree of physiological and chromosomal effects, is regarded as most effective. But induction of heritable changes determines the efficiency of a mutagen. For most of the crops optimum doses for mutation induction are reported. When such published information is not available it should be obtained through experimentation. Usually the dose is preferred which kills 50 to 70 per cent (LD_{50} to LD_{70}) of seeds or developing seedlings (Briggs and Knowles, 1967) or GR_{50} (influences growth adversely of at least fifty per cent population). With ionizing radiations a dose, which causes survival of fifty per cent (LD_{50}) or growth of 50 per cent (GR_{50}) is a good treatment. In actual practices one dose lower than and one higher than LD_{50} or GR_{50} should be employed.

Gaul (1964), however, is of the view that a dose of physical or chemical mutagen, which kills over ninety per cent of the treated populations, would be desirable. He observed that mutation rate increases with high doses far exceeding LD_{50}, the increase being linear or even exponential. Using extremely high doses of ethyl methane sulphonate (EMS) and X-rays he observed that 97 and 99 per cent lethality of seeds could increase mutation rate two to three times as compared to the LD_{50} dose.

In case of chemical mutagens determination of LD_{50} value is complicated by difficulty in determining the effective termination of treatment. In such cases graded series of dilutions of treatment durations on the basis of reports in published literature should be tried (Sharma and Chopra, 1989).

17.6 HANDLING THE TREATED MATERIAL

The untreated seed or other plant part is denoted by *M*. The first generation after treatment is denoted by M_2 and so on. In order to ensure recovery of desired mutants surviving population of M_1 is of critical importance. The M_1 plants obtained from germinating seeds are usually reduced in vigour and those seedlings which survive till maturity may be sterile. Only 30 per cent of the total number of mutants of *Pisum* manifested high fertility so that the mutants could be propagated directly by seeds (Gottschalk and Wolff, 1983). Remaining

70 per cent represented in similar proportions lethal, sterile and weakly fertile genotypes. The plant material treated with mutagen is normally sown directly in the field. In case of cereals, the seeds of which possess more than one primordia at the time of treatment, close planting is recommended so that only one tiller grows from each seed. It is the first tiller in cereals, which has shown maximum number of mutations (Briggs and Knowles, 1967). But care should be taken not to raise too dense a population because plants carrying mutation are frequently less competitive.

Out crossing should be prevented in case of M_1 generation. This is more important in case of self-incompatible crop species and monoecious crops.

Because most of the mutations are recessive, M_1 plants will very rarely manifest mutations. Those M_1 plants should be discarded which exhibit reduced fertility because sterility is almost always associated with chromosome changes. Harvesting in M_1 is accomplished plant wise.

Large M_2 has to be raised because the frequency of useful mutants is very low. For example, 12 lakh plants were raised by Yamaguchi and Yamashita (1979) of barley in M_2 from which seven mildew resistant mutants were selected. Three blast resistant mutants of rice were isolated from 51530 plants grown in M_2 (Yamasaki and Kawai, 1968). From their experience with peas and barley, Gottschalk and Wolff (1983) concluded that only about 0.5 per cent of all the experimentally produced mutants are of interest from breeding point of view. The true proportion may be higher if the micro-mutations are considered in an adequate way. M_2 is raised as single plant progenies. Single row of each selected M_1 plant, usually comprising of 10 to 12 plants may be raised. Because mutants are to be identified in this generation space planting is desirable. Selection begins in M_2. Hence the M_2 generation should be grown under conditions that will permit expression of the character. For example, if resistance is to be incorporated in pepper against *Phytophthora*, M_2 seedlings are inoculated by dipping the roots in the inoculum before transplanting in field/pots. Screening may start after 10-12 days (Ram, 1998). Daskalov (1986) has reviewed this topic. About 70000 to 1 lakh plants are to be grown in M_2. Jaranowski and Micke (1985) have discussed about mutation breeding in peas. Gottschalk and Wolff (1983) have discussed about induced mutation breeding of several crops for wide range of characters, e.g. alteration of flower shape, colour and function, ripening time, disease resistance, drought and heat tolerance, shattering resistance etc. Enough M_3 plants of selected M_2 plants should be grown. Selection may continue in M_4 also after which selections which appear phenotypically uniform within rows and are promising may be entered in coordinated programmes.

Promising mutants could be utilized in one of the three possible ways, as suggested by Gottschalk and Wolff (1983).

(i) Selected mutants with favourable properties are directly developed into varieties,
(ii) Prospective mutants are utilized indirectly, and lastly,
(iii) Different favourable mutants could be crossed with each other.

17.7 VARIETIES DEVELOPED

Mutations, both spontaneous and induced, have been utilized for development of new varieties in cereals, oilseeds, pulses and vegetable crops.

NP 111 is a spontaneous mutant of wheat from variety *NP* 4 (Jain, 1989). It is glabrous glumed. *NP* 836, a mutant from a tipped awned variety *NP* 799 was released for cultivation in 1955. Sharbati Sonora and Pusa lerma with amber coloured kernels were obtained after induction of mutation in Sonora 64 and Lerma Rojo 64, respectively. Both the latter are red kernelled. These were released in 1967.

Barley mutants *RDB* 1, PL 56 and *DL* 253 were released for cultivation. RDB 1 was obtained as a mutant from *RS* 17 and *PL* 56 from *C* 164. Mutant *RDB* 1 was crossed with various tall barley varieties and semi-dwarf varieties Rajkiran and *RD* 103 were developed. Former is resistant to nematodes and latter has better grain weight and more seeds per ear than *RDB* 1.

GEB 24 is the earliest Indian rice variety obtained as spontaneous mutant in traditional cultivar Konamani of TamilNadu (Sharma, 1989). *T* 141 is a popular rice variety of Orissa. It was exposed to X-rays and semi-dwarf, fertilizer responsive mutant was obtained which was released as Jagannath. Sattari a variety which matures in seventy days was selected from gamma ray irradiated progeny of *NSJ* 200 × Padma.

Seeds of groundnut variety Spanish improved were irradiated with X-ray doses ranging from 25 to 75 Kr. Frequency of mutations was higher at 35 to 45 Kr. But promising mutants were obtained from 75 Kr. Varieties *TG* 1 to *TG* 6 were developed from this material. These were superior than parental material as well as best check variety for pod yield and several other characters. *TG* 1 (Vikram) was released for cultivation (Arunachalam, 1989). At Ludhiana *Brassica* varieties *RL* 81, *RLM* 198 and *RLM* 514 emanating from mutation breeding were released for cultivation. A new variety *RLM* 619 was also developed (Rai, 1989).

Rajasekharan and Shanmugavelu (1984) irradiated bitter gourd variety MC 103 with gamma rays and from M_2 a promising selection was isolated. It was stabilized and released as MDU 1. CO 10 Avarai, a high yielding variety of *Dolichos lablab* was isolated from gamma ray irradiated (24 Kr.) bushy variety CO 6 (Sivasubramanian *et al.*, 1989). French bean variety emanated from irradiated material of *EC* 1906, and released as Pusa Parvathi. In this case dry seeds (with 10 per cent moisture contents) were exposed to X-rays. In case of tomatoes, *S* 12 was developed via mutation breeding. Similarly Pusa Lal Meeruti was developed from improved Meeruti following gamma ray exposure @ 15 to 30 Kr. S12 was obtained from mutated population of Sioux (Peter, 1998).

Reference

Ahmad, S. and M.B.E. Godward (1981). Comparison of a radioresistant with radiosensitive cultivar of *Cicer arietinum* L. — II Differences in the number of chromosome aberrations at the same dose. *Environ.Exp.Bot.*, 21: 143-151

Allard, R.W. (1960). *Principles of plant breeding*. John Wiley and Sons, Inc. New York, pp. 444-454.

Al-Rubeai, M.A.F. and M.B.E. Godward (1981). Genetic control of radio sensitivity in *Phaseolus vulgaris* L. *Environ.Exp.Bot.*, 21: 211-216.

Arunachalam, V. (1989). Oilseeds: Groundnut. In *Plant breeding — theory and practice*. Ed. V.L. Chopra, Oxford and IBH Publishing Co. Pvt.Ltd., New Delhi, pp. 139-158.

Auerbach, C. and J.M. Robson (1947). The production of mutations by chemical substances. *Proc.Roy Soc.*Edinburgh, B., 62: 271-283.

Briggs, F.N. and P.F. Knowles (1967). *Introduction to plant breeding*. Reinhold Publishing Corporation, NewYork, pp. 301-316.

Brock, R.D. (1964). Induced mutations affecting quantitative characters. FAO-IAEA symposium on *The use of induced mutations in plant breeding. Radi.Bot.Supp.*

Daskalov, S. (1986). Mutation breeding in pepper. *Mutation breeding review* No. 4, p. 26.

Gardner, E.J. and D.P. Snustad (1984). *Principles of genetics*. 2nd edition. John Wiley and Sons, NewYork, pp. 273-319.

Gaul, H. (1964). Mutations in plant breeding. *Rad.Bot.*, 4: 155-232.

Giles, W.F. (1941). Cauliflower and broccoli — What they are and where they came from. *J.Roy.Hort.Soc.*, 66: 265-277.

Gottschalk, W. and G.Wolff (1983). *Induced mutations in plant breeding*. Springer Verlag, Berlin.

Horne, F.R. (1954). Winter cauliflower: History and breeding in South West. *Sci.Hort.*, 11: 128-139.

Jain, K.B.L. (1989). Cereals: Wheat and Barley. In *Plant breeding — theory and practice*. Ed. V.L. Chopra. Oxford and IBH Publishing Co. Pvt.Ltd., New Delhi, pp. 77-99.

Jaranowski, J. and A. Micke (1985). Mutation breeding in peas. *Mutation Breeding Review* No.2, pp.23. Int. Atomic Energy Agency, Vienna.

Kuckuck, H. and R. Peters (1979). Induzierte mutationen in selbstfertilen und inzuchtfesten kulturformen aus der kreuzung *Secale cereale* L × *Secale vavilovii*. Grossh. Und ihr Verhalten in Kreuzungen mit panmiktischen Sorten und setbststerilen spontanmutationen. *Z.Pflanzenzucht*, 82: 97-115.

Lawarence, C.W. (1964). Radiation induced polygenic mutation. FAO-IAEA Symposium on the use of induced mutations in plant breeding. *Rad.Bot.Suppl.*

Muller, H.J. (1927). Artificial transmutation of the gene. *Science*, 66: 84-87.

Natrajan, A.T. and M.S. Ramanna (1966). Chromosome breakage and induced mutagenesis in barley. *Indian J.Genet.*, 26A: 287-297.

Nayar, N.M. (1989). Vegetatively propagated crops. In *Plant breeding — theory and practice*. Ed. V.L. Chopra. Oxford and IBH Publishing Co.Pvt.Ltd., New Delhi, pp. 275-283.

Nayar, N.M. and T.R. Dayal (1970). A method of screening high frequencies of induced mutations in potato. *Z. Pflanzenzucht*, 63: 155-160.

Nilan, R.A.; E.G. Sideris; A.Kleinhofs; C.Sander and C.F. Konzak (1973). Azide — a potent mutagen. *Mutation Res.*, 17: 142-144.

Okuno, K. and T. Kawai (1978). Genetic analysis of induced long-culm mutants in rice. *Jpn.J.Breed.*, 28: 336-342.

Peter, K.V. (1998). *Genetics and Breeding of vegetables*. I.C.A.R., New Delhi.

Poehlman, J.M. (1987). *Breeding Field Crops*. AVI Publishing Co., Inc. West-Port, Connecticut, USA, pp. 109-126.

Rai, B. (1989). Brassicas. In *Plant breeding — theory and practice*. Ed. V.L. Chopra. Oxford and IBH Publishing Co.Pvt.Ltd., New Delhi, pp. 159-170.

Ram, H.H. (1998). *Vegetable breeding*. Principles and Practices. Kalyani Publishers, Ludhiana.

Rao, R.N. and A.T. Natarajan (1965). Mutagenicity of some alkyl alkane sulfonates in barley. *Mutation Res.*, 2: 132-148.

Sharma, S.D. (1989). Cereals: Rice. In *Plant breeding — theory and practice*. Ed. V.L. Chopra. Oxford and IBH Publishing Co.Pvt.Ltd., New Delhi, pp. 101-120.

Sharma, R.P. and V.L. Chopra (1989). Mutation breeding. In *Plant breeding — theory and practice*. Ed. V.L. Chopra. Oxford and IBH Publishing Co.Pvt.Ltd., New Delhi, pp. 409-431.

Sharma, S.K. and B. Sharma (1979). Pattern of induced mutability in different genotypes of lentil (*Lens culinaris*, Medik.). *Z.Pflanzenzucht*, 83: 315-320.

Singh, B.D. (1993). *Plant breeding. Principles and methods*. Kalyani Publishers, Ludhiana, pp. 595-631.

Sivasubramanian, P.; R.S. Annappan; S. Iyemperumal; S.R. Sreerangaswamy and M.Pandiyan (1989). CO10 dual purpose lab-lab. *Madras Agricultural Journal*, 76: 411-412.

Srieenivasan, T.V. (1989). Sugarcane. In *Plant breeding — theory and practice*. Ed. V.L. Chopra. Oxford and IBH Publishing Co.Pvt.Ltd., New Delhi, pp. 253-274.

Swaminathan, M.S. (1967). The origin of macro-from micro-mutations and factors governing the direction of micro-mutational changes. *Indian J.Genet*., 26A: 29-41.

Yamasaki, Y. and T. Kawai (1968). Artificial induction of blast resistant mutations in rice. Rice Breeding with induced mutations, pp. 65-73, IAEA, Vienna.

Yamaguchi, I. and A. Yamashita (1979). Resistant mutants of two-rowed barley to powdery mildew (*Erysiphe graminis* f.sp. *hordei*). Tech.News, 20, Inst.Rad.Breed.Ohmiya, Japan.

18

Distant Hybridization for Plant Improvement

Usually recourse to generate more desirable genetic variability is through hybridization of two or more varieties of same crop species. Hybridization involving varieties from same species is known as intra-specific hybridization. The chromosome number and the structure of chromosomes do not differ in two varieties of same species. The chromosomes of two species mostly differ with respect to gene contents only. In the absence of incompatibility, male or female sterility and some lethal factors, the F_1 hybrid between two varieties of same species is completely fertile. In the F_2 generation also there is no sterility and maximum genetic variability, as per Mendelian expectations, is observed if large population is grown. But situations arise when some beneficial gene (or genes), say for example gene for resistance to Yellow Vein Mosaic Virus of okra, is not available in the varieties of cultivated species. Under these circumstances a breeder has to search for beneficial gene(s) in related wild species or weedy species or other species of same genera. Hybridization involving plants from two species of same genera is designated as inter-specific hybridization. The species is defined as interbreeding natural population which is genetically different and reproductively isolated from other such groups. Because of restrictions on free genetic exchange among individuals from different populations genetic differentiation takes place sooner or later. Consequently, success of inter-specific hybridization is not predictable in advance. As will be discussed latter there may be impediments to obtain crossed seed and if crossed seed is obtained to raise F_1 hybrid and further if F_1 hybrid plant grows it may or may not be fertile. Even the magnitude of genetic variability that one would expect according to Mendelian theory in the F_2 generation, may be disappointing and not congruent with expectations. Only a fraction of the expected genetic variability appears in the F_2 population that is grown. In the extreme cases the desirable gene may not exist among the various related species from same genus. The breeder has then no alternative, but to search related genera for the desired gene. If such gene is found in another genera from same family, the plants from two different genera are crossed to produce what is designated as inter-generic hybridization. Prediction about the success of inter-generic hybridization is difficult.

In general inter-specific and inter-generic hybridization are designated as '*distant hybridizations*' or '*wide crosses*'. Not all the inter-specific crosses and most of the inter-generic hybridizations confer success in achieving the object. But recently developed Plant Genetic Engineering enables identification of the nucleotide sequence of DNA for desirable gene, isolation and its multiplication (cloning), transporting it through an appropriate vector to the cells of desired plants, which ultimately integrates into the chromosome of the plant. This gene expresses itself in the plant. Such plants are called transgenic plants. Sexual mating is essential for exchange of genetic material, but plant genetic engineering has not only shattered the sexual barriers but can accomplish integration of beneficial gene(s) across the kingdoms (See Section 15.4). Nevertheless, it may have its limitations.

One of the prime objects of turning to wide crosses by the plant breeders is to incorporate some beneficial gene, such as gene for disease or insect resistance, gene for abiotic stress or some quality aspect. Reader should recall that inter-specific hybridization has played significant role in the evolution of several crop species.

Agriculturally and horticulturally important types were developed through inter-specific hybridization. The modern strawberry (*Frageria x ananassa*) was actually evolved from the segregation products of an eastern North American species, *Frageria virginiana* and a Chilean or Pacific Coast wild species *F. chiloensis*. The large fruit size of Pacific Coast species has been combined with fruitfulness, quality of flesh and excellent flavour of eastern North American species in the modern strawberry. As will be delineated subsequently genes for disease resistance have been incorporated in cultivated species of several crops from wild or related species through distant hybridization.

Inter-specific hybridization can be made to achieve new character expression not present in either parent. Some times novel types emerge from wide hybridization. Unusual leaf forms were obtained in *Viola* and unusual carpel and sepal numbers in *Antirrhinums* and *Paeonia* following distant hybridization (Stebbins, 1950). Actually inter-specific hybridization or inter-generic hybridization may be more useful in ornamental plants than field crops, vegetable crops and orchard crops. Novelty is more fascinating in ornamental plants. But consumers are very much meticulous about the quality aspects of fruits, such as flavour, taste etc. Same is the case with field crops and vegetable crops. Hence breeder will usually not like to disturb the existing genetic constellation by introducing large number of genes from alien species.

The genus *Chrysanthemum* has more than 160 species, but commonly they are called *Chrysanthemum morifolium* (now the name has been changed to *Dendranthema grandiflora*). Presently grown Chrysanthemums are the result of 2500 years of continuous repeated cycles of hybridization and selection from several species of *Chrysanthemum* that grew wild in China and Japan. Improvement was started in Japan, of the imported Chinese types, by crossing them with the then existing Japanese varieties (Dutta, 1996). Three species of gladiolus were being grown in England before 1730, out of which *Gladiolus byzantinus* was introduced from Constantinople. In 1837, H. Bedinghaus of Belgium crossed parrot gladiolus, *G. natalensis*, with *G. oppositiflorus* making development of present day gladioli (Mukhopadhyay, 1995). In 1833, 'Ramosus' hybrids were produced in Holland by crossing *G. insignis* with *G. oppositiflorus* (Floribundas). The 'Nanus' were evolved by crossing *G. insignis* with *G. scullyi*. According to Emsweller *et al.* (1937) wide crosses, including inter-generic hybridization, have been frequent in ornamentals, especially among roses, orchids and lilies. Distant hybridization has played an important role in orchard crops like apple, plums, cherries, grapes and various kinds of berries. Inter-specific hybridization has been important in vegetatively propagated crops, such as potato and sweet potato.

From the studies with *Avena sativa, Pennisetum americanum, Sorghum bicolor* and *Hordeum vulgare*, Frey *et al.* (1984) inferred that crosses of these cultivated crop species with their wild and weedy relatives lead to increase in yield because these relatives possess genes that can increase yield potential. They suggested that if wild relatives are used in breeding programme to improve cultivated cereals, the base population should contain 3 to 12.5 per cent wild germplasm to allow the greatest short-term progress. Further, to take advantage of specific "nicking" ability among germplasm sources different cultivated parents should be used in each back-cross.

Distant hybridization has played significant role in the evolution of crops like wheat, oat, cotton, tobacco and sugarcane.

Distant hybridizations are also made for curiosity and aesthetics. Inter-specific crosses are effected by the scientists to explore the relationship of one species with another. Such studies are also carried out to learn about the origin of allopolyploid species. Goodspeed and Clausen (1928) concluded from their studies that *Nicotiana tabacum* is an amphidiploid of *N. sylvesteis* and *N. tomentosa*.

While discussing the causes of heterosis genetic diversity was suggested as one of the causes. One would expect the F_1 hybrids of inter-specific hybridization to be most heterotic, because the two parents are much diverse. The F_1s between distant hybridizations may not be heterotic for characters associated with fitness, but they may manifest heterosis for certain characters, particularly the growth rate of plants. Dobzhansky (1950) refers to heterosis for fitness as "Euheterosis" and to heterosis for character which does not confer fitness upon the F_1s as "luxuriance".

Further it was mentioned that the magnitude of heterosis depends upon differences in the gene frequency. Hence one would expect that the magnitude of heterosis increases with degree of differentiation between two populations, with the limitation that F_1 of inter-specific cross is not sterile. According to Falconer (1960) error in extending our conclusion about differences in gene frequencies to wide crosses arises from the fact that we assume absence of non-allelic interaction. The existing genetic differentiation between wild populations has mostly arisen by evolutionary adaptation to local conditions. If the two populations are adapted to two different situations, their F_1 generation may be adapted to neither of the situations. Hence it will be less fit than either parent. Suppose genes *A* and *B* are selected in one population because when they are present together they increase fitness of the population, but either of the gene alone does not. In similar way genes *a* and *b* in another population are selected for similar reasons. With respect to fitness both the genes interact, i.e. *A* and *B* (and also *a* and *b*). But if these genes are fixed throughout the two populations, and become part of their constant genetic structure, variation arising from non-allelic interaction will disappear. When these two populations, adapted to different situations, are crossed the F_1 hybrid will be adapted to neither and hence will be less fit than populations *AABB* or *aabb*.

18.1 DIFFICULTIES ENCOUNTERED DURING DISTANT HYBRIDIZATION

Because the two species to be crossed, whether they belong to same genera or different genera, may differ with respect to number of chromosomes, may differ from each other with respect to chromosome aberrations, may possess factors in nucleus or cytoplasm which are not favourable to alien genomes and several other still unexplainable causes, the F_1 hybrids may be sterile, partially sterile or may be fertile.

Hence achieving success with distant hybridizations starts from pollination, fertilization, and seed setting to F_1 hybrid plant. As will be elucidated later the F_1 hybrids may be sterile. The genetic variability that is anticipated as per Mendelian theory is not obtained in F_2.

There are several examples where the F_1s from inter-specific crosses are fertile. The genus *Lycopersicon* is naturally subdivided into two sub-genera, viz., red fruited sub-genus *Eulycopersicon* and green fruited sub-genus *Eriopersicon. Lycopersicon esculentum* and *L.pimpinellifolium* belong to sub-genus *Eulycopersicon*. The F_1 hybrid of *L. esculentum* × *L. pimpinellifolium* shows high degree of fertility and some taxonomists regard the parents as one species (Peter, 1998). According to Rick and Butler (1956) there is no barrier to hybridize these species and secure recombination.

There are several factors which are impediments to the success of distant hybridization. A few examples will give an idea about the complexity.

18.1.1 Failure of fertilization

It has been observed in *Datura* that pollen tube bursts in the style of another species and hence fertilization is not accomplished.

Pollen grains were unable to germinate on the stigma even upto 72 hours after pollination in case of crosses of *Cucumis melo* with *C. figarei* or *C. meeusii* or *C. zeyheri* (Peter, 1998).

Fertilization may not take place because the pollen is unable to traverse the long style and reach the female gamete in time. This was observed in cross between maize and *Tripsacum*. The style of maize, the reader will recall, consists of long silk like structure emerging from ears. When *Tripsacum* is used as pollen parent in inter-specific cross the pollen is unable to traverse the great length between stigmatic surface and the ovary (Mangelsdorf and Reeves, 1931). Kalloo (1998) reported that the pollen tube of *Solanum melongena* failed to reach the ovary of *S. khasianum* in inter-specific cross *S. khasianum* × *S. melongena*. Inter specific cross *S. bulbocastanum* × *S. veerucosum* failed due to inhibition of growth of pollen tube of latter in the style of *S. bulbocastanum*.

18.1.2 Presence of Undesirable Genes

Babcock and Collins (1920) described hybrids between *Crepis capillaries* (n = 3) and *Crepis tectorum* (n = 4) which died after developing to cotyledonary stage. Hollingshead (1930) renewed the attempts to secure viable hybrids using different strains of parental species. One hybrid population was obtained of which half matured normally and half were inviable. Investigations indicated that *Crepis tectorum* was heterozygous for a lethal factor (*Ll*) which halted the development of hybrid to which it was transmitted at cotyledonary stage. Progeny of *Crepis tectorum* were all normal but on crossing various plants to *Crepis capillaries* it was evident that progeny were segregating for a lethal in 1: 2:1 ratio. Plants homozygous for non-lethal (*LL*) allelomorph were viable, those with *ll* gave inviable hybrids and those heterozygous for this locus gave half viable and half inviable progeny. In genus *Crepis* Babcock and Navashin (1930) reported that 19 out of 103 inter-specific hybrids were abnormal or weak, in most cases dying before reaching flowering stage. East (1935) found 14 inviable types in 77 hybrid combinations in *Nicotiana*.

Oka (1964), on the basis of research carried out for a decade, presented evidence that both haplontic and diplontic sterilities between *Oryza sativa* varieties and occurrence of weak F_2 plants could be explained by duplicate or complementary genes. Oka (1953*a*, *b*) explained that the gametic sterility was governed by a set or sets of duplicate genes (gamete development genes), which in double recessive condition lead to inviability of gamete. The fertility in the hybrids of cultivated varieties and *O. perennis* was interpreted to be due to double dominant genes for sterility.

Brown and Menzel (1952) reported that most of the inter-specific hybrids involving a wild American cotton species *Gossypium gossypioides* died at various stages. Inter-specific hybrids of another American wild species, *G. davidsoni* with *G. barbadense, G. hirsutum, G. stocksi* and *G. thurberi* die in cotyledonary stage. Silow (1941) attributed the deaths of inter-specific hybrids of these *Gossypium* species to the action of lethal genes.

In several inter-specific and inter-generic crosses the failure to secure viable seed was attributed to very early death of embryo. Sears (1944) studied the genetics of lethality of hybrids between *Triticum monococcum* and *Aegilops umbellulata*. He concluded that two alleles were present in *Triticum monococcum* which acted as dominant lelthals with *Aegilops umbellulata* but without ill effect on *T. monococcum*. The two alleles differ in earliness with which they cause death. L^e and L^l cause early and late death, respectively. Gene *l* was responsible for viability. Acccoding to him varieties of *Triticum monococcum* are either $L^e L^e$ or $L^l L^l$ and *Triticum aegilopoides* carries genes *ll*.

Mercy *et al.* (1974) failed to secure F_1 hybrid seed from inter-specific cross *Cicer arietinum* × C. *soongracium*, which they attributed to pollen-pistil incompatibility. On the other hand, Bassiri *et al.* (1987) discovered that failure to get seed from inter-specific crosses of *Cicer arietinum* × *C. reticulatum* and *C. arietinum* × *C. cuneatum* was due to slow pollen tube growth or collapse of fertilized ovules. Ahmed *et al.* (1988) believed that crossability barrier was due to factor(s) operating after fertilization.

Two recessive genes (kr_1, kr_2) located on chromosomes 5*B* and 5*A*, respectively, have been found to facilitate crossing between wheat and rye. The dominant genes, Kr_1 and Kr_2 reduced crossability (Sandha and Upadhyay, 1986).

18.1.3 Differences in The Number of Chromosomes

It has been observed that the inter-specific cross fails in one direction but could be successfully accomplished in other direction. Usually a cross is more successful when the female parent carries larger number of chromosomes. Seeds of F_1 hybrid between tetraploid wheat species (4*x*) and hexaploid species (6*x*) are usually shriveled, which fail to germinate. However, the seeds of F_1 hybrid of inter-specific cross hexaploid × tetraploid are smaller in size but normal with respect to germination. According to Thompson (1930a) in cases in which the two species differ in chromosome number the sterility may be attributed to death of gametes with intermediate chromosome number or to their inability to function normally owing to unbalanced condition of chromosomes. Thompson (1930b) attributed normal seed development, when hexaploid wheat species was taken as female parent, to chromosome balance of endosperm cells. In case of 6*x* × 4*x* cross the cells of endosperm contained 42 chromosomes from female parent and only 14 chromosomes from pollen parent. But in 4*x* × 6*x* cross only 78 per cent chromosomes were from female parent. Abnormal development of endosperm in 4*x* × 6*x* cross may be due in part to quantitative relationship between chromosome numbers of endosperm and embryo, but unbalance of dosage appears to be responsible for some difficulty in some crosses.

18.1.4 Embryo Abortion

Seeds from large number of distant crosses are not fully developed and are shrunken due to poorly developed endosperm. Such seeds are poor in germination or may often fail to germinate. The embryos developed in distant crosses suffer from post-fertilization development blocks. This restricts normal development of hybrid seeds. The developing embryos are heterotrophic i.e. depend on outside source of organic nutrients, and the nutritional supplies are provided by endosperm. In the inter-generic crosses, endosperm disintegrates and results in collapse of embryo. Such embryos may abort, collapse shortly following zygote formation and such seeds are not viable. The hybrid embryos can be rescued from collapse by allowing them to grow on an artificial medium (Chopra and Sharma, 1989).

An inter-generic hybrid of *Eruca sativa* ($n = 11$) and *Brassica campestris* ($n = 10$) was produced by embryo rescue (Agnihotri *et al.*, 1990a). *Brassica spinescens* exhibits resistance to *Alternaria* blight. An inter-specific hybrid of *Brassica spinescens* with *B. campestris* was obtained by embryo rescue (Agnihotri *et al.*, 1990b). Plants were multiplied by micropropagation.

In tomato it was observed that *Lycopersicon peruvianum* can be crossed with *L. esculentum* but the embryos abort (Peter, 1998). *L. peruvianum* is a source of resistance to *Verticillium* wilt and leaf curl virus. *Lycopersicon hirsutum* carries resistance to *Fusarium* wilt and belongs to sub-genus *Eriopersicon.* The inter-specific hybrid *L. hirsutum* × *L. esculentum* results in seed set and fruit formation but the F_1 is self-incompatible.

18.1.5 Sterility of F_1 Hybrids

The inter-specific hybrids, when made between two species, which have been differentiated from each other, show hybrid sterility. Although the F_1 hybrid may attain flowering the plants will be either partially or completely sterile. Sterility can be observed in intra-specific crosses of certain genera but it is not as pronounced as seen in crosses between different species.

There are probably several causes of hybrid sterility but most of them arise from some kind of genetic unbalance. If the unbalance affects the metabolism in early stages of development of F_1 hybrid or the later vegetative stages of the development, hybrid inviability results.

However, when genetic unbalance influences gametogenesis hybrid sterility results. The sterility of F_1 hybrids could be divided into two types, viz., chromosomal and genic.

When the chromosomes contributed by two species lack structural homology it will lead to sterility. The reader is aware of the fact that inversion heterozygotes, translocation heterozygotes do not produce 100 per cent fertile gametes. They are deficient in the chromosome segments or exhibit various configurations during meiosis in F_1. Most of the gametes produced by individuals with chromosomal aberrations are non functional. Difference in the number of chromosomes per genome contributed by the two parents can also lead to chromosomal sterility. There may not be complete homology among chromosomes from two parental species and hence meiosis will not be normal. If the number of chromosomes differ in two species, there will be univalents, for excess number of chromosomes, assuming that the two species have some common number of chromosomes with homology. But if the chromosome number of F_1 hybrid is artificially doubled (with colchicine) then the amphidiploid is fertile. This evidence shows role of chromosome number as the cause of sterility.

Perfect synapsis in the cells of F_1 hybrid does not ensure its fertile nature. There are examples of F_1 hybrids where chromosomes from two species manifest loose pairing at first metaphase but are completely sterile. Sterility associated with completely normal or near normal pairing between chromosomes from two species is ascribed to genic sterility. Stebbins (1950, 1958) presented evidence for wide spread sterility due to heterozygous structural differences which are too small to influence pairing between chromosomes. Stebbins (1950, 1958) calls such small structural differences in chromosomes in heterozygous condition as '*cryptic structural hybridity*'. Hybrid sterility which results from genic causes does not involve chromosomes. Genic sterility may be caused by disturbances in synapsis at meiotic division. Cultivated rice of the world can be broadly divided into three groups on the basis of geographical adaptation and morphological characters, viz., *indica, japonica* and *javanica*. In addition to morphological, physiological differences there is large amount of hybrid sterility between *indica* and *japonica* groups (Chandraratna, 1964). There is regular occurrence of partial sterility in wide inter-varietal crosses within *Oryza sativa* species. This is of frequent occurrence when *japonica* and *indica* groups are crossed. This has been explained on the basis of (*a*) recombination and segregation of specific genes, i.e. genic sterility, (*b*) chromosome abnormal irregularities and cytoplasmic influences (Chandraratna, 1964; Chang, 1964; Henderson, 1964).

Although some F_1 hybrids from inter-specific crosses may exhibit normal development and normal or near normal fertility some plants in F_2 generation may be quite abnormal or sterile. This could be attributed to segregation of complementary gene system (Muller, 1940). However, Stephens (1949, 1950) is of the opinion that degeneration in F_2 results due to deficiencies and duplications of the chromosomes, in case of cotton. Each of the duplication or deficiency is so small that synapsis is affected. In other words, the degeneration results from cryptic structural hybridity.

18.2 Segregation in Distant Hybridization

When F_1 is heterozygous for n gene pairs it was shown in chapter 5 that there will be 3^n types of genotypes in F_2 generation resulting from varietal crosses. In case of intra-specific hybridization the parental species do not differ from each other with respect to number of chromosomes. There is not much differentiation which will restrict association of homologous chromosomes at zygotene stage in meiotic cell division. In the absence of genes causing lethality or sterility of F_1, normal seed is produced. F_2 also exhibits genotypic variability among the segregates, which results from recombination and independent assortment.

However, when plants belonging to different species are crossed or from different genera are hybridized results deviate from these expectations. Provided such a cross could be successfully accomplished and F_1 is also fertile, tremendous diversity of types shall appear in F_2 and subsequent generations. But the population does not even approach to free recombination of parental values. Instead recombination is found to be restricted to a narrow segment of total, running from combinations more or less like the one parent to combinations more or less like the F_1 to combinations more or less like the other parent (Anderson, 1939).

An experiment was started with *Nicotiana langsdorffi* and *N. alata* in 1930 at the John Innes Horticultural Institution to study the actual recombinations obtained in second generation as compared with the total imaginable recombinations of all the characters of both parents. This experiment was carried further at the Missouri Botanical Garden in the following year and at the Bussey Institution of Harvard University from 1931 to 1933. Edgar Anderson (1939) was a staff member of

Arnold Arboretum. For this study two morphologically dissimilar species were chosen, viz., *N. langsdorffi* and *N. alata*, which yield hybrids with high degree of fertility. The experiment was confined to the study of flowers because as shown by East (1913) flower size is less affected by environmental variables than is the size of vegetative parts. *N. langsdorffi* and *N. alata* differ strikingly in floral characters. Corolla tube is long (5.8 to 8.9 cm) in case of *N. alata* but short in *N. langsdorffi* (1.7 to 2.1 cm). The corolla limb is also wide in *N. alata* (2.5-4.0 cm) as compared to *N. langsdorffi* (0.6-0.9 cm). Further the corolla limb is deeply lobed in *N. alata* but scarcely lobed in *N. langsdorffi* (Figure 18.1).

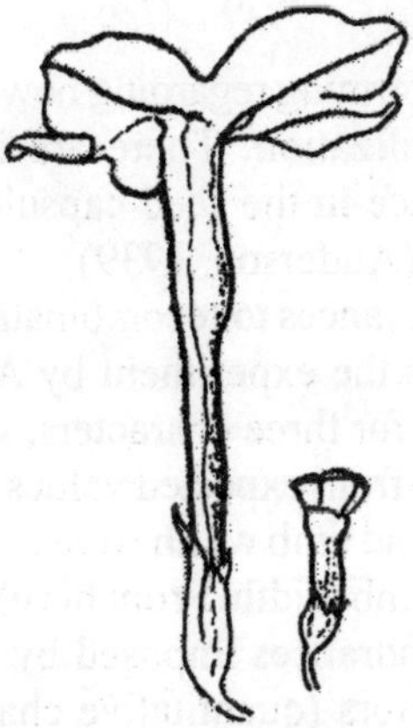
A B

Fig. 18.1: Representative flowers of *N. alata* and *N. langsdorffi* (Adapted from Anderson, 1939)

From the observations Anderson (1939) concluded that actual character recombinations in F_2 were less than 1/64 of the total imaginable recombinations. According to him the hindrances to character combinations could be due to four reasons, viz., gametic elimination, zygotic elimination, pleiotropy and linkage. Indication of the amount of gametic elimination can be obtained by examining the pollen of F_1 plants. Avery (1938) made a cytological study of the F_1 between *Nicotiana alata* and *N. langsdorffi* and reported an average of 44 per cent of good pollen as judged by its morphology and staining reactions. Bulk of the pollen according to her, was non-viable due to gross irregularities in chromosome distribution, a small percentage may well be due to formation of gametically inviable recombinations of the two species. It is expected that such combinations would be found more among extreme recombinations than among those which are close to the original parental types.

Gametic elimination also occurs due to selective fertilization by the F_1 pollen. Precise information is available for only one locus in germplasm, the self-sterility alleles having been the subject of detailed study. *N. langsdorffi* is homozygous for self-incompatibility alleles (S_fS_f). At this locus in *N. alata* there is a series of self-incompatibility alleles designated as S_1, S_2, S_3 etc. The pollen bearing any one of these alleles will not grow on the stigma of plant bearing the same allele (see chapter 3).

There is no precise informing regarding how many recombinations are eliminated after fertilization. There are indications that such eliminations can take place in the seed capsule, in aging seeds and shortly after germination (Anderson, 1939).

One of the severe hindrances to recombinations may be caused by linkage. It was evident in the experiment by Anderson (1939). The expected recombinations for three characters, viz., tube length, limb width, and lobing differed from expected values in F_2. It was predicted that genes for tube length and limb width were tightly linked and lobing was away from gene for limb width. From his observations Anderson (1939) concluded that hindrances imposed by linkage of characters governed by multiple factors (quantitative characters) were severe. And recombination in crosses between species and races was limited in second generation to a relatively insignificant fraction of total recombination.

It is interesting to note that male gametogenesis is more easily upset by chromosomal or genic disharmonies than female gametogenesis (Allard, 1960). Hence propagation of inter-specific F_1 hybrids often depends upon back-crossing the F_1 as seed parent to either parental species (used as pollen parent). Even when the F_1 is fertile and can be self-pollinated, the offspring may still be more like back-cross type rather than true F_2 segregates.

18.3 HOW TO OVERCOME SOME BARRIERS TO ACHIEVE DISTANT HYBRIDIZATION

In section 18.2 several hurdles were pointed out which tend to reduce the possibility of achieving success with distant hybridizations. Nevertheless, due to extensive studies by various scientists there are some effective means to overcome the barriers which obstruct the possibility of distant hybridizations.

18.3.1 Shorten the Length of style

In case of *Zea mays* and *Tripsacum* cross, the pollen grains of latter are unable to traverse the long distance between stigma and ovary since they are not used to it. If the length of style is shortened success may be possible in crossing *Tripsacum* with maize.

18.3.2 Removal of Stigma Before Pollination and Placing a Drop of Sugar

The F_1 hybrids of *Capsicum frutescens* × *C. annuum* range from partial fertility to complete sterility. Radhakrishnan *et al.* (1977) developed a technique which resulted in seed set in this inter-specific cross. The upper part of gynoecium, comprising of stigma and some style, was excised and a drop of 5 per cent sucrose solution was applied to the cut surface before pollination. This was successful in attaining the objective. An interesting fact was that fruit set was considerably lower when *C. annuum* was used as female parent. Similar results were obtained by Nair *et al.* (1979) in a cross of *C. annuum* × *C. frutescens*.

The species of *Solanum* belonging to section *Tuberarium* and occurring in Mexico have been of interest to the potato breeders because several of them possess resistance to late blight caused by *Phytophthora infestans*, de Bary. Mexican species also form a polyploid series comprising of diploids ($2n = 24$), triploids ($2n = 36$), tetraploids ($2n = 48$), pentaploids ($2n = 60$) and hexaploids ($2n = 72$) like the species of *Solanum* found in South America. A study of causes of failure of seed setting in crosses *S. pinnatisectum* × *S. bulbocastanum* and *S. pinnatisectum* × *S. lanciforme*, indicated that the pollen of respective male parents did not germinate on stigma, even after 72 hours from the time of pollination (Swaminathan, 1955). This incompatibility was overcome by removal of a small portion of style of *S. pinnatisectum* and placing a drop of an agar-sucrose-gelatin medium on decapitated surface of the style (Swaminathan, 1955). This medium was prepared by dissolving 0.5 g of agar and 2.5 g of sucrose in 25 cc of distilled water to which 0.5 g gelatin was added. Pollen grains of many species of *Solanum* were found to grow well in this medium.

18.3.3 Application of Growth Promoting Substances to Pistils of Female Parents

Crane and Marks (1952) successfully crossed diploid and tetraploid varieties of pear by treating the ovaries with β-naphthoxyacetic acid.

18.3.4 Pollination at Different stages

Several times early pollinations, before the stigma becomes receptive, shall lead to seed set. Bud pollination has been suggested in cole crops to achieve seed set. This technique is useful in those cases where growth of pollen tube in style of female parent is slow.

18.3.5 Grafting of Female Parent on Male Species

Pissarev and Vinogradova (1944) observed that embryos and endosperm began to develop after pollination of wheat flowers by rye pollen, but in most cases further growth of organs ceased completely. To overcome this difficulty they developed embryo implantation technique. Embryo together with scutellum of wheat was cut from dry caryopsis, which was then transferred to the rye endosperm, whose embryo had been primarily cut out. 'Graft embryo' was fixed in place by means of a paste prepared from a flour of same genus to which endosperm belonged. Hall (1954) grafted the wheat embryo, according to the technique developed by Pissarev and Vinogradova (1944), in the rye endosperm. The graft kernels were placed in sterilized soil in small pots. The emasculated flowers of the wheat grafts were pollinated by rye pollen. The results showed that crossabiltiy between wheat and rye can be increased if wheat plants derived from embryos which were transplanted on to rye endosperm were used as female.

Singh and Hymowitz (2001) successfully produced an hybrid between *Glycine max* ($2n = 40$) and *G. tomentella* ($2n = 78$). They noticed that hybrids between *Glycine max* and wild soybean species are difficult to produce. Gynoecia usually dropped within 3 days after pollination. This drop of gynoecia was stopped by spraying a growth hormone mixture (100 mg *GA*3 plus 25 mg *NAA* plus 5 mg kinetin per one litre distilled water). This spray was applied once a day for about twenty days on developing pods. Pods were physically removed about 20 days after pollination and cultured. Putative F_1 plants were identified on the basis of morphology and cytology. F_1 plants were relatively slow growing. F_1 plants had $2n = 59$ chromosomes and were completely sterile. In order to restore seed set, actively growing shoots of F_1 (scion) were grafted on soybean (root stock) and internodes of scion were wrapped with cotton which was socked in 0.2 per cent colchicine. Slow growing shoots with thick leathery dark green leaves were retained. Rapidly growing shoots were removed because they were most likely not doubled chromosomally. Amphidiploid soybean carried $2n = 118$ chromosomes, flowered profusely and occasionally set one seeded pod.

18.3.6 Doubling the Chromosome Number

Diploid species of *Solanum* have been crossed with tetraploid *Solanum tuberosum* by doubling the chromosome number of diploid species and subsequently crossing the two at tetraploid level. Using this technique crosses have been made between diploid *Solanum polyadenium* and tetraploid *S. tuberosum* (Stelzner, 1949). Similarly crosses between diploid *S. chacoense, S. kesselbrenneri* and *S. rybinii* have been accomplished, rather freely, with some *S. tuberosum* (Swaminathan, 1951).

Some tetraploids, some times, do not cross normally between themselves, but do so when chromosome number of one of them is doubled. Thus crosses between frost resistant *S. acaule* and *S. tuberosum* was effected fairly easily when doubled plants of *S. acaule* ($2n = 96$) were used as female parent (Lamm, 1943).

18.3.7 Use of Mixed Pollen

Beasley (1940) found that certain inter-specific crosses of cotton failed because only a few seeds per boll were produced and there was abscission of bolls. When he pollinated initially with 6 to 12 pollen grains from female and subsequently with excessive amount of pollen from male parent bolls were retained.

18.3.8 Unilateral Incompatibility

In some inter-specific hybridization programmes success can be attained only in one direction and reciprocal differences are noticed between species with regard to crossability.

Caera rubber (*Manihot glaziovii*) is a closely related plant to tapioca (or cassava, *M. utilissima*). Inter-specific crosses between these two species were attempted to introduce some wild genes into cultivated tapioca and evolve a new type (Abraham, 1957). Thousands of crosses were made between these two species employing, *M. glaziovii* as female and tapioca as male. But they failed to give fruitset. Nevertheless, reciprocal cross using tapioca as female parent and caera rubber as male was successful, although seed setting was very poor, just 1 per cent. The F_1 plants were quite different from tapioca and caera rubber. Unilateral incompatibility has been reported in *Solanum, Lycopersicon* and *Capsicum* (Kalloo, 1988). In these vegetable species pollen tube inhibition in style was the main reason for unilateral incompatibility.

18.3.9 Embryo Rescue

Parthasarathy and Kedarnath (1949) reported that inter-specific cross between *Sesamum orientale* and an Argentina species *S. radiatum* was made but hybrid seedling could not be obtained as although the pollen of *S. orientale* germinated on stigma of *S. radiatum* and pollen tube traversed down through the style and fertilized the ovules, due to collapse of endosperm and consequent starvation of embryo, shriveled seeds were obtained which did not germinate.

Corchorus capsularis and *C. olitorius* are of economic importance. *C. olitorius* can tolerate diseases and insect pests more as compared to *C. capsularis*. Numerous efforts were made to combine the characters of these two species by crossing but no success was achieved. The reason for failure of the crosses was due to abortion of hybrid embryos before attaining maturity (Sikka, 1957).

Several cases of such embryo abortions have been witnessed by the breeders engaged in distant crossing.

Hybrid embryos from crosses between distantly related species are often weak and inviable due to inadequate nourishment by the endosperm. Embryos from distant hybridization suffer from post-fertilization developmental blocks which prevent formation of normal seed. The developing embryos are heterotrophic, i.e. depend on outside source of organic nutrient, and the nutrition is supplied by the endosperm. In the distant hybridization endosperm disintegrates and leads to collapse of embryo. Consequently, viable seeds are not produced. The hybrid embryo can be excised in aseptic condition from the female plant and immature embryo from the ovule can be grown on an artificial medium. Aseptic excision of immature embryo from the ovule, before the abortion of endosperm, is called 'embryo rescue'. According to Chopra and Sharma (1989) excised embryo can be cultured by three different methods, viz., embryo culture, embryo implantation and ovule culture.

Embryo in the initial stages, i.e. proembryo, is excised from the female plant and transferred to a solid agar medium or liquid medium with appropriate nutrition required for its normal growth. After development of plantlets they are transferred to sterile soil for growth into mature plants.

Liabach (1929) crossed *Linum perenne* with *L. austriacum* but was unable to get success. He removed the developing seed from unripe capsules, about 14 days after pollination and placed in a vial on cotton wool moistened with cane sugar solution. He was able to obtain plants from these embryos. Since then exçised embryo technique has been used to achieve seed from inter-specific crosses.

Embryo culture depends for its success on precise nutrition medium, which differs from one species of plant to another. Alternatively the aseptically excised embryo could be transplanted into a new endosperm, a technique called 'embryo implantation'.

Embryo culture was successfully employed for developing *Triticale*, an inter generic hybrid between *Triticum durum* and *Secale cereale* (Sandha and Upadhyay, 1986). It was also used by Agnihotri *et al.* (1990*a*, *b*).

18.3.10 Developing Amphidiploids

Whenever, the two species differ in number of chromosome sets, i.e. level of ploidy, the F_1 if at all obtained, may be sterile because the cells of such F_1 lack homologous chromosomes. Consequently pairing can not take place at zygotene. Resulting gametes are therefore inviable. Doubling the chromosome number, say with colchicine, will lead to an amphidiploid which will be fertile (Chapter 16).

References

Abraham, A. (1957). Breeding tuber crops in India. *Indian J. Genet.*, 17: 212-217.

Ahmed, F.; A.E. Slinkard and G.J. Scodes (1988). Investigations into barriers to inter-specific hybridization between *Cicer arietinum* L. and eight other annual *Cicer* species. *Plant Breed.*, 100: 193-198.

Allard, R.W. (1960). *Principles of plant breeding*. John Wiley and Sons, Inc.New York, pp. 423-443.

Agnihotri, Abha; Vibha Gupta; Malathi Lakshmikumaran; K.R. Shivanna; Shyam Prakash and V. Jagannathan (1990a). Production of *Eruca* × *Brassica* hybrids by embryo rescue. *Pl.Breed.*, 104: 281-289.

Agnihotri, Abha; Malathi Lakshmikumaran; K.R. Shivanna; and V. Jagannathan (1990b). Embryo rescue in inter-specific hybrids of *B. spinescens* × *B. compestris* and DNA analysis. In *Progress in plant cellular and molecular biology*. Eds. H.J.J. Nijkamp, L.H.W. van del Plas and J. van Atrijk: 271-274.

Anderson, Edgar (1939). Recombination in species crosses. *Genetics*, 24: 668-698.

Avery, Pricilla (1938). Cytogenetic evidences of *Nicotiana* phylesis in the *alata* group. *Univ.Calif.Publ.Bot.*, 18: 153-194.

Babcock, E.B. and J.L. Collins (1920). Inter-specific hybrids in *Crepis* I. *Crepis capillaris* (L.) Wallr. × *C. tectorum* L. *Uni.Calif.Pub. Agric. Sci.*, 2: 191-204.

Babcock, E.B. and M. Navashin (1930). The genus *Crepis*. *Bibliogr. Genet.*, 6: 1-90.

Bassiri, A.; F. Ahmed and A.E. Slinkard (1987). Pollen grain germination and pollen tube growth following *in vivo* and *in vitro* self and inter-specific pollinations in annual *Cicer* species. *Euphytica*, 36: 667-675.

Beasley, J.O. (1940). Hybridization of American 20 chromosome and Asiatic 13 chromosome species of *Gossypium*. *J.Agric.Res.*, 60: 175-181.

Briggs, F.N. and P.F. Knowles (1967). *Introduction to plant breeding*. Reinhold publishing Corporation, New York, pp. 317-333.

Brown, M.S. and M.Y. Menzel (1952). Additional evidence on crossing behaviour of *Gossypium gossypioides*. *Bull.Torrey.Bot.Club.*, 79: 285-292.

Chandraratna, M.F. (1964). *Genetics and breeding of rice*. Longmans, Green and Co. Ltd., London, p. 389.

Chang, T.T. (1964). Persent knowledge of rice genetics and cytogenetics. *The International Rice Research Institute Technical Bulletin*, 1.Los Banos, Laguna, Philippines, p. 96.

Chopra, V.L. and R.P. Sharma (1989). Innovative approaches for crop improvement. In *Plant Breeding, Theory and Practice*, Ed. V.L. Chopra, Oxford and IBH Publishing Co., New Delhi, pp. 453-466.

Crane, M.B. and E. Marks (1952). Pear apple hybrids. *Nature*, 170, 1017

Dobzhansky, Th. (1950). Genetics of natural populations. XIX. Origin of heterosis through natural selection in populations of *Drosophila pseudoobscura*. *Genetics*, 35: 288-302.

Dutta, S.K. (1996). *Chrysanthemum*. National Botanical Research Institute, Lucknow.

East, E.M. (1913). Inheritance of flower size in crosses between species of *Nicotiana*. *Bot. Gaz.*, 55: 177-188.

East, E.M. (1935). Genetic relations in *Nicotiana*. I. Compatibility. *Genetics*, 20: 403-413.

Emsweller, S.L.; P. Brievley; D.V. Lumsden and F.L. Mulford (1937). Improvement of flowers by breeding. *U.S. Dept.Agric. Yearbook*, 890-998.

Falconer, D.S. (1960). *Introduction to Quantitative Genetics*. Oliver and Boyd. Edinburgh.

Frey, K.J.; T.S. Cox; D.M. Rogers and P. Bramel Cox (1984). Increasing cereal yields with genes from wild and weedy species. In *Genetics: New Frontiers, Proceedings of XV International Congress of Genetics*, Vol.IV. Eds. V.L. Chopra *et al.* Oxford and IBH Publishing Co., New Delhi, pp. 51-68.

Goodspeed, T.H. and R.E. Clausen (1928). Inter-specific hybridization in *Nicotiana* VIII. The *Sylvestris-tomentosa-tabacum* triangle and its bearing on the origin of *Tabacum. Uni. Of Calf.Pub.in Bot.*, 11: 245-256.

Hall, O.L. (1954). Hybridization of wheat and rye after embryo transplantation. *Hereditas*, 40: 453-458.

Henderson, M.T. (1964). Cytogenetic studies at the Louisiana Agricultural Experiment Station of species relationship in *Oryza*. In "Rice *Genetics and Cytogenetics*". Elsevier Publishing Co., Amsterdam, pp. 103-110.

Hollingshead, L. (1930). A lethal factor in *Crepis* effective only in an inter-specific hybrid. *Genetics*, 15: 114-140.

Kalloo, G. (1988). *Vegetable breeding* Vol.I. C.R.C.Press, Florida, USA, pp. 137-162.

Lamm, R. (1943). Notes on octoploid *S. punae* plant. *Hereditas*, 29: 193-195.

Liabach, F. (1925). Das Taubwerden von Bastardsamen und die kunstliche Aufzuchtfruh absterbender Bastardembryonen. *Z.Bot.*, 17: 417-459.

Mangelsdorf, P.C. and R.G. Reeves (1931). Hybridization of maize, *Tripsacum* and *Euchlaena*. *J.Hered.*, 22: 329-343.

Mercy, S.T.; S.N. Kakar and T.M. Varghese (1974). Embryology of *Cicer arietinum* and *Cicer soongracium*. *Bull.Toreey Bot.Club.*, 100: 26-30.

Mukhopadhyay, A. (1995). *Gladiolus*. ICAR, New Delhi, p. 83.

Muller, H.J. (1940). *The New Systematics*. Ed. J.Huxley. Oxford University Press, New York, pp. 185-268.

Nair, K.M.D.; M.G. Rao and B.G.S. Reddy (1979). New inter-specific chilli strains of Karnataka. *Current Research*, 8: 158-160.

Oka, H. (1953a). Phylogenetic differentiation of the cultivated rice plant. VI. The mechanism of sterility in intervarietal hybrid of rice. *Jap.J.Breed.*, 2: 217-211.

Oka, H. (1953b). Phylogenetic differentiation of the cultivated rice plant. VIII. Gene analysis of inter-varietal hybrid sterility and certation due to certain combinations of gametic-development genes in rice. *Jap.J.Breed.*, 3: 23-30.

Oka, H. (1964). Patterns of inter-specific relationships and evolutionary dynamics in *Oryza.Symp.Rice Genet. and Cytogenet.*, Elsevier Amsterdam, pp. 71-90.

Parthasarathy, N. and S. Kedarnath (1949). The improvement of Sesamum crop in India. *Indian J.Genet.*, 9: 59-71.

Peter, K.V. (1998). *Genetics and breeding of vegetables*. ICAR, New Delhi.

Pissarev, W.E. and N.M. Vinogradova (1944). Hybrids between wheat an *Elymus* — Comptes Rendus (Doklady) de TAcad.des Sciences de TURSS Vol.XLV, No. 3.

Poehlman, J.M. (1987). *Breeding Field Crops*. AVI Publishing Company, Inc.Westport, Connecticut, USA.

Radhakrishnan, K.P.; S.T. Mercy and M.K. George (1977). Crossability studies and analysis of incompatibility in three species of *Capsicum*. *Agricultural Research Journal*, Kerala, 15: 124-127.

Rick, C.M. and L. Butler (1956). Cytogenetics of tomato. *Adv.in Gen.*, 8: 267-382.

Sandha, G.S. and M.K. Upadhyay (1986). Progress in Triticale Research. In *Twentyfive years of Coordinated Wheat Research, 1961-86*. Wheat Project Directorate, IARI, New Delhi, pp. 235-252.

Sears, E.R. (1944). Inviabilty of inter-generic hybrids involving *Triticum monococcum* and *T. aegilopoides*. *Genetics*, 29: 113-127.

Sikka, S.M. (1957). Breeding new varieties of fibre crops. *Indian J.Genet.*, 17: 180-196.

Silow, R.A. (1941). The comparative genetics of *Gossypium anomalam* and the cultivated asiatic cottons. *Jour.Gen.*, 42: 259-358.

Singh, B.D. (1993). *Plant breeding — principles and methods*. Kalyani Publishers, Ludhiana, pp. 661-686.

Singh, R.J. and T. Hymowitz (2001). Exploitation of the wild perennial *Glycine* species for improving soybeans. In Proceedings *Harnessing the soy potential for health and wealth*, Ed. Bhatnagar, P.S. India Soy Forum, 2001. March 17-18, 2001. The Soybean Processors Association of India, Indore, pp. 58-61.

Swaminathan, M.S. (1951). Notes on induced polyoploids in the tuber bearing *Solanum* species and their crossabiltiy with S. *tuberosum*. *Amer.Pot.Jour.*, 28: 472-489.

Swaminathan, M.S. (1955). Overcoming cross-incompatibility among some Mexican diploid species of *Solanum*. *Nature*, 176: 887-888.

Stebbins, G.L. (1950). *Variation and evolution in plants*. Columbia University Press, New York, p. 643.

Stebbins, G.L. Jr. (1958). The inviability, weakness, and sterility of inter-specific hybrids. *Adv. In Gen.*, 9: 147-215.

Stephens, S.G. (1949). The cytogenetics of speciation in *Gossypium*. I. Selective elimination of the donor parent genotype in inter-specific back-crosses. *Genetics*, 34: 627-637.

Stephens, S.G. (1950). The internal mechanism of speciation in *Gossypium*. *Bot.Rev.*, 16: 115-149.

Thompson, W.P. (1930a). Causes of differences in success of reciprocal interspecific crosses. *Amer.Nat.*, 64: 407-421.

Thompson, W.P. (1930b). Shrivelled endosperm in species crosses in wheat, its cytological cause and genetical effects. *Genetics*, 15: 99-113.

Bibliography

Allard, R.W., 1960. *Principles of Plant Breeding*. John Wiley and Sons, New York.

Bassett, M.J., 1986. *Breeding Vegetable Crops*, (Edited). AVI Publishing Co., Connecticut., U.S.A.

Chopra, V.L., 1989. *Plant Breeding—Theory and Practice*. Oxford and IBH Publishing Co. Pvt. Ltd., New Delhi (Edited).

Kalloo, G., 1988. *Vegetable Breeding. Vol. I, II and III*. CRC Press, Florida, U.S.A.

Mukherjee, B.K., 1995. The Heterosis Phenomenon, Kalyani Publishers, Ludhiana.

Poehlman, J.M., 1987. *Breeding Field Crops*. 3rd edition, AVI Publishing Co. Inc., Westport, Connecticut, U.S.A.

Russell, G.E., 1978. *Plant Breeding for Pest and Disease Resistance*. Butterworth. London.

Seshadri, V.S., 1989. "Vegetable Crops". In *Plant Breeding—Theory and Practice*, Ed. V.L. Chopra. Oxford and IBH Publishing Co. Pvt. Ltd. New Delhi. pp. 391-408.

Singh, B.D., 1993. *Plant Breeding—Principles and Methods*. Kalyani Publishers, New Delhi.

Singh, H.B., Swarup, V. and B. Singh, 1998. *Exploitation of hybrid vigour in Vegetables*, I.C.A.R., New Delhi.

Swarup, V., 1977. *Breeding Procedures for Cross-Pollinated Vegetable Crops*. I.C.A.R. Publ. New Delhi.

Van der Plank, J.E., 1968. *Disease Resistance in Plants*. Academic Press, New York.

Index